# Why We Snap

# WHY WE SNAP

## Understanding the Rage Circuit in Your Brain

## R. Douglas Fields, PhD

DUTTON
— est. 1852 —

# DUTTON
—• est. 1852 •—

An imprint of Penguin Random House LLC
375 Hudson Street
New York, New York 10014

Copyright © 2015 by R. Douglas Fields

DUTTON—EST. 1852 (Stylized) and DUTTON are registered trademarks of Penguin Random House LLC.

Names: Fields, Douglas.
Title: Why we snap : understanding the rage circuit in your brain / Douglas Fields.
Description: New York : Dutton, 2016. | Includes index.
Identifiers: LCCN 2015016464 | ISBN 9780525954835 (hardback) | ISBN 9780698194311 (ebook)
Subjects: LCSH: Anger. | Violence. | Neurosciences. | BISAC: SCIENCE / Life Sciences / Neuroscience. | MEDICAL / Research.
Classification: LCC RC569.5.A53 .F54 2016 | DDC 616.89/142—dc23
LC record available at http://lccn.loc.gov/2015016464

Printed in the United States of America
1   3   5   7   9   10   8   6   4   2

Set in Minion Pro with Mercury Text G1
Designed by Daniel Lagin

*For my family, Kelly, Morgan, Dylan, and Melanie*

# Contents

# PART ONE

# ENRAGED

# 1

# Snapping Violently

Rage is a short madness.                    Horace, *Book 1, Epistle ii,* line 62

Y ou mustn't say things about Melanie," he warns her.

"Who are you to tell me I mustn't?" she snaps back, vibrating in anger. "You led me on. You made me believe you wanted to marry me!"

"Now, Scarlett, be fair," he pleads, trying to calm her fury. "I never at any time—"

"You did! It's true! You did." She cuts him off. "I'll hate you till I die!" she screams. "I can't think of anything bad enough to call you!"

Sobbing in rage, she suddenly slaps her lover across his face. As he retreats she grasps a vase and hurls it across the room. The delicate porcelain shatters against the wall.

Later the jilted woman sobs desperately as the second man in her lovers' triangle walks out on her: "Oh, Rhett! Rhett, Rhett! Rhett . . . Rhett, if you go, where shall I go? What shall I do?"

He faces her calmly and delivers these enduring words: "Frankly, my dear, I don't give a damn."

*Gone with the Wind*, the 1939 film classic based on Margaret Mitchell's novel, captures the paradoxical moment of snapping that is familiar to us all, but inexplicable. Why smash a treasured vase? Why slap a lover across the face? The immediate aftermath brings regret and shame, and upon reflection bewilderment. The explosive impulse of destruction is driven by a powerful righteous rage, overwhelming but pointless.

Who has not lost self-control in a blind rage, smashing a dish—or worse? We all wish to believe—need to believe—that we are in control of our behaviors and actions, but the fact is that in certain instances we are not. Something unexpected in our environment can unleash an automatic and complex program for violence, destruction, and even death—all of it an unconscious pre-established program.

Rage explodes without warning. Overpowering judgment, compassion, fear, and pain, the fiery emotion serves one purpose—violence, both in words and actions. While this human response has been vital to our survival since our species evolved, rage simultaneously puts one's life at risk. And it seems there is no escaping the rage circuit once it has been activated. So if rage is an automatic reflex, are you really in control of your fate? That flare-up with your partner or child or friend or even a complete stranger can change your life in an instant, forever.

Despite the essentially peaceful lives most of us lead most of the time, killing is programmed into the human brain. This is because, as with most animals, individuals in the natural world must be able to defend themselves and their offspring. Moreover, carnivores must kill other living creatures for food. These behaviors are hardwired in the brain, not in an area where consciousness resides but instead deep in the core of the brain where other powerful impulses and automatic life-sustaining behaviors (feeding, thirst, and sex) are programmed. Each of these behaviors, just like the complex rage behavior, is automatic once triggered. The question is, what triggers this deadly switch for violence and killing?

Late one summer night in a torrential downpour, my daughter and I threaded our way through the dark cobblestone back alleys of Paris, hungry and lost. Like most scientists, I travel the world to lecture and collaborate with other scientists, and I almost always travel alone. This night my seventeen-year-old daughter was with me. The springtime of proms, graduation ceremonies, and anxious anticipation of leaving high school behind had cleared a momentary opportunity for a father and daughter to share time together. It was wonderful seeing Kelly's eyes open to the world. Soaking wet, we leaped over puddles and escaped into a steamy one-room restaurant. No one spoke English. Kelly applied her high school French to order from one of the three frantic middle-aged women who shared the burden of all the cooking and serving.

Suddenly in exasperation the woman jabbed at the menu, scolding Kelly. She had not ordered a glass of wine for herself. The idea that anyone would enjoy a fine dinner without the requisite glass of wine was unthinkable. For Kelly, underage for drinking alcohol in the United States, this was a revelation. Not everywhere in the world is necessarily the same as the place in which you were reared.

After Paris we traveled to Barcelona for my next lecture at an international meeting of neuroscientists. The morning before the meeting began we made a quick visit to the Gaudi cathedral. Ascending the steps out of the dingy subway station smelling of concrete dust and sweat, we emerged into the brilliant Barcelona sun. The crowd of passengers pressed upon us in a gray blur.

Suddenly I felt a sharp tug at my pant leg. As if swatting a mosquito I slapped the zippered pocket above my left knee. My wallet was gone!

My left arm shot back blindly. In a flash I clotheslined the robber as he pivoted to hand my wallet to his partner and flee down the steps. As if swinging a sledgehammer I hurled him by his neck over my left hip and slammed him belly first onto the pavement, where I flattened him to the ground and applied a head lock.

Splaying my legs for hip control like a wrestler pinning an opponent I yelled for help. Fifty-six years old, 130 pounds, with wire-rimmed glasses and graying hair, I have no martial-arts training, no military experience, no background in street fighting. Drawing on junior high school wrestling moves from forty years ago, I found myself applying an illegal choke hold. The street-smart hoodlum struggling in my arms was in his late twenties or early thirties.

"Police!" I shouted. "Call the police! I've got him!"

There was no reply . . . no gasps of shock from the dense crowd . . . no one was coming to my aid. Instead, from my perspective on the ground all I saw were men's feet closing in around me in a tight circle. They were all part of the gang. Oblivious to being hunted as prey, we assumed that the crowd was the normal throng of passengers bumping and jostling through the Barcelona Metro system.

The muscled man beneath me struggled to break my grip. With his neck in the crook of my left arm I cinched with all the force my biceps could produce, cutting off blood to his brain and air to his lungs. Bending his head back I torqued his spine backward painfully, tipping his face skyward. His eyes and mouth opened wide in shock, pain, and fear. The wallet popped free as he tossed it toward his accomplice and grasped furiously at my arm to break my stranglehold.

"That's my wallet!" I yelled.

A woman's hand shot between the thicket of legs. Instantly I recognized it as my daughter's. She had been cut off by the gang that had stalked and trapped us, encircling me silently like a pack of wolves. Captain of the Ultimate Frisbee team, Kelly dove through the air in an arc to deflect the disk inches from an opponent's grasp in a full-on layout onto solid concrete. She intercepted the pass in midair and tipped the wallet into the palm of my outstretched right hand. Reading the eyes of an accomplice fixating on my BlackBerry spinning on the pavement, she lunged again and beat him to the prize.

With my wallet retrieved and realizing that I was horribly outnumbered, I released the thief and bounced onto the balls of my feet as he scurried backward on his butt like an injured crab escaping. "Crazy man!" he gasped.

Looking into the eyes of the half-dozen muscular thugs surrounding me, I tried to discern if he choked out those parting words to deflect suspicion or if he meant it as a threat.

Now what?

A massive surge of adrenaline fueled my twitching muscles and nerves to levels of raw power I had never felt before. I was now struggling not to pick up the next hoodlum squaring off with me, hoist him over my head, and hurl him into his accomplices, knocking them down the steps of the Metro station like bowling pins. It was not a question of whether I could execute the superhuman feat. I had no doubt that I could do it. Rather, I was trying *not* to do it, simply because this might not be my best option. At least, not yet.

Suddenly a middle-aged, well-dressed Spaniard stepped casually between me and the attackers and with flicking shooing motions of his fingers he said, "He no crazy—go." Without breaking stride he descended the steps into the Metro station. As he passed me he smiled and said, "*Bueno*—good—go now." In passing he had defused the situation to its best possible outcome: a draw. The band of robbers scattered into the Metro station like rats down a sewer, leaving my daughter and me standing there stunned, my wallet clenched in a death grip in my right hand.

Unfortunately, that was not the end of it. The gang pursued Kelly and me throughout the city for the next two hours. They were not after my wallet anymore. I had humiliated and beaten up a member of their gang. They wanted revenge.

We tried every trick to elude them—fleeing into tourist shops and through noisy restaurants, cutting through back alleys, abruptly crossing streets to reverse course, changing clothing, and when they got too close, leaving the sidewalk and running down the center of the boulevard, weaving through oncoming cars. At one point we stopped traffic to jump into a taxi in the middle of a three-lane boulevard, but they had cell phones and wherever we went they sent increasingly menacing tattooed thugs with steroid-bloated biceps to intercept us. As we dodged the gang of robbers, we witnessed them casually pick wallets from two more tourists. I even snapped photographs of them doing it—a stupid mistake, as it turned out, because their lookout on my side of the street caught me doing it. The unshaven goon came running up the sidewalk jabbering Russian into a cell phone and extending a video camera in an unconvincing attempt to pass as a tourist. We fled down a side street. As they closed in on us, we were forced to jump into a taxi and escape to a small town an hour away.

In the cab, my daughter asked in a tone filled with shock and disbelief, "Where did you learn to do that? I looked over and saw you swinging some guy around by the neck. I couldn't figure out what was happening."

I laughed in nervous relief.

A 170-euro cab fare later, we were broke but safe.

Now my daughter is convinced that my job at the National Institutes of Health is just a cover for my real job as a spy.

My daughter's and my experience in Barcelona with the pickpocket gang was the inspiration for this book. Where, indeed, had I learned to do what I had done? How, with the lightning-quick reflexes of someone snatching a fumbled coin from the air, had I unleashed such a flurry of moves on my attacker without conscious thought? Had I contemplated the situation, I never would have attempted such a thing. No amount of money is worth being injured or killed. I could have been kicked in the head by the gang as I struggled with their comrade on the ground and left to die brain damaged and in a coma. Or they could have easily held my daughter at knifepoint and used her as ransom. I never imagined that I would or could react this way. Yet it worked.

One fifty-six-year-old tourist and his daughter had defeated a gang of criminals.

This unconscious explosion of violence to protect my daughter and myself is the same behavior that is triggered inappropriately in so many everyday instances of sudden and regrettable violence. We need to understand this unconscious neural circuitry and recognize what trips it.

In the aftermath of the attack I found myself wondering: Does everyone have this latent unconscious ability for rage waiting to be unleashed, or is this relatively uncommon? Would another person react differently—rather than fight, become a helpless victim, run, or negotiate? Why? Which strategy is the best in which situation? Would I always react this way if caught unawares in similar circumstances? Whatever our response might be to sudden threats like this one, is there any way to control it? Can the inherently meek, if they exist, be taught to fight back (to overcome or ignore their meekness, as it were), or are our individual reactions to such threatening situations preprogrammed?

As a neuroscientist, I wonder how this unconscious reflex was possible. Without even seeing the robber in my peripheral vision I had snared the person by the neck and thrown him to the ground. How had my brain taken in all that information while my attention was fully occupied by the enormous challenges of negotiating my way through a strange new environment? Somewhere deep in my brain I must have been taking in this situational information unconsciously and guarding against the threat to which I was consciously oblivious. It had been a blind snatch. I did not really know who I had grabbed by the neck until the instant I saw my wallet being tossed to his fellow gang member, but the fact is, I found myself on the ground not with an eighty-year-old lady innocently walking in my blind spot: It was indeed the bad guy.

Until I found myself on concrete in combat, I had simply witnessed my violent reflexes unleash themselves. Clearly, in many dangerous situations the process of conscious thought would be too slow. As when recoiling from a hot stove, our unconscious protective reflex kicks in before we feel the burn consciously, revealing a deeply embedded automatic lifesaving reaction that through millions of years of evolutionary struggle has been preprogrammed into our DNA. But snatching your hand back from a burning-hot stove is a simple reflex. Responding appropriately to a perceived threat within a fluid social environment is far more complex—yet your life or the life of friends and family will depend on executing the proper split-second reflex for either rage or retreat. The rage reflex can unleash furious, uncontrollable anger or, as in this instance, trigger a rage of intense and purposeful violence devoid of anger. In either case the response is automatic and apparently unchecked by rational thought.

The power of rage gives a petite woman strength to lift a car off the ground to free a trapped child. It is the stuff that drives a US Marine, 180 degrees against all normal instinct, to run into a hail of bullets to save a comrade in jeopardy. But sometimes this automatic lifesaving rage reflex embedded in our brain by evolution clashes against the modern world. "I just snapped," the remorseful man confesses tearfully after having strangled his girlfriend in a fit of rage. Rage can ignite a crowd, resulting in sudden mob violence. The triggers can be small or large, individual or collective. The results can be devastating.

We must understand the biology of how the animal instinct inside us works in order to appreciate how rage arises. We must learn to control rage if possible, and to exploit it when necessary to save our lives. How much of this propensity toward rage is genetically predetermined and how much is learned? Precisely what is it in any given situation, and in an individual's personal history, that will trigger rage? Would I have reacted the same way to the pickpocket had I been traveling alone rather than with my daughter? Does the tendency to unleash rage reside latent in everyone, or is it only programmed into a few? How do men and women differ with respect to rage triggers? As individuals and societies we need to examine the beast within us and confront, in the context of modern society, the biological roots of rage.

## Losing It

"We like it here," the woman said. Her shaggy brown-and-white mongrel stretched its leash to sniff a tree. "There are some older couples—a lot of older couples—some families with teenage kids and some with younger-aged kids like mine."

Her two daughters, one with bright-blue extensions woven into her curly black hair, both of them in dresses, fidgeted behind her.

"It's very diverse ethnically and culturally," she continued. "I like that."

Earlier, one of the woman's neighbors, sixty-seven-year-old Ray Alfred

Young, emerged from his home. As he latched the front door of his tidy brown-brick townhome, he could not have imagined that he would not be returning.

Dappled sunlight splashed the windshield of his Toyota Corolla as he drove through the quiet streets lined with towering red oaks shading residents from the blazing July heat. Retired after thirty-seven years of service to the federal government in the Department of Labor, Young was heading to the US post office two and a half miles from his home in the Dumont Oaks neighborhood of Silver Spring, Maryland. Completely bald, Young had battled throat cancer, losing his teeth and part of his jaw to the disease, but he remained active in his church and community.

"He is the kind of neighbor who checked your mail when you were away and checked your house while you were on vacation," said Mary Anne Darling, a special-education teacher who lived next door to Young for the past ten years.

"He was always very polite to me," said Evan Schluederberg, another neighbor commenting after the bloody events.

The post office on the corner of New Hampshire Avenue always seemed busy. The line of customers often stretched nearly to the door, and indeed there was a line as Young entered. He stood waiting his turn with the rest of the afternoon crowd when he saw a man cut in line. In fact, a postal worker had directed the person away from the counter to complete his paperwork and then motioned him back.

Stunned witnesses described what happened next. Young waited in the vestibule of the post office for the fifty-eight-year-old man who he thought had cut in line. He started yelling and arguing with the man; then he reached into his pocket and pulled out a butterfly knife and began stabbing the man over and over. Young sank its four-inch blade into the man's chest two times and then two more times into the man's left shoulder blade. The victim fought barehanded to defend himself from the deadly blade, suffering slashes to his forearm and biceps.

Two female postal employees rushed to the fight and blasted Young with pepper spray. Young recoiled, his eyes and nose burning, before escaping outside, where he darted to his car and retrieved a baseball-bat-sized club from the trunk—then stormed back to beat the victim, who was bleeding profusely and suffering serious injury. Shocked bystanders shouted at Young, warning that police were on their way. Young retreated to his Toyota Corolla and drove off.

Police stopped the car at an intersection less than a mile from the shady streets of Dumont Oaks. Quickly the scene was surrounded by multiple police cars. Young gave up without a fight. Soon he sat on the curb in his yellow shirt and khakis, his hands cuffed behind his back, his chin sunk to his chest, and his eyes still smarting from the pepper spray while officers searched his car.

"I've never seen him be aggressive or volatile in any way," said Darling after Young's arrest for attempted murder.

"He seemed just like a standard, normal dude," his neighbor Schluederberg reflected. No one, not even Young's lawyer, claimed that Young was mentally ill.

"He's as stable a citizen as you could ask for," Young's defense attorney, Gary Courtois, told Judge Eugene Wolfe at Young's bond hearing. Young, dressed in a blue prison jumpsuit, appeared in Maryland District Court via closed-circuit TV. He was unable to see his brother and son seated in wooden pews behind the thick glass knee wall separating spectators from the attorneys' oak desks. Seated at the imposing bench in his black robe, Judge Wolfe read through documents in a bright-yellow file folder. According to Montgomery County police, Young had no previous history of violence.

The judge, a scholarly-looking black man with graying hair, denied the request to reduce the $500,000 bond. Looking up at Young's defense attorney through his silver wire-rimmed glasses he said, "I am not willing at this stage to put him back out on the street knowing what I do about this case."

How can this violent break with reality be explained? A sixty-seven-year-old man suddenly departing from a life of civility, snapping uncontrollably in an instant and viciously attacking and attempting to murder a man who was a decade younger than him because he thought, mistakenly, that the man had cut in line. It made no sense; not even to the attacker afterward. "Maybe I should have stayed home. I'm too old for this stuff," Young said soon after he was arrested.

We tend to ignore this subject of snapping violently not only because we have become numbed to it but because this hostile behavior is so disconcerting. On a societal and personal level, we are all too familiar with this seemingly irrational rage response. No matter where you live, the daily papers and news media are filled with similar instances where "normal" law-abiding individuals with no history of violence suddenly "snap" and attack violently. Often the rage is triggered inexplicably by the slightest provocation. There are countless horrific examples from the national news, but the incident I just described happened at my local post office, only two miles from my home. This shows just how pervasive and commonplace rage attacks are.

This behavior is so common on the highways we have a name for it: road rage. Also from my local newspaper: A thirty-two-year-old woman from the Alexandria area of Fairfax County was arrested Monday night and charged in what authorities described as a road-rage incident in which a woman was killed. The woman was charged with felony hit-and-run in the incident, in which she ran over and crushed to death a twenty-one-year-old woman after a verbal altercation broke out and matters escalated. "Never get out of your vehicle," Officer Don Gotthardt, a county police spokesman said. The woman surrendered herself later that night at a police station.

This rage response is not exclusive to men or to women, or to the aged or young. It is inexcusable and perplexing, but not incomprehensible. These incidents of snapping in rage merely *seem* incomprehensible because we have become numbed to them and we avoid the subject. On a certain level we would rather dismiss snapping as pathology, but such thinking is wrong. Psychopathic homicides, which are driven by mental illness, grab attention because these acts are rare. The commonplace blind rage attacks between spouses, coworkers, and complete strangers are the cause of far more aggression and violence than that caused by the mentally ill or psychopathic killer.

Two-year-old Angelyn was on an outing with her extended family at Tysons Corner Center shopping mall in Fairfax, Virginia, on November 29, 2010. Mall surveillance cameras captured the unimaginable act of rage at this mall near my home.

"It's chilling now, seeing it and knowing what's going to happen," attorney Raymond Morrogh said during closing arguments. "She was intent on completing this horrible act."

In the footage, the family, including Angelyn's diminutive fifty-one-year-old grandmother Carmela dela Rosa, can be seen leaving the mall and exiting across an elevated pedestrian bridge. Suddenly dela Rosa sweeps

Angelyn up in her arms and throws her over the edge. Shoppers saw Angelyn hit the ground forty-four feet below.

"It looked like a jacket had fallen over the edge of the bridge," one witness described. Another witness driving by said she saw it out of the corner of her eye and thought at first that it was a bird.

The girl's mother and the rest of the family ran down six flights of stairs to Angelyn. Carmela dela Rosa stayed on the bridge, leaning over the edge and watching the horror she unleashed on her family unfolding below.

Kathlyn Ogdoc, dela Rosa's daughter and Angelyn's mother, testified at trial how the last thing she heard on the ambulance ride to the hospital was her daughter crying before the EMTs sedated her child. Angelyn died at Fairfax Hospital twelve hours later.

"She'll never be able to go to kindergarten. I can't teach her how to put makeup on. We can't teach her how to drive. She won't be able to grow up," Angelyn's grief-stricken mother testified at trial against her own mother.

In statements to police, dela Rosa said she threw the toddler over the edge of the walkway because she was angry at James Ogdoc, her son-in-law, for conceiving Angelyn with her nineteen-year-old daughter out of wedlock, preventing Kathlyn from meeting new people and exploring the world. That anger grew into an all-consuming rage.

The jury rejected the defense's plea of Not Guilty by Reason of Insanity. Psychologist Dr. Stanton Samenow, who evaluated dela Rosa while she was in prison, said that the hatred she harbored for her son-in-law never subsided.

"She was basically angry at the world and her place in it. . . . She decided to take it out on the child," prosecutor Morrogh concluded.

Samenow testified that dela Rosa, in interviews with Fairfax County police in the hours immediately after the crime, didn't inquire once about the condition of the child. Instead she asked, "What's going to happen to me next?" and later, "Is this going to be out in the public?"

After an anguished trial that could have no real winners, dela Rosa was convicted of first-degree murder. Judge and jury rejected the lesser charge of second-degree murder, concluding that the act had been committed with malice.

"I'm very sorry for what I've done. I'm sorry to James and Kat," dela Rosa said in court, referring to her granddaughter's parents before her sentence was delivered.

"That a grandmother would do something like this to a granddaughter is almost incomprehensible," Judge Bruce White said before sentencing the woman to thirty-five years in prison.

It would be comforting to dismiss this tragedy as an aberration of mental illness rather than what the jury concluded it was: a violent act of fury. Morrogh observed, "It is a natural instinct to want to think she was insane when she did it because it's easier to admit than acknowledge evil exists in ordinary families."

"This isn't insanity," he said. "This is depravity."

Human behavior, once the domain of education, religion, and psychology, is now becoming comprehensible as neurobiological science delves deeper into the circuitry of the brain. In the late 1920s, Walter Rudolf Hess, an ophthalmologist in Zurich, Switzerland, who gave up his practice to conduct brain research, surgically inserted a fine-gauge wire deep into the brain of a cat. After allowing the cat to recover from anesthesia Hess flipped a switch to energize the electrode. Sometimes this caused the cat to cock its head or twitch its muscles. Probing deep into the middle of the brain (the diencephalon), Hess found that electrical stimulation of specific spots in that area evoked changes in "autonomic" brain functions, which maintain the internal stability of the body such as heart rate, dilation of blood vessels, and respiration. While he was probing this region of the hypothalamus, the cat suddenly sprang into vicious attack mode. Its eyes dilated, its hair stood on end, its back arched, its teeth and claws were bared, and it began snarling and striking out at the nearest object.

Using this method in the 1960s, John P. Flynn and his colleagues at Yale University added one more critical element to the experiment: a rat. When Flynn activated the same spot in the hypothalamus, the docile laboratory cat subject, indifferent to the rat in its cage, suddenly became enraged as Hess had shown, but the rage was directed *toward* the rat, which it killed immediately. Stimulating this spot had not simply evoked a ragelike emotional state; the cat's fury was directed toward another living creature. This was the same type of behavior triggered in sixty-seven-year-old Ray Young. Somehow the line-cutting incident at the post office tripped this circuit deep in the unconscious part of his brain. Driven by these deep-brain neurons in his hypothalamus, Young was forced into actions focused irrevocably on killing.

Flynn also found something remarkable that Hess's experiments done without a rat in the cage would have failed to detect: If he moved the electrode very slightly to stimulate a nearby spot in the hypothalamus, the cat suddenly sprang into a quiet, stealthy attack, without exhibiting rage or anger. Instead, the cat coldly and meticulously stalked and attacked a test object or killed a prey animal placed in its cage. The cat would cease the attack, whether cool-tempered or enraged, when the investigator stopped stimulating the electrode.

Unlike cocking the head or twitching a muscle, these attacks are enormously complex behaviors, tightly focused on another animal in the environment, involving the release of many hormones, profound alterations in sensory and motor function, the initiation of complex behaviors to kill, and overwhelming the cognitive ability of the animal to any purpose other than attack. All of this somehow linked to that tiny knot of neurons at the core of the brain where conscious thought cannot penetrate: the beast within.

## Misfiring Neural Circuits of Violence

Releasing the rage response to defend against attack or to obtain food in the wild makes sense, but murdering your granddaughter on an impulse to spite your son-in-law? Running over a twenty-one-year-old woman because of her annoying driving? Knifing a man to death—for the man surely would have died if the postal workers had not intervened—because he appeared to cut in line?

Such furious reactions might seem like aberrations, but these acts are epidemic in modern society. Add to these extremely violent reactions the commonplace experience of witnessing someone losing it at an airline counter or at a bar or in a car on a crowded highway, and the need to understand how the rage circuit works becomes obvious. This book offers a new understanding based on the latest research in my and others' labs. In the following chapters we will uncover the brain circuitry responsible for the rage response and explore how these circuits are tripped and consider whether modern society inadvertently causes misfires of the rage response embedded in our brains.

However misguided my reflexive response to the pickpocket in Barcelona was, the underlying reasons that it happened are not hard to see. The loss of all our money in a foreign country would have left my daughter and me without means of obtaining food, lodging, or transportation. But the point is that the "decision" to trip this reflex was made automatically and unconsciously in a fraction of a second. It had to be so. But the snap reflex means that certain stimuli in the environment of which I was not consciously aware had tripped the circuits in my hypothalamus to compel me to risk my life in a street fight.

So too in each of the examples I've given thus far can we see the factors that triggered the rage responses. The grandmother's actions, while horrible and misguided, were motivated to protect her daughter's welfare. Road rage is understandable from the perspective of protecting one's own territory. This urge to protect our territory is critical for life in the wild, as in modern times, but whether it is an intrusion into the space surrounding our vehicle or an airline seat tipped back into our row, the question is whether the sudden explosion of rage to physically engage in combat with the intruder is rational, or instead a misfiring of automated brain circuits of threat detection. Do sensory stimuli while driving a car trigger that violent urge irrationally because our brain evolved millions of years before man invented the wheel or middle seats on airplanes? Ray Young acted out of a motivation to maintain public order. All social animals must do so to some degree, or there is no social structure, but at some point the situation tripped a self-protective response as if he were in a mortal fight with an aggressor.

"He must have felt threatened," Attorney Courtois told the judge at Young's bond hearing, trying to account for how a peaceful, sane retired person could have suddenly attempted to murder a stranger in the post office.

"I was defending myself," Young explained, giving his distorted perspective. "His wife or someone pepper-sprayed me."

Understanding how the rage response circuit in the hypothalamus becomes activated helps explain why some cases of extreme violence occur, but it does not excuse them. Sometimes these are criminal and profoundly cruel acts, but none of the ones I've recounted was the result of insanity, or even the result of conscious reasoning. This explains the apparent paradox of a seemingly stable and nonviolent individual snapping suddenly and committing horrible violence.

This violence is fundamentally different from the brain circuitry underlying such crimes as assault and robbery. These and other "scheming" crimes originate in the cerebral cortex, where consciousness arises. The hypothalamic rage response is also distinct from the neurobiology underpinning the mass murders committed by someone like Jared Loughner, who gravely wounded congresswoman Gabrielle Giffords outside a Tucson, Arizona, shopping center on January 8, 2011. Seen grinning in his mug shot, Loughner appeared mentally deranged. Adam Lanza, who sickeningly massacred schoolchildren at Sandy Hook Elementary School, as well as his own mother, was by all accounts a twisted, mentally ill person who was receiving psychiatric treatment and medications long before the attack. These are broken minds. Understanding the criminal behaviors committed by insane individuals is currently impossible, even as we endeavor to heal them. Such insanity-driven murder is not within the scope of this book.

A gruesome mass murderer in Norway, Anders Behring Breivik, who massacred seventy-seven innocent students and bystanders in his far right-wing outrage over immigration, is a case of rage triggered by social conditions and politics. Psychiatrists and Norwegian courts ruled that Breivik was not insane, but rather his mass slaughter of innocents was driven by a righteous sense of rage. Indeed, Breivik's greatest fear was that the court would judge him insane. For Breivik, the guilty verdict was a triumph, which he greeted with a smirk and a clenched-fist salute to the court. He then apologized at his sentencing for having not killed more people, but Judge Wenche Arntzen abruptly cut him off before sentencing him to prison for twenty-one years, the maximum possible under Norway's law. This was a crime of rage.

Could it be that subconscious neural circuitry and behaviors designed for human life in the distant past are susceptible to triggering violence in everyone? Could these circuits of rage be prone to misfiring in response to stresses and encounters in modern life that did not exist when these survival mechanisms were laid down over millions of years of evolutionary struggle? I believe the answer is yes.

While it is possible to see the neurobiological origins of Ray Young's rage response, for example, not everyone in the post office responded the way that he did to the same situation. Why? Are certain people more prone to snapping in this way? Indeed, there is evidence to support the conclusion that the suppressors and triggers of the rage response differ in different people. Both genetic factors and environmental experience contribute to how easily one reacts and which of many different types of stimuli provoke the rage response. Many forms of rage are recognized in people and in animals, including territorial aggression; dominance; and predatory, defensive, sexual, social, and disciplinary rage. Each one emerges in response to very different yet specific sets of stimuli playing against different internal patterns of brain activity in the individual that ultimately link to the hypothalamus. New research is showing that differences in genes that control certain neurotransmitters and hormones lower the threshold for aggressive behavior, reduce fear, and increase impulsivity in some people.

On the other hand, Ray Young had no prior history of violence or hostility. This must mean that rather than being a genetic predisposition, a complex set of factors in his body and stimuli in the environment combined to release the ordinary inhibitions on the deadly explosion of aggression. Could factors in Young's recent experience or in his background have intersected with the sensory input transmitted during the unexpected occurrence of a man cutting in line so as to impinge upon the hypothalamus and trigger the rage response circuitry?

Consider some subtle details: It was not a penknife that Young drew out of his pocket. It was a butterfly knife. The weapon, originating in Asia for defense and knife fighting, has a blade concealed within a split handle, so that it can be rapidly deployed single-handedly with the flick of a wrist. A second weapon (a club) was stashed in the trunk of Young's silver Toyota Corolla.

At the preliminary hearing the defendant, clad in prison garb, appeared in the courtroom on closed-circuit TV from the jailhouse. Young is a big man. Despite his age and his ongoing battle with cancer, Young retained his imposing 190-pound burly longshoreman's physique. He stood behind the wooden podium calmly with his hands stacked one upon the other, fingers to wrist, palms down, and his elbows jutting out symmetrically and motionless throughout the entire proceeding. His statuesque posture evoked the Buddha in lotus position, but with palms down for strength and grounding rather than upward in reverent meditation. Young's presence on courtroom TV appeared to be a case of bad casting in contrast to all the other defendants who preceded him. Their loose-jointed "gangster" walk and fidgety, irreverent body language on camera portrayed them as battle-tested gang members, drug pushers, and robbers in a B-grade cops-and-robbers movie. One-third of them relied on a foreign-language translator to communicate with the judge.

"He's as stable a citizen as you could ask for," I recalled Young's defense attorney, Gary Courtois, telling Judge Eugene Wolfe previously. To me he did indeed seem "just like a standard, normal dude," as his neighbor had stated.

Has the hectic pace and stress of the twenty-first century put these explosive behaviors on a hair trigger that can inadvertently discharge as a result of constant bombardment of the senses by culture, the media, feelings of anxiety, alienation, and the stresses of modern life?

The lifesaving circuits of rage engraved in our brain by evolution now clash with the transformational changes in the modern environment. Humans are an organism evolved from the open plains of Africa. These neural circuits were formed and tuned for survival in the natural world. To understand the paradoxical everyday human tragedies that fill the daily papers, you must add to this situation the availability of drugs—both legal psychotropic drugs for treating mental illness and behavioral problems, and such things as steroids for treating disease, as well as powerful street drugs that the human brain never encountered prior to the last few decades—and we see the human nervous system grappling with an internal and external environment that it was never designed to handle. Finally, add on the most dangerous component, without which such aberrations in the past would have been relatively benign behavioral oddities: the availability of dangerous new implements of destruction, including automobiles, sophisticated knives, propaganda via the Internet, and firearms that can amplify the power of an individual to levels well beyond the ability of the strongest and bravest among us to combat with bare hands. None of these current realities of human life existed during the millions of years it took to evolve the human brain.

Coded into our DNA through eons of battle for survival of the fittest, the circuits of aggression reside latent in the hypothalamus of everyone. New methods and new information are revealing how they work. Let's take a closer look at these circuits, the biological roots of rage.

# 2

# Neurocircuits of Rage

I'm enormously interested to see where neuroscience can take us in understanding these complexities of the human brain and how it works, but I do think there may be limits in terms of what science can tell us about the meaning of good and evil . . .

**Francis Collins, email to author on May 21, 2015**

There is a beguiling popular notion that a "reptilian brain" lies at the core of the human brain. Its cold, lizardly logic governs the most basic survival functions of life; among these are feeding, sexual behavior, and self-defense. Aggression, dominance, territoriality, and ritual displays are purportedly governed by this neural tissue, a vestige of our long-distant reptilian ancestors lurking inside us. The violent outbursts and unconscious reflex to attack or defend to the death are programmed by the automatic "doomsday" neural circuits in our "lizard brain."

Layered on top of this primal lizard brain is purportedly an early-mammal brain. "If you imagine the lizard brain as a single-scoop ice cream cone, the way you make a mouse brain out of a lizard brain . . . [is] to put a second scoop on top of the first scoop," writes neuroscientist David Linden in his excellent book *The Accidental Mind*. This second scoop of the neural ice cream cone is a more refined and complex gelato in contrast to the plain vanilla first scoop of the reptilian brain.

Superimposed on top of that second scoop is the modern primate brain.

This is the superior outermost layer of neural tissue (neocortex) found in primates and to a lesser extent in other higher mammals. This cerebral mantle provides us with higher-level cognitive ability, abstraction, tool-making mastery, language, self-awareness, and the capacity to control the reptilian urges within ourselves and in our civilizations. This hot-fudge topping on the ice cream cone (cerebral cortex) is what makes us human. An essential aspect of this "triune brain" theory is that these three brains compete with one another, explaining how someone can be overtaken suddenly by irrational impulses or rage. Those with weakened "mammalian brains" can commit acts of violence and crime propelled by unchecked primitive urges.

This colorful confection, however, is simply not true. You no more have a lizard brain curled up inside your skull than you have lizard scales armoring your skin. The popular idea of the triune brain originated back in the 1960s, a period before the word "neuroscience" even existed. Our understanding of how the brain worked was abysmally primitive back when TV was just progressing from black-and-white pictures to color. The first neurobiology department in the country began at Harvard University in 1966. At that time anatomy, physiology, psychiatry, and neurology all dealt with brain function, but each was considered a different area of science. The concept that brain science could be a separate and coherent scientific discipline distinct from all others was novel. Psychiatrist Dr. Paul MacLean conceived the triune brain theory based on little more than stitching together comparative anatomy and evolution into a Frankenstein scheme.

As a psychiatrist working at Yale University in the late 1940s, Paul MacLean had become interested in the brain's control of emotion and behavior. Like Flynn and Hess before him, MacLean used electrodes to stimulate different parts of the brain in conscious animals to induce aggression and sexual arousal. Based on these experiments he pinpointed the center of emotion in the brain as the limbic system. The limbic system is located at the center of the brain and includes the hippocampus, amygdala, and other structures. MacLean proposed that the limbic system had evolved to control the fight-or-flight response to danger and to react to emotionally pleasant and painful sensations. This is indeed one of the prime functions of the limbic system, but it also participates in several other functions.

In the 1960s MacLean proposed that an even more primitive center, the reptilian brain, secluded at the core of the human brain, controlled the most basic bodily functions such as breathing. The lizard brain, he argued, resided in what anatomists term the brain stem, which is where the brain begins to swell from the top of the spinal cord inside the skull. These functions of the brain stem were never in doubt. The anatomy of the brain stem—that is, the connections leading into and out of it—and the immediate lethal result of damaging it, revealed its function vividly. But it was new to refer to this part of the human brain as the brain of a lizard.

MacLean proposed that the outermost layer of the brain, the neocortex, was the crowning achievement of evolution and thereby responsible for reasoning and all that is uniquely human. This is not a matter of scientific debate, but the idea that the human brain was formed by evolutionary accretion, beginning not with fish or amphibians but with the age of reptiles, bypassing birds, skipping to mammals, and then ascending to nonhuman primates and humans was a novel perspective. The human brain and mind, according to MacLean's view, were revealed much like layers of past civilizations uncovered in an archeological dig.

To neuroscientists the triune brain was nothing more than an attempt to conceptualize, simplify, and popularize some general aspects about the brain's neuroanatomical framework, but nonscientific audiences and the media latched on to the concept of the triune brain as though it were a neurological truth, a Rosetta stone to understanding human behavior and emotion in terms of neural circuitry. Never mind that popularizers of the triune brain frequently managed to confuse what the three brain parts even were, most often referring to the "second brain layer" emotional limbic system in the human brain as the reptilian brain (brain stem) at the brain's core that actually controls breathing and other automatic bodily functions. The pervasive error is seductive, because raging dinosaurs and angry crocodiles snapping into vicious attack are compelling allegorical images for primitive human emotions unleashed.

Still, the public and the media latched on to the concept, and the pseudoscientific assertion of the triune brain has entered into common usage and suffused our culture.

From *Dexter*, a TV series about a serial killer:

**DEB:** *How do you know he'll kill again?*

**DEXTER:** *An alarm is going off inside my lizard brain.*

"The greatest language barrier," MacLean wrote in a *New York Times* article in 1971, "lies between man and his animal brains; the neural machinery does not exist for intercommunication. . . ."

Nothing could be further from reality. But it is easy to see how this mistake arises from the dramatic and complex aggressive behaviors that can be triggered by using an electrode to stimulate an appropriate spot in the brain. The problem here is that the question motivating the brain-stimulation experiments is itself faulty. A search made by probing electrodes into brain tissue carries a hidden assumption that there *is* such a cluster of neurons somewhere inside the brain that can overrule free will and take command of primitive human behaviors and emotions. But the fact that rage can be triggered by stimulating a particular spot in the brain does not necessarily lead to this conclusion any more than concluding that the power that propels an automobile engine comes from its electric battery. Although you can launch the complex systems of the automobile engine into action by closing the circuit on the battery, you would be mistaken to conclude that the battery itself provides the automobile with the force to move. Delivering an electrical shock to a particular spot in the brain may simply tap into circuitry in a complex network that can span large areas of the brain that affect multiple systems necessary for the behavior.

The human brain is a unified, highly interconnected, complex system. It is distinct in countless ways from the brain of any other animal, just as the brain of each species has evolved independently through its own line of ancestry. New techniques and modern thinking recognize that brain function typically involves widespread communication through circuits spanning many different neural systems and regions of the brain. Some vital aspects of this circuitry may be more heavily concentrated in particular spots in the brain, but this does not necessarily mean that a particular brain region is "responsible" for generating the behavior that is evoked by stimulating it. As cognitive tasks become more complex, more areas of the brain become linked into operation to produce the behavior. In contrast to the lizard brain explanation for human rage, aggression, and defense, the reality is a bit more complicated, and more interesting.

## Losing the Lizard Tale

When the sneaky pickpocket's fingers brushed my leg in Barcelona, an electrical signal shot up my nerves to my lizard brain and it instantly sprang into action to attack the threat. The signals never reached the mammalian chocolate-fudge syrup cortex of my ice-cream sundae brain, so the response was entirely automatic and unconscious.

Hardly.

The nerve endings in my leg are not wired directly to my lizard brain in such a way that, much like laboratory electrodes stimulating the hypothalamus of a cat, the touch of my leg triggered the lightning-fast jujitsu moves that enabled me to ensnare the bandit by his head and throw him to the ground before any signals had reached my conscious awareness in the hot-fudge sundae of my cerebral cortex. No. A quick look at the wiring diagram of the human nervous system shows that nerve endings in the skin connect to the spinal cord, not to the hypothalamus.

You've no doubt heard of the fight-or-flight response, that adrenaline rush that boosts heart rate and sets muscles twitching to prepare you to run for your life or fight to the death. Indeed, this physiological response was launched in my body as well as in my daughter's during my fight with the pickpockets and during our subsequent chase through the Catalonian city. The enormous strength that infused my body when I fought the urge to pick up the leader of the gang and hurl him into his accomplices flowed directly from this lifesaving physiological response. The fight-or-flight response gave us the strength and mental sharpness to escape our pursuers with ceaseless energy during the two hours we eluded the gang hunting us down like prey, and it gave us heightened mental power and intense focus to outwit them, but the fight-or-flight response cannot explain what triggered my throwing the robber to the ground in a split second. In the milliseconds from the time I felt a tug at my pocket, I had grabbed him by his neck. The fight-or-flight response is a hormonal response, driven by an injection into the bloodstream of adrenaline (epinephrine), which is secreted from the adrenal glands attached to the kidneys. The power of adrenaline that is released in a life-threatening situation is transformative, elevating the body and mind to unparalleled levels of performance, but any process that is initiated by a substance injected into the bloodstream is far too slow to explain how I had reacted.

Are the nerve endings in my leg endowed with the ability to discern the light touch of a pickpocket's fingers as a personal attack requiring an immediate violent protective response? How could nerve endings in my skin possibly be equipped with the logic and wisdom to decide to launch an attack on a villain that could potentially risk my life in combat in a Barcelona street fight? Clearly the lizard-brain tale of how humans react in rage is fantasy.

So what did happen? And what happens inside the minds of those overtaken by violent criminal attacks when afterward the bloody perpetrator offers the bewildering, pitiful explanation: "I just snapped"?

## Toward the New Neuroanatomy

To begin to understand what happened to me in Barcelona and to countless other people every day, you need to leave the outdated notions of the 1940s–60s behind and enter the twenty-first century of neuroscience. Innovative techniques are uncovering a wealth of information and understanding, little of which has reached the public or even college textbooks, let alone courts of law. In preparation for this new research in later chapters, a basic foundation in neuroanatomy will be helpful.

If you are already familiar with the anatomy of the human brain, this section can be skipped entirely. Neurosurgeons, for example, will have no problem understanding descriptions of brain anatomy as we consider the new research on how the brain detects and responds to threat. The names for all the anatomical structures (usually in Latin) will roll off their tongue with ease, but for others the strange names, such as *precentral gyrus*, *insula*, and *parahippocampal gyrus*, are tongue twisters—as meaningless as the names of villages and cities in an unfamiliar foreign country. As with learning any language or reading a map, meaning develops with familiarity and usage as information, interrelationships, and associations with various names and places accumulate. But there are shortcuts.

Think of a map of the human cerebral cortex (the surface of the brain) as if it were a map of the United States. Imagine looking at your brain in profile, as if viewed through your skull by someone standing on your left side. Now envision sketching out your brain in profile as a simple two-dimensional map. The front of the brain would be on the left (west), and the back of your brain would be on the right (east).

Different spots in the brain are specialized for different functions, just as

New York City is the financial center of the country and Hollywood is the center of movie production. So it is with the brain. The visual cortex, which would be concentrated around New York City (the back of the brain), is the brain center where visual information is processed. Input from the eyes is relayed through a series of connections to this area.

Keep in mind that much the way New York City is, at least to many, both the literary and financial center of the country, individual brain regions can and often do participate in processing different kinds of information and controlling behaviors for more than one distinct function.

The prefrontal cortex, which would be located near San Francisco (the front of the brain), controls higher-level cognitive functions such as attention, decision-making, and other "executive" functions that require integrating multiple kinds of sensory input and various types of information from many other brain regions.

Just as New York City is the financial center of the United States, what makes it so is the broad network of connections it has to other parts of the country. Wall Street cannot function in isolation, and the same is true of the brain. Different regions of the brain are specialized for different functions, but the pattern of connections with other brain regions is what makes each brain center work as part of a system. The brain's anatomy must be viewed as a schematic, just like an electronic circuit diagram or a road map.

Thus, there are names for major "roads" through the brain as well as "cities." (Cities in this analogy are concentrations of neurons, and roads are the connections between them over which electrical impulses are sent.) It is important to keep in mind when interpreting experiments on the brain that the operation of the New York Stock Exchange can be disrupted either by a local problem in New York City or by a break in the network of connections between Wall Street and financial centers around the country. Roadways in the brain are called *tracts*. These are bundles of nerve fibers (axons) connecting groups of neurons across distant regions of the brain.

Brain tissue is not homogenous like putty; it is filled with lumps and holes and layers of cells in a characteristic anatomical arrangement. Tight clusters of neurons are called *nuclei*; for example, the *nucleus accumbens*, an important relay point in the brain's reward circuit, gives us the sensation of pleasure and achievement. The surface of the human brain is highly wrinkled and folded, a consequence of the great expansion in surface area of the human cerebral cortex over the course of our evolution. A hill between two furrows on the cerebral cortex is called a *gyrus*, and most of them are named, just as hills and valleys have names on a topographic map.

Oftentimes it is useful to refer to a larger region of a geographical area rather than individual cities—for example, the West Coast, or the Deep South. The geographical regions have distinct characteristics because of their local features and their relationships with other parts of the country. The Pacific Northwest is rainy; the Deep South is humid; there are "grain belts" and "rust belts" and politically defined "red" and "blue" regions; all of them extremely useful in discussing and understanding the country and how it operates. Neuroanatomy is the same, and the concept of different brain regions is no more complicated than that. For example, the *occipital* lobe of the brain (Latin for "back of the head"), would be the region east of the Mississippi. The *temporal* region would be located around Texas and New Mexico. I'd previously mentioned the function the occipital region is noted for—vision. The temporal region is important for memory and auditory processing.

As you might expect, there are strong networks of connections from the temporal lobe to the prefrontal cortex to draw upon information stored in memory to inform the higher-level executive functions carried out in the frontal lobes. Just as in most transportation and information systems, the connections between regions typically go in both directions. Consider, for example, that a memory must become associated with other memories and contain multiple sensory modalities (sights, smells, sounds, emotions). These must become blended together to give any memory its context and meaning. It is information returning from the prefrontal cortex to the temporal lobe, and in particular to the *hippocampus*, that ties all of these elements together to make a rich memory with context, sequence, and layers of meaning. For example, your memory of your mother's face is not a static snapshot; it is a multifaceted concept with rich meaning and emotional significance. The prefrontal cortex and hippocampus working together make this possible.

Of course, this geographical analogy has its limitations. The brain is a three-dimensional structure and it is a "paired" organ with symmetrical left and right halves. Regions deep inside the brain that connect to the cerebral cortex are important, as are connections between the left and right brain hemispheres, but we can build on this foundation gradually rather than try to master brain anatomy all at once. The description thus far has also left out the "little brain" or *cerebellum*, which is a multi-lobed structure situated at the back of the brain like a woman's hair bun. The cerebellum is important for coordination of movement and many other functions, to which we will soon return.

An important concept for the purposes of understanding why we snap is that consciousness arises from neuronal activity in the cerebral cortex. We lose consciousness if electrical activity in large areas of the cortex is suppressed or altered: during deep sleep, when the cortex fires in slow synchronous oscillations (waves); during epileptic seizures, when electrical activity in the cortex fires erratically in high-frequency waves and discharges; and while in a coma or following anesthesia, when electrical activity in the cerebral cortex is greatly suppressed.

The rest of the brain beneath the cerebral cortex operates under the radar of our awareness, as the hypothalamus does in controlling automatic bodily functions necessary for survival. Unless operations in the deep brain send some output to the cerebral cortex, we have no conscious awareness of their operation. Many emotions arise from signals sent to the cortex from the deep structures in the brain that control bodily function automatically. Hunger results when neurons unconsciously monitoring blood-sugar levels detect a drop below the optimal range, then send an alert to the cerebral cortex in the form of a strong urge to eat something. Similarly, thirst is the conscious sensation generated by deep brain structures (in the hypothalamus) that monitor the body's water content. More pertinent to the subject here, fear, anger, craving, and the pleasure of reward all involve unconscious information processing in deep brain structures, such as the amygdala and limbic system, sending their output to the cerebral cortex.

The old saw that we use only 10 percent of our brain overlooks how much of the brain is at work below our conscious awareness. Complex motor skills—for example, typing on a keyboard, riding a bike, driving a golf ball—all take place unconsciously. Although the motions of our voluntary muscles (for example those in our limbs) are under conscious control, skilled performance in most motor tasks could not be coordinated consciously, because conscious control is too slow. This is why it takes effort and time to learn to ride a bike, drive a car, play a guitar, or type on a keyboard. When the movements required for these tasks were new to you and you had to evoke each precisely placed and timed movement by conscious control, your performance was awkward and very slow. Over time, those movements became automated and unconscious. This involved encoding the motor programs to

execute the skill in deep brain regions and non-cortical regions. The cerebellum and deep brain region called the *striatum* are where our learned motor
skills are stored and executed. Usually these motor programs are set into
motion under conscious control, but not always, as we have already seen.
Control of behavior and unconscious brain functions by the prefrontal cerebral cortex is termed "top-down" control by cognitive neurobiologists. Control of function and behavior by information from the unconscious brain to
the cerebral cortex is termed "bottom-up." In threat detection and snapping
in rage, both top-down and bottom-up control are involved.

Another important concept in appreciating how the cerebral cortex
operates is that different sensory inputs target different specific regions of the
cortex. As mentioned, vision is processed in the occipital region (New York)
and sound in the temporal region (Texas/New Mexico). However, vision,
hearing, and all other sensory perceptions are very complex processes. To see,
you must be able to recognize light and contrast, shapes, colors, 3-D space,
and movement, and then attach significance and meaning to the objects to
form a comprehensible scene. Patterns of ink on paper must eventually evoke
meaning, memories, and emotions in order for you to read them. To accomplish this, information has to be processed in one module of the cerebral cortex for one purpose (visualizing a letter, in the case of reading), and then the
output of that information processing is sent to a different cortical module,
which operates on this input to produce a more complex analysis (for example, where in the word the letter appears determines its sound and the word's
meaning, and this information is fed into higher-level brain modules for
understanding the significance of the sentence and plot of the story).

As a rule, information processing of sensory input gets more complex as
information is relayed to regions toward the front of the cerebral cortex.
Here the complex information from many different senses comes together
for higher-level analysis. Referencing back to our map analogy of the cerebral cortex, information processing flows in increasing complexity from the
East Coast (the back of your head) to the West Coast (the front of your head).

Remember that these connections are typically two-way highways, with
information echoing back from "higher" to "lower" cortical modules. Interestingly, consciousness is believed to arise from the recurrent interactions
flowing back and forth among cortical regions. Consciousness, then, is only
an echo of what has already occurred without any awareness—and this is the
key to understanding threat detection and why we snap. Subliminal input,

for example, is information that is presented too rapidly to provoke any conscious awareness. We are constantly taking in subliminal information about our environment, and this dictates not only what we perceive but how we respond.

That is a first course in what is a feast of new information on the neurobiology of rage and aggression gained from new methods of research and discoveries made in only the last few years. These discoveries include astonishing new developments in how to surgically control the circuits of rage. But let's look first at some of the history of interventions in the brain's operation.

The infamous prefrontal lobotomy surgery developed by Portuguese neurologist Egas Moniz, who received the Nobel Prize in 1949 for his work, pacified the most violent psychiatric patients into docile beings (with sometimes awful side effects). The excessive use of this surgery, including its use to control aggressive behavior in children and adolescents by American neurologist Walter Freeman and his followers in the 1950s, put an end to the practice, but other psychosurgeries, for example cingulotomy, replaced it. This brain surgery is the removal of the *cingulum*, which is the highway of nerve fibers connecting the cerebral cortex to the limbic system deep in the brain. The surgery was developed by neurosurgeons Eldon Foltz, Lowell White, and Thomas Ballantine. Severing connections in the cingulum interrupts circuits in the rage response, relieving severe panic disorders and other psychiatric conditions that cannot be treated successfully by other measures. These surgeries could be quite effective and without serious side effects in many cases, but by the 1980s most neurosurgeons had abandoned psychosurgery. The social pressure against it and legal repercussions when things went wrong were too much to bear.

Brain stimulation became a fashionable alternative to psychosurgery. The pioneering brain-stimulation research of Walter Rudolf Hess in cats had only been the beginning. Many researchers pursued the stimulation approach to understand rage and aggression, most notably the celebrated Spanish doctor José Manuel Rodriguez Delgado. Delgado attracted something of a cult following among science-fiction enthusiasts and alarmed some members of the public by the frightening prospect of "mind control" through radio-activated brain stimulation. Delgado is famously remembered for a dramatic demonstration in the bullring, where he used a radio-controlled device rather than a *muleta* (red cape) to control the raging animal and save himself from being gored to death. He brought the charging bull to a halt with the flick of a switch, delivering a shock to a specific spot in the *toro*'s brain. Delgado had devoted his scientific career from the 1950s to the 1970s to understanding brain circuitry through an experimental approach using electrical stimulation of the brain. The circuits he activated in the hypothalamus, amygdala, and limbic system are now understood as circuits involved in aggression and rage. By stimulating the appropriate spot he could turn a cat into a hissing and clawing predator or a stealthy, stalking killer or an antisocial bully prowling its cage looking for fights with subordinate animals. Stimulating the lateral hypothalamus could even make the enraged cat attack the friendly experimenter if he approached the cage. The parallels to human behavior were, and remain, clear. Delgado found that deep-brain stimulation in monkeys could elicit complex and purposeful attack behaviors but that these responses were influenced by the animal's gender, sexual interactions, and social dominance within the colony.

Today, *transcranial stimulation* can be used on humans to stimulate or inhibit neural activity in targeted spots in the brain without the need of implanting electrodes. A powerful electromagnetic pulse can be delivered through the skull and focused on a targeted region inside the brain. This technique, sometimes coupled with scans of the brain using fMRI (functional magnetic resonance imaging), is revealing answers to mysteries about how the unconscious brain detects threats, responds in fear, and motivates behavior. Together these experiments show that rage and aggression are not the product of specific neurons in particular spots in the brain; these complex behaviors are dependent on neural activity sweeping through vast areas of the brain—the *human* brain, not some embedded lizard brain.

So because these circuits that operate beneath the level of conscious awareness are complicated, that complexity might suggest they are not easy to manipulate or trigger. We tend to expect nuanced responses from such a sophisticated mechanism. Yet the most profoundly alarming thing about the rage response is that something in our environment can instantly wrench us from whatever we may be doing and set us on a path to violence. A simple, direct, and immediate response. But what are the triggers of rage in our environment? Are they so rare, so complex?

# 3

# What Are the Triggers?

Nothing in life is to be feared, it is only to be understood. Now is the time to understand more, so that we may fear less.

**Widely attributed to Marie Curie**

What is so baffling about suddenly "losing it" is that the eruption of anger and destruction is sometimes triggered by the slightest provocation. Ferocious anger erupts without forethought, unleashed automatically before or beyond the power of your rational mind can act to keep it in check. Sometimes the violent behavior is dangerous or even life-threatening. The aftermath is often regret and bewilderment.

The outcome of snapping in rage is often blatantly counterproductive: a broken dish now useless. One's most cherished china or highly prized automobile can be sacrificed in a fit of purposeless rage, and sometimes it is not an inanimate object that is demolished but rather a relationship or a person. In the worst cases the results are cruel, tragic, or criminal. How does one explain shaken-baby syndrome, for example? With the perspective of hindsight, a fit of rage is likely to be seen as senseless by the person committing the act. Ray Young in the post office attack, Carmela dela Rosa throwing her granddaughter to her death, and the woman mentioned in chapter 1 who ran over another driver in a fit of rage all regretted their horrific violent actions shortly afterward. Some, such as Ray Young, went to prison apologizing for what he had done and still grappling to understand it, completely mystified about what had happened "to him."

Such furious reactions can be conveniently dismissed as aberrations of flawed individuals, but these acts are epidemic in modern society. Consider the case of Oscar Pistorius, the South African double amputee dubbed "Blade Runner" who competed in track using carbon fiber prosthetic legs against able-bodied athletes in the 2012 Olympics. His rival and winner of the 400m championship, Kirani James of Grenada, traded numbers from his own jersey with Pistorius immediately after the race and embraced him. Every other member of the field joined in embracing Pistorius at the finish line in a public and powerful display of respect and admiration. Pistorius was beloved the world over as an inspiration. This all changed in a flash and soon he was on trial for murder. Did he shoot his beautiful girlfriend Reeva Steenkamp with a spray of bullets in a fit of rage or by mistake? The fact that this question can be raised at all reveals the universal implicit understanding that this beastly propensity toward rage is deeply ingrained in human nature, and that seemingly anyone can suffer a rage attack with a horribly remorseful outcome.

Staff Sgt. Robert Bales walked out of his army outpost in the Kandahar province of Afghanistan in the middle of the night and massacred innocent civilians. After shooting indiscriminately with an M4 rifle twenty-two men, women, and children—most of them in their homes—and then setting their bodies on fire, Bales was later as perplexed by his behavior as anyone. This model soldier, athlete, and past president of his high school class wept with sincere apology and regret at trial, where he was sentenced to life imprisonment after pleading guilty to the gruesome mass murder. Father of two, Bales apologized to the relatives of his victims at his trial, but he could not explain why he committed the killings. "I am truly, truly sorry to all the people whose family members I have taken away," he said. "I have murdered their families."

NFL star Marc Edwards, a childhood friend of Bales, played football with him on their high school team, and he later went on to earn a Super Bowl ring with the New England Patriots. Testifying on the stand as a character witness, Edwards said that he could not fathom how the young man he knew could become the killer of women and children in Afghanistan. "It didn't make any sense to me," he said.

Bales's commander in Iraq said that he was a great leader and "stood out and had a real positive attitude."

Bob Durham, who lived next door to Bales growing up in Nebraska, said Bales was like another member of the family. Durham broke down in tears on the stand describing Bales's compassion and how, as a teenager, Bales helped him care for his developmentally disabled son.

Haji Naim, an elder in one of the two small villages Bales attacked, flew several thousand miles to appear in person in court and describe the murderous crimes. "This bastard stood right in front of me," Naim said. "I wanted to ask him, what did I do? And he shot me [in the face]."

Afghan witnesses testified that Bales ignored their pleas for mercy, including boys and girls who shouted, "We are children!"

In considering mitigating circumstances that might aid his defense and help us understand the horror, the defense originally considered the possibility of PTSD, substance abuse, and potential brain trauma from repeated concussions. In the end, all of these "rational" explanations were dismissed and none of them were introduced at trial in Sergeant Bales's defense. Explaining what had happened that night, defense attorney John Henry Browne concluded, "I don't think anybody with a rational mind could say Bob Bales didn't snap."

Anyone immediately comprehends the explanation of "snapping" and committing a terrible rage attack, but this "explanation" provides no understanding at all. What are the mechanisms driving this explosive human behavior that causes a person to suddenly lose control in a brutal attack of violent rage?

These are extreme examples, and the Bales case is multilayered, but nearly all of us have lost it in a burst of rage triggered by some slight provocation, leaving us perplexed and regretful. Snapping can erupt in many ways: an angry outburst of profanity, banging a fist on the table, scolding a child out of proportion to the incident provoking the anger, or slamming a door in disgust. Even Jesus Christ snapped in a fit of rage when he saw the moneychangers desecrating his house of worship. Completely out of character, the man of peace violently upended the merchants' tables and chairs, exclaiming, "My house shall be called the house of prayer, but you have made it a den of thieves!" (Matthew 21:13 KJV). Jesus grabbed some cords and physically whipped the moneychangers out of the temple (John 2:13–15 NIV). Sure, he had good cause, but ideally he might have gone about it another way and still achieved his worthy objective. "Know this, my beloved brothers: let every person be quick to hear, slow to speak, slow to anger, for the anger of man does not produce the righteousness of God" (James 1:19–20 ESV). Sometimes we get angry for good reason. Other times, anger gets us.

Road rage is probably the most easily relatable example of "snapping" in this way. Some 30 percent of US drivers admit to experiencing road rage, but Leon James, professor of psychology at the University of Hawaii, says that he believes that *everyone* experiences it. Road rage ranges from belligerent horn honking, verbally assaulting another driver, and delivering insults with obscene hand gestures to aggressive high-speed vehicular jousting intended to induce a wreck or a collision by slamming on brakes, cutting lanes, running the other car off the road, pursuing another vehicle to threaten the driver, and, in the extreme, pulling a gun and just shooting the other driver.

What precipitates such sudden rage on the road? Frustration and anger accompany the actions, but they are not the *reason* for the violent behavior. This is why a "Sunday school" moralistic approach to preventing rage is so ineffectual. It is not the emotion itself that causes the behavior, so efforts directed at suppressing the emotions can be futile. Sure, stay calm, don't lose your temper, love one another, treat everyone as you would like to be treated . . . perfectly rational and laudable principles of behavior, but rage does not erupt from the rational mind.

Dismissing rage attacks as the product of morally or mentally defective individuals is contrary to the preponderance of evidence, the frequency of the attacks, and the broad spectrum of people, some just like you, who experience them.

If attempting to understand and control the rage response by focusing on suppressing anger and frustration is ineffectual, what approach will bring insight? The emotions of anger and frustration, like hunger, fear, or sleepiness, are indeed powerful drivers of human behavior, but to understand and modify the behaviors it is necessary to identify *why* one is hungry, sleepy, angry, or frustrated. If we understand why these emotions erupt in specific situations, we can more readily recognize what is happening in ourselves and others and see the rage response for what it is—a deeply ingrained biological process, critical to our survival, that is entirely automated but can be influenced by the rational, conscious mind. Again, these impulses to undertake life-threatening violent action were once absolutely necessary for our survival. So too are the emotions of anger and frustration. It makes perfect sense that human beings, just as other animals, have these hardwired capabilities programmed in our unconscious mind. The conscious mind acts far too slowly when confronted by sudden danger or other situations where violent action is the necessary response.

Modern life is so alien to the environment and lifestyle human beings experienced in our distant past when our neural circuits evolved and were sharpened. The rage response equipped our ancestors to cope with survival in a threatening environment. But when these protective circuits misfire in modern society, irrational attacks on vehicles are made that were in some sense intended for mastodons. One of the most memorable scenes from the BBC television show *Fawlty Towers* has John Cleese in the role of Basil Fawlty going berserk and beating his uncooperative car with a tree branch, yelling, "I'm going to give you a damn good thrashing!"

If the trigger for a violent act is apparent and the violent reaction unleashed is obviously necessary for biological survival, the action will be seen as justified. In such cases the behavior is not identified as a rage response or "snapping," it's considered quick thinking. What makes road rage and other cases of rage attacks so perplexing is that the trigger is often hidden, and it can remain elusive even long after the act.

Consider the case of Trayvon Martin, the seventeen-year-old shot to death in February 2012 by George Zimmerman, who was patrolling his neighborhood with a handgun in search of potential vandals and robbers. What made deliberation so difficult for the jury was the trigger for the action. If Zimmerman's actions were triggered in response to actions taken by Martin that placed Zimmerman's life at risk in a mortal battle, a jury of Zimmerman's peers would accept the action and the killing as regrettable but not criminal. On the other hand, if Zimmerman had pursued Martin and provoked the altercation, Martin would be judged as victim and Zimmerman the criminal. The question of whether the jury got it right in this difficult case we can set aside; the point is that the *behavior* itself is not at issue. What is at issue in characterizing a behavior as a sudden rage response or not is the specific *trigger* that unleashed that behavior.

The triggers can be very subtle. Consider that road rage erupts frequently even in professional racecar drivers. This seems peculiar because the things that set off road rage on the highways—cutting lanes, overtaking another driver, impeding the advance of a faster car trailing you—are the norm, the nuts and bolts of racing. Yet a professional racecar driver can be seen throwing off his helmet in a fit of anger, storming over to another car, and going fisticuffs with the driver who spun out in front of him, ruining his chances of winning.

"I remember one race in particular," a professional racecar driver told me in relating a specific instance of road rage in competition:

> My career had gone into the Dumpster. My credibility had been questioned. I was racing in a smaller league. I would be so angry at times when these other competitors weren't doing what I thought they should be doing. My frustration with having to be back in that arena—the lesser arena—and that guy . . .
>
> This one time it was night. We were doing a twenty-four-hour race and I had passed him. I think he came out of the pits and I was clearly faster, but maybe he was fresh and full of it, but I just could not shake him. I passed him and he just would not leave me alone and I just wanted to be in the rhythm. It was at night and I was in a good rhythm and running hard and this guy just wouldn't let it go.
>
> I remember myself just getting madder and madder and I think ultimately I pulled away from him but you know it took a long time. I'm thinking, I'm so much better than this guy and why is this guy . . . ?

Circumstances beyond present events on the road were compounding the situation. The driver who spoke to me had suffered a serious accident in competition with the best drivers in the United States and Europe, and the accident had been caused by a mechanical failure. Such a failure in the machine one trusts their life to, beyond one's ability to prevent, will shake anyone's confidence. At this elite level of racing where every competitor is pushing to the limit to win, even a fraction of a second's hesitation or sliver of doubt may allow another person to pass. "It's a hard thing to overcome. In time you overcome it," he said, but the accident was still fresh. Moreover, he was dealing with health issues and physical pain, even some personality conflicts among his crew.

"I remember being frustrated with the world. I was falling back."

I asked him if his road rage in this instance enabled him to increase his lead on the other driver. "Yeah, I think so, but it took its toll and I had an accident after that. It is embarrassing to admit, but it is true."

Sound familiar?

The important point here is to identify what triggered the rage response and appreciate an important component in this volatile mix: a person's internal state and external stresses change the threshold at which the trigger is tripped. The driver continued:

> A lot of these times when you are in the middle of the event [road rage] it is too late. If you stood a hundred feet above and you looked down on yourself and you say, "Really? You are actually gonna pull a gun and shoot somebody because they got too close to you with their car?" I mean, there's *way* more going on here. That's not what made you mad. You are already flipped-out.
>
> The whole world was unraveling and there wasn't a damn thing— despite what felt like heroic efforts—it was just going to unravel. The cognizant, conscious mind when it is working correctly would go, "This one race is not a make-or-break thing. What *is* make-or-break is if you have another accident." The cognizant mind would say, "Fix this problem. Stop, fix this problem." It is a very difficult thing to admit.
>
> If I had the kind of maturity I have now, I would have sought help. The smart thing would have been—"Like, OK, you've got to rehabilitate. You've got to get yourself one hundred percent." But what I did is I went racing because I could. I hurt [after the injury], but I could get around . . . it was ridiculous.

It is interesting, in this respect, to recall how Ray Young, who was convicted of knifing a person for apparently cutting in line at the post office, was battling cancer. Similarly, my pickpocketing incident in Barcelona happened while my daughter and I were dealing with some chronic stresses, which I will describe later. But for now, let us focus on the immediate nine triggers.

## The Nine Triggers

It would appear from the daily headlines that countless situations can provoke violent, often deadly, rage, making it impossible to know how such varied tragic events get triggered: "Man Shoots Girlfriend in Head, Then Kills

Self in Southwest Houston," "Disgruntled Employee Goes on Deadly Shoot-ing Rampage at Lumber Company," "Alleged Car Thieves Killed by Irate Mob," "Texas Man Shoots Wife Dead and Wounds Her Lover after Catching Them in Bed," "Mother Shoots, Kills Son's Armed Attacker," "Barroom Brawl Erupts over Insult" . . . there is no end to the headlines. They fill the daily newspapers and broadcast media. Similar altercations that do not make the news keep law enforcement officers busy. But it is not the case that these rage attacks are set off by so many different situations that they are incom-prehensible. Setting aside cases of pathology, the normal human brain will not engage in violent behavior without very specific provocation. I propose that nearly all of the array of possible provocations can be reduced to only nine specific triggers.

These triggers of rage can be remembered by using this mnemonic: LIFEMORTS ("life/deaths"; "deaths," in this case, in French). The triggers are listed briefly here and then analyzed in greater depth in chapter 6. The triggers could be lumped together and split apart in several ways, just as the infinite variety of colors can be reduced to only nine basic colors (red, orange, yellow, green, blue, purple, brown, black, and white) for the purposes of rec-ognizing and categorizing. For the important purpose of being able to quickly identify the trigger in any provocative situation, I have collected these provocations of rage into nine major categories that can be recalled quickly with this LIFEMORTS mnemonic. Knowing this mnemonic can change your life. In a dangerous situation, it could save it.

If you learn to recognize these triggers you can understand why a person snapped in a specific situation. No matter how misguided the response might have been, it will no longer be a mystery beyond comprehension. If you can recognize which of these triggers is igniting your sudden rise in anger or frustration, you can quickly disarm the rage response. Sometimes it is fully appropriate to unleash "the beast within," because fundamentally all of these triggers exist to release violent behavior to save your life. The trick is to rap-idly identify the trigger or triggers in a fluid situation, and ask yourself if this is indeed a potentially life-or-death situation or whether the trigger designed for life in the jungle misfired in the modern world. When encountering potential rage in others, the ability to recognize the triggers will help you in understanding and reacting to it by avoiding inflaming the situation, and possibly defusing it. A lot of things can make a person angry, but if you perceive that the source of a person's sudden anger springs from one of the LIFEMORTS triggers, you will instantly recognize that you are in a potentially violent, even deadly, situation.

**Life-or-limb.** Almost anyone, and most animals, will defend themselves in what is perceived as a life-or-death attack. If an individual is about to die—or merely suffer serious injury—fighting back immediately and vigorously makes perfect sense from a biological perspective. There is nothing left to lose if your life is truly on the line.

**Insult.** Insults will easily provoke a rage response. Perceived insults often precipitate barroom brawls. They underlie family feuds, and in the not-so-distant past insult instigated duels to the death between gentlemen. The reason for this can be illuminated by extrapolating insults between people to a broader biological perspective.

In the animal world, violent interactions are one of the most common ways of establishing dominance. Among many social species this violence can become standardized, such as head-butting between bighorn sheep or bloody battles between male elephant seals. Such battles can result in serious injury or death. The use of violence to establish dominance is widespread throughout the animal kingdom, from fish to chimps; even social invertebrates, such as insects and marine snails (the keyhole limpet, a few peaceful-looking inches in length!), use physical violence to establish dominance over other members of the species. So deeply ingrained is the quick use of violence to establish dominance and the willingness to snap into a deadly attack, the behavior persists even in domesticated dogs encountering an unfamiliar dog, and it persists in us. Verbal insults are the human equivalent of head-butting—a means of challenge and establishing dominance. Bear in mind that it is the perception of insult that triggers the rage, even if the action was not intended as an insult. A hardworking employee may feel insulted by being skipped over for a promotion. . . .

**Family.** Animals will protect their offspring and family members against attack or other threat. Evolutionary success is determined by passing on an individual's genes to the next generation. Protecting offspring, and even siblings and parents who closely share your genes, increases the odds that your genes will be passed on. We see this in a mother bear protecting her cub or when people are killed when they come between an adult moose and her calf. What father or mother would not protect and willingly sacrifice their own life for their children? This is one of the most basic of all instincts in animals as well as people.

**Environment.** Most animals will protect their own environment, their territory and home. The reasons are clear: home and territory provide the basic necessity for survival. Many social animals—cats, dogs, birds of prey, and people—are fiercely territorial. They establish and patrol the borders of their territory relentlessly. The concept of private property, signs that say INTRUDERS WILL BE SHOT, or the right to kill a stranger invading one's home all illustrate this fundamental right of people and other creatures to kill to protect their environment. Border disputes between neighbors are a common trigger for violent attacks that are instigated by such trivial matters as an encroaching tree or a shortcut across their property.

**Mate.** Violence to obtain and protect a mate is the rule of the jungle. Many species, such as wild horses and seals, use violence to acquire and maintain mates or even harems. The Darwinian drive to pass on genes from the fittest individuals is the bedrock underlying the ready willingness to fight to the death over mates. Think: "All is fair in love and war." At the same time, violence between males and females frequently occurs within intimate sexual relationships. This is the trigger for violence in domestic disputes, infidelity, and attacks by jealous lovers.

**Order in society.** This drive exists among other social animals but it is highly developed in human beings, because our species is so utterly dependent on social order for its survival. The accepted use of violence to maintain social order differs from violence to establish dominance among individuals in a group. The orderly operation of society is the purpose of this trigger of rage—not dominance among rivals. Violence is used to enforce the rules of society, to assure fairness, and to correct transgressions. Imprisonment, fines, firing someone from a job, revoking a professional license are all modern practices of forcing people to comply with the rules of orderly society through violent action: the forceful removal of liberty and things of value with the intention to harm and punish the individual who transgresses. Rage attacks frequently break out in response to a perceived social injustice. This trigger often ignites mob violence.

**Resources.** Animals in the wild must fight for their food. Violence will be used to obtain it and to retain it against theft. Likewise, human beings are intolerant of theft and they will react with violence to prevent it. In human society, money and other forms of valuable property are equivalent to food, because these valuables can be transformed into food, housing, and territory.

**Tribe.** Humans, requiring a tight social structure for survival, will fiercely defend their own tribe. This altruistic behavior is much less common in the animal world, but it does exist. When prehistoric tribes of human beings assembled for their survival in nature, an encounter with another tribe was likely to result in competition, the loss of resources, or the triggering of any of the other LIFEMORTS events leading to violence. Thus human beings have always been wary of others. An us-versus-them imperative rules. Throughout history we see human beings divided by tribe, country, or religion attacking and defending against one another. The Great Wall of China, medieval castles, forts and stockades in the Western Frontier ("Indian country"), the remains of stone walls and moats in Italy, throughout Europe, and Japan, and the tight and clearly defined borders of modern countries that will be defended to the death in war are all derived from this requirement to protect the tribe that is an immutable characteristic of our species. Groups of individuals specialized for using violence to protect the tribe are formed as armies or militia in nearly all societies. Tribalism is what drives inner-city gangs, and tribalism is the basis for racism and war. Avoiding this trigger is essential for peace.

**Stopped.** Animals will struggle violently to escape restraint, even to the extent of gnawing off their paw if caught in a trap. Humans are no different. Backpacker Aron Ralston amputated his own right arm to free himself after falling into a crevice and getting it trapped under a rock in April 2003. The ordeal is captured in his book *Between a Rock and a Hard Place* and the film *127 Hours*. In the right circumstance this is a rational and lifesaving response. Being restrained, cornered, imprisoned, or impeded from the liberty of pursuing one's desires will trip this trigger of rage. The accompanying emotion is frustration. The Stopped trigger is what motivates individuals or groups to seek liberty through violent action, such as revolution and war. It also motivates revenge against those who are perceived as having impeded a person's progress, as illustrated recently by the rogue Los Angeles police officer Christopher Dorner, who went on a shooting rampage after being fired. Struggling violently to escape impediment and restraint is natural, but artificial situations presented in the modern world that did not exist in the distant past can misfire the Stopped trigger of rage. The anger and frustration that builds in waiting in long lines (or when someone suddenly cuts in line) and similar situations in traffic on the road can set off this trigger.

> This trigger can also result from a feeling of oppression, which is in effect the perception by an individual or groups of individuals that they are being prevented from enjoying their rightful benefits in society. Any circumstance that leads an individual to feel cornered falls into this category of trigger to violent rage, as can illness if it is perceived as limiting one's right to enjoy and fully participate in their rightful place in society.

———

Any one or more of these LIFEMORTS triggers can initiate an automatic rage response. Whether it does or doesn't depends on additional factors specific to the situation, the individual's state of mind, and other contributing stresses and influences. Frequently a situation will present multiple LIFEMORTS triggers simultaneously or cumulatively. This, not surprisingly, increases the likelihood of violence.

Take my reaction to the pickpocket in Barcelona as described in chapter 1; several LIFEMORTS triggers were pulled when the thief snatched my wallet, including Life-or-limb, Family, Environment, and Resources. The loss of resources was the prime trigger (Resource trigger). Without money for food, shelter, and a place to stay (Environment) my well-being was very much at stake. Add to this Family, as my daughter's welfare was also threatened and she was dependent on my reaction to the situation. Finally, Life-or-limb: once the situation had exploded into a street fight, my life and limb were certainly in peril. It was him or me at that point. Had my daughter not been accompanying me, there would have been one less trigger, and that could have made all the difference in how I responded instinctively. Family is also what triggered my daughter to come to my aid, leaping through the air to intercept my wallet and recover my BlackBerry while I struggled with the thief.

Now consider why the gang of pickpockets reacted by chasing my daughter and me through Barcelona for the next two hours after I had defeated and nearly strangled their pickpocket and derailed their coordinated intentions to rob us. Insult: the pickpocket had been insulted by my beating him up, and indeed the gang's credibility had been threatened. Tribe: tribalism, the essence of any gang, was then evoked to chase us down and seek revenge.

Now it is possible to understand why driving an automobile in traffic is such a potent instigator of people snapping violently. Driving sets off so many of these triggers. Several of them are, in reality, relics from the past that are mistakenly tripped by the artificial experience (in biological terms) of driving an automobile. We tend to view the area around our vehicle while driving as territory. When a vehicle cuts in front of us, this can pull the Environment trigger to defend that territory. Not allowing someone to pass is motivated by the same defense of territory. But when traveling in a car, it is time, not territory, that matters (assuming the other vehicle is not threatening our safety). Whether a vehicle is in front or in back of your car on the freeway makes no real difference in terms of the purpose of your activity in driving: to be transported to a destination rapidly. Either way, it amounts to a difference of a few seconds, which is of no real consequence. Yet if viewed as territory, it's easier to see how road rage can provoke a shooting if another driver encroaches.

This irrational response arises because the brain circuits did not evolve at a time when we had motor vehicles. Back in the jungle, as is the case today, if a number of people are running, people *do* cut in and out and pass one another all the time and it provokes no angry reaction. The fastest person is free to run to the lead. In the case of a footrace, your brain perceives that you are running over territory, not *with* territory as is the virtual illusion when driving a car.

The triggers of road rage can include: Life-or-limb, when another driver commits some action that nearly causes you to have a car accident. Insult, such as flipping the bird, honking aggressively, swearing at another driver. Family, as an altercation on the road can be influenced positively or negatively by whether family members are in the car with you. If a driver feels that another driver's action has nearly put his family in jeopardy, the first driver may be more inclined to seek revenge. Conversely, awareness that family is threatened can restrain a driver from acting in rage. How many cars have you seen with that BABY ON BOARD sticker or similar? Environment; "He cut into *my* lane!" Order in society, when someone fails to yield, cuts in line, passes around others who are patiently waiting to merge due to a lane closure. An individual is not following the rules, thus making him the target of righteous violence. Tribe, in which many road rage incidents are precipitated by particular makes of vehicles or loud music booming from another vehicle being considered obnoxious. These attributes identifying the vehicle's driver as a member of a different group—redneck, hick, lowrider, tree hugger, rich guy, low-class—or member of a different race or nationality. Stopped, as obviously being bound in traffic is not much different from being bound by any other method that once physically restrained our early ancestors.

No wonder the artificial behavior of driving an automobile provokes people so easily to rage. Driving is a veritable complex booby trap of the nine triggers. There are additional important factors that enter into road rage because of the artificiality of the behavior—namely, inability to fully regard the automobile as another human being and the diminished ability to communicate nonverbally with the person in the other vehicle, in contrast to the rich unconscious vocabulary of body language. If people accidentally bump into each other on foot, instant apologies erupt, but on the road the same accidental bump provokes rage.

What triggered the rage described by the racecar driver during his twenty-four-hour race was primarily Stopped (but also perceived by the driver as Insult). The driver felt that his dominance was being challenged; he felt oppressed, cornered, and unable to sustain his rightful place in the world because of circumstances beyond his control, including restrictions imposed by health. The driver also simply felt his livelihood was being stolen. This is a common trigger of the rage reaction in many different situations. His rage had little to do with the car pressing on his tail—that's what a race is.

All of the triggers of road rage—passing, cutting in lane, following closely, and so on—are encountered routinely by professional racecar drivers in competition. If professional drivers responded to them with sudden rage as happens on the highway, all road races would degenerate into destruction derbies. Professional racers adjust their response to these triggers of rage so that they are not tripped, but circumstances in racers' internal and external environment can lower that threshold, allowing the triggers to be tripped.

Interestingly, the external circumstances that modify the threshold for pulling the triggers are the same LIFEMORTS triggers of the sudden rage attack, only experienced over a prolonged period of time rather than suddenly by a specific event. This cocks the trigger and sets it off with the slightest provocation when it is hit again suddenly in a later encounter. Consider the racecar driver recounting the contributing pressures that lowered his threshold for road rage in competition:

> So here I am in this race, but here I am as my career is going downhill. But I am physically and mentally not right and totally frustrated because I have tasted it [success of being world champion] and I know I have the ability . . . I mean, when I was on track nobody could touch me.

Any major-league batter struck by a fastball confronts the brain's strong protective response, making him gun-shy the next few times at bat, and his batting average is likely to dip. While getting beaned by a hardball hurled at 95 mph by a professional pitcher must really hurt, it is rarely fatal.

"First of all," the racecar driver explained, "it really almost took me [the serious crash on the racetrack], and that feeling of one minute you're doing something and in the next minute (*pause*) . . . When that car left the ground I knew this was a serious thing—and it was. I was lucky to survive it. That car was just demolished. It left the ground at a hundred and eighty-five miles an hour. It flew. It shouldn't have failed. It shouldn't have failed like that."

These stresses of personal injury and loss of confidence, as difficult as they are to grapple with, were compounded by others. "I had to fight through so many things. Little things, but they added up," he said, speaking about simultaneously dealing with difficulties within his team of coworkers and with personal stresses in his family.

The evidence suggests that ongoing stresses were weighing on Sergeant Bales's life too. Bales told the jury that he had struggled with anger issues that worsened after his third military deployment to the Middle East. He said trivial things like dirty dishes in the sink would make him angry, and that he was "mad at myself for being mad." He said he started flying off the handle when soldiers made small mistakes. He also started taking steroids and drank alcohol at times in the looser environment that prevailed at the Special Forces outpost. Bales said the fear was constant: "In my mind, I saw threats everywhere." Bales was struggling with family stresses and financial difficulty at home, telling his superior that he had "bad kids, an ugly wife, and was not anxious to make it back home to see them."

Recognizing the nine triggers of rage and understanding the compounding circumstances that press on them helps us comprehend how someone can snap and commit a violent crime, but it does not excuse the actions any more than understanding that a bank robber commits his crime for financial gain, but it does help illuminate why these perplexing crimes occur. Acting on that understanding could help prevent them. I would like to think that if Bales (and his superiors) understood why he was "mad at [himself] for being mad," a model soldier, father, and past president of his high school would not be spending the rest of his life in a prison cell, and twenty-two innocent people would be unharmed and living out their lives in Afghanistan.

If the LIFEMORTS triggers can be recognized, tragedy can be averted.

These extreme examples show that snapping violently over minor provocations is the cause of much violence and even murder, but more often snapping in everyday life leads to angry and thoughtless verbal outbursts. The angry outbursts and actions can have lasting and regrettable consequences on relationships, damaging or destroying professional or family connections. Almost everyone snaps at one time or another, with the possible exception of the late Mr. Fred Rogers of infinite patience on the public television children's show *Mister Rogers' Neighborhood*. Even Pope Francis admits to snapping.

"It's true that we can't react violently," the pontiff says, "but, for example if Dr. Gasbarri here, a great friend of mine, says a curse word against my mother, then a punch awaits him. It is normal. It is normal."

The pope was speaking in the immediate aftermath of the 2014 murders by Islamic extremists in Paris in response to the controversial cartoons of the Prophet Muhammad published in the satirical French magazine *Charlie Hebdo* (in this case, Insult being the trigger to violence).

Professional and amateur sports is an excellent place to witness this unconscious neurocircuitry of rage working for and against athletes. "Players when they get frustrated can hijack themselves and end up doing something that they didn't want to do. Whether that is throwing at a batter, whether that's going into the dugout after they strike out and throwing a Gatorade container down, kicking a water cooler, breaking a bat—whatever it is, those kind of immediate responses do happen. Typically players will see that as something that they shouldn't have done. They will wind up apologizing for it and move on," Cleveland Indians team psychiatrist Charlie Maher explained to me. But these same neurocircuits of rapid response are crucial to athletic performance, enabling players to operate instantaneously and appropriately in the stressful arena of competition.

We are in the fifth inning, July 6, 2012, and the Cleveland Indians are in deep trouble. The score is 6–2 with the Tampa Bay Rays at bat facing Indians pitcher Justin Masterson, with only one out. The Indians send left-handed relief pitcher Nick Hagadone to the mound. Hagadone has a 95-mph fastball and he started the season strong. Recently, though, he has fallen into a slump, giving up fifteen runs in the last seven games. This opportunity on the mound will bring Hagadone's ugly streak to a sudden end, but not in the way anyone could have imagined. By the time the inning was mercifully ended, the Rays had increased their lead to bury the Indians 10–2.

Hagadone left the pitcher's mound frustrated. Reaching the clubhouse he slammed his fist into the door, fracturing his pitching arm. The injury required surgery to insert a metal screw to mend the broken bones, and at least eight months to heal. His season was over. The Indians immediately optioned him to the minor leagues, and placed him on the minor-league disqualification list so that they would not have to pay him while he was sidelined. Hagadone's major-league contract worth $480,500 was immediately sliced to $78,250 in the minor league.

Indians manager Manny Acta said, "I think Nick learned his lesson. A big part of this game is learning how to control your emotions."

Hagadone suffered embarrassment and indignity, but everyone can relate to him. People gripped by frustration frequently suffer sudden, self-inflicted, unintended injury to their toes or fists in an explosion of aggression taken out on an inanimate object.

"I think we all shared Nick's frustration, and we're certainly disappointed with the reaction," Indians general manager Chris Antonetti said. "We wish he would have handled it differently. But he's very remorseful, he's sorry that it happened, and he's been very accountable for it."

I asked the Cleveland Indians team psychologist, Charlie Maher, if Hagadone might be one of those hotheaded athletes with anger-management issues. Not in the least, he said: "Actually, this was very, very surprising, because of how he is. This player is someone who I know very, very well. He is someone who would at home be very laid back, very easygoing.

"That was a situation that happened in a game. He snapped, so to speak. He's a perfectionist. He wants everything to be right. When he is locked into the moment, this guy is very, very good. He can pitch. He can move on to the next pitch. He doesn't think about it."

These are the rare qualities required of a relief pitcher. Someone who can step into the game at the most stressful moment and extract the team from their dire state with a precision performance under pressure.

"He got too caught up in what other people were thinking about him, and he let that get to him," Maher continued. "He is more of a controlled kind of individual, and he's very good. When he is ready to pitch he'll do a good job. Here, he just snapped."

I mentioned to Dr. Maher that if this player did not care intensely about doing well, this never would have happened. In a world where it sometimes seems that many people don't care very much about much of anything, Hagadone needs to appreciate how his intense passion to excel is extraordinary, and this is one of the keys to his success as a professional athlete. Maher agreed: Hagadone snapped as a result of activating the Insult trigger of rage, in this case from feeling his personal status diminished in the eyes of his colleagues and fans. The Stopped trigger was also a contributor. "He was, like, trapped," Maher said, expressing how he believes Hagadone must have felt, "and that's how he got out of the trap."

Setting aside the extremes of rage murders and terrorists and high-performing athletes, let's consider this rage circuitry at work in the brains of everyday people in everyday life. Most parents have experienced abrupt rises in anger toward their own children, whom they love dearly and desire to raise properly, and a parent may lash out verbally or even physically in ways that they immediately regret. Married couples, coworkers, shopkeepers, or anyone who deals with the general public, and many others in daily life understand what snapping is and that it needs to be avoided in most cases. But the "anger management" approach is frequently not sufficient. Consider the following comments posted on various websites or printed in magazine articles on subjects ranging from parenting to marital relations to anger-management services.

"I have a bad temper and my son is very stubborn and very jealous, lately every time he behaves wrong I just start yelling at him which makes me feel bad later. Will that have a deep impact on him? Time outs doesn't seems [*sic*] to work with him, he just won't listen to me."

This concerned mother was not only worried about her snapping but was more deeply troubled that her angry reactions could have long-term harmful effects on her son. The moderator's reply confirmed that her angry responses were indeed harming her son's development. The mother likely already knew this. Her question was cloaking a deeper one: How do I stop snapping at my child?

A father posting on the same website: "When I come home from work, I find myself snapping at my son. How can I stop doing it?" He signs himself "Too Tired Dad."

The moderator confirmed what the father (and most of us) already know: It is easy to bring the stresses of work home and have anger released on family members over trivial incidents.

Professional advisers typically respond with the following helpful advice to parents:

1. Understand what is causing the underlying stress.
2. Try to understand that at the child's current age, he or she has limited ability to cope with life.
3. Always reprimand your child in private. Your statements as a parent should enable your child to maintain a sense of dignity.
4. Do not use your child as a target when you are frustrated at work or in some other area of your life.
5. Take a deep breath and count to ten.

All of these are helpful suggestions. The problem is that most of them can be ineffective in the moment of snapping. Such advice tips are really just restatements of what snapping is, and further elaboration on why snapping should be suppressed. The problem is that this rational approach does not disrupt the rage circuitry effectively. The concerned father and mother are already well aware of the problem. Their strong desire to avoid these behavioral responses is why these people went to the trouble to bare their souls online seeking help.

As helpful as anger-management advice can be, there is a need for a supplemental approach to these practical techniques in order to suppress anger. On a PBS news report, a veteran standing in court charged with committing a violent crime after returning to civilian life shot back at the judge, "Even me being in anger management . . . what good is that? Because I snapped!"

"In the moments where I feel like I'm going to snap, as much as I feel like I need help, I don't really need advice," observes Ann, who writes the blog *PhD in Parenting*. "I do not need someone telling me how great time outs are. I do not need someone telling me that I was disrespectful to my child and that if I just focused better on connecting with her that these problems would not arise. I do not need someone saying that I expect too much of myself or that I expect too much of my children. I do not need someone telling me that we need strict consequences for misbehaviour. I need a hug. I need empathy. I need help. I need a break. I need space. I need time to think. Once I've had that, I need one-on-one time with the kids to reconnect with them and I need a few days of calm for us to get back to normal and leave our stresses behind."

What people really want is to understand *why* they snap. What is it? A husband named Chris posts online seeking to understand why he snaps at his wife:

> It seems like I snap at her at times and she gets upset and starts crying. Why do I do this? I don't "mean" to do it. Today I was restless as my mom died a week ago. I wasn't happy doing anything. She wanted to take me to the movies but I didn't want to go. Then I snapped at her about some silly thing and she got upset and left in the car. I don't know what to do.

Chris clearly grasps that underlying stresses are part of the snapping response, but he wants to understand how and why. Understanding is the first step to managing anything.

This concerned wife is asking the same question as the others seeking help with snapping:

I snapped out on my husband this morning. I did it last week, too. Needless to say I'm disappointed with myself. This morning I was bugged because we missed the recycling truck. Yes, it's his job (so to speak). He normally takes it out in the morning. His schedule has changed so he missed it. I bugged out about it; yelled, had a fit, etc. But why should I give a flying fig? If it's so darn important I should take it out myself the night before. Another thing, last night I dumped my bag out all over the living-room floor. I was looking for my keys and I had too much cr*p in my bag, so I dumped it and walked out the door to go get whatever bologna I wanted to go get.

Normally, I'd point to hormones, being tired, etc. If I'm chemically off I can feel it. The thing is there wasn't any emotion that made sense to me. Sure, there's anger involved I guess, but I question that because it didn't linger . . . at all. I can't even recall it. It seems like anger on the surface and just a pile of numb underneath.

On top of everything our anniversary is tomorrow. Meh, who knows. My cousin is getting married this Sat and he plans to have a special dedication for my brother. Maybe this is some weird grief thing I'm not in touch with.

So, what do you do when your spouse snaps out on you (if they ever do) and what do they do with your snap outs (if you ever snap)? Then, how can it be made up to the snappee?

The current behavioral counseling approach does not provide an answer. What she wants to know is, what is the biology underpinning this odd and undesirable automatic behavior?

She has perceptively pinpointed two of the factors, stress and hormones, and she is trying to identify the triggers. Why did her husband's failure to take out the recycling set her off? It makes no sense to her.

Another woman asks:

It seems like lately whenever i am with my husband i am snappy and short with him. I raise my voice and become really irritated very easily which is very unusual. I have gone over everything in my head which could contribute: He is working alot, i am responsible for EVERYTHING at home, i have switched birth control twice in the past 3 months, i am on phentermine to lose weight, it is getting close to when i am supposed to start my period, just stress in general, our son is teething so therefore very fussy. I don't know what is really the cause but i want to stop doing it. I feel terrible when i do and it has been pushing him away which is not good. Does anybody have some advice for me?

What this woman and the others are seeking is an understanding of the brain's unconscious and automated circuits of rage, and the LIFEMORTS triggers that instantly engage them. A common misconception is that snapping behavior is caused by anger. In fact, anger is an emotional response, and a lifesaving one in the right situation, provoked by the brain's threat-detection circuitry.

In an article entitled "Men and Anger Management" published on WebMD, an expert writes, "Anger is a very powerful emotion that can stem from feelings of frustration, hurt, annoyance, or disappointment. It is a normal human emotion that can range from slight irritation to strong rage."

The problem with this perspective is that from a neuroscience view it mixes up the *triggers* of rage with the *response* of rage. The emotions—anger, disappointment, annoyance, hurt, frustration, jealousy—are the responses evoked in the brain by specific situations or circumstances (triggers) that communicate beyond conscious words. These powerful feelings motivate behavior to address an immediate threat in your environment. These emotions are the *result*, not the *cause*, of snapping. This is why advice to suppress anger and the other emotional responses is often ineffective. The horse is already out of the barn, so to speak.

Applying techniques to manage emotional responses is the present approach to addressing snapping, as in this example from an anger-management resource:

## SOME SIMPLE STEPS YOU CAN TRY

- Breathe deeply, from your diaphragm; breathing from your chest won't relax you. Picture your breath coming up from your "gut."
- Slowly repeat a calm word or phrase such as "relax," "take it easy." Repeat it to yourself while breathing deeply.

- Use imagery; visualize a relaxing experience, from either your memory or your imagination.
- Nonstrenuous, slow yoga-like exercises can relax your muscles and make you feel much calmer.

Practice these techniques daily. Learn to use them automatically when you're in a tense situation.

Again, these are all very useful techniques, but still, they are predicated on an approach that might be likened to "putting out the fire, rather than preventing fires by understanding what starts them." The longing for this missing insight is what comes through loudly in all of the experiences shared by the people crying out for help through advice columns and websites.

All the wife wanted to know is *why* she felt a sudden rise of anger because her husband forgot to take out the recycling. In the context of the brain's automated circuitry of sudden aggression and anger, her husband's action threatened her dominance—the Insult trigger of aggression. If she could perceive this, she would see that this was clearly a neurological misfire of a normally very important brain function. Her husband meant no insult. The emotion of anger is useless in this instance, as there is no need for any aggressive response.

If we think back to the quote earlier from Pope Francis, we see how he fully recognized, and he was trying to teach others, that insult is a powerful provocation to violence. It is necessarily a part of being human. Some insults are more likely to cause a violent response than others, such as insulting a person's mother or their religion. In the LIFEMORTS context, disrespecting one's mother or a person's religion pushes two rage buttons simultaneously with the Insult button: Family and Tribe. It does not matter at all if the insult is unintentional. If a trigger is pulled, the violent response is released.

Likewise, in many of the previous incidents, additional factors made the person unusually touchy, causing them to erupt in rage at seemingly trivial incidents, for example, chronic stress, being tired, or hormonal, dietary, or pharmaceutical factors. It is helpful to understand why, in the biological sense, the triggers of rage need to be on a hair trigger under these situations. The lowering of the threshold for a snap response is an important component of the mechanism. Consider that people commonly become irritable when they are hungry. Low blood sugar is the result of inadequate food, which in the primal sense is an immediate threat to survival. Facilitating aggression during hunger in our hunter-gatherer days would have promoted survival by boosting a person's ability to act aggressively to hunt down and kill prey. That neurological relic of our past can be counterproductive in the modern world of fast-food restaurants, but we are stuck with it. It is helpful to understand this, however, if you are a waiter. You should not delay serving bread and drinks to hungry diners as soon as they are seated.

## Stress Pressing on Triggers

The incident of Cleveland Indians pitcher Nick Hagadone suddenly snapping and slamming his fist into the clubhouse door is less bewildering if you are aware that he had been laboring under the chronic stress of enduring a seven-game slump in his performance leading up to the incident. Chronic stresses affect many people in modern life. Work-related stress, for example, frequently increases volatile reactions, sometimes with career-changing consequences that are as sudden and debilitating as Hagadone's fractured pitching arm. Today one's career has replaced hunting and gathering as the means of putting bread on the table, but the same threat-detection brain circuits are at work providing for ourselves and family in our career today as were operating in the brains of our hunting and gathering ancestors of prehistoric times.

"Yes, I have seen firsthand the effects of chronic stress on whistleblowers," says attorney Jason Zuckerman, principal of Zuckerman Law, a Washington, DC, firm that represents whistleblowers nationwide. Whistleblowers are people who reveal wrongdoing within a government agency or private enterprise. Whistleblowers are glamorized in movies, but the reality is much different. They typically suffer unrelenting threats and reprisals from their employer. Most whistleblowers end up broke, out of a job, emotionally destroyed, and often psychologically or physically ill from enduring the chronic stress and battling the injustice of their employer's actions against them.

I interviewed one employee in a government agency who was stripped of all resources and personnel and sent to an isolated room in the basement of the multistory building. I found him alone in a windowless room with yellow-painted cinder-block walls, sitting at a metal desk with a black telephone and a computer monitor. There was nothing else in the room. He no longer had any job function, and no interaction with other employees. He talked about the difficult struggle to withstand the constant stress of isolation and injustice. He was consumed by the fight to get his old job back, by his legal expenses, and with devising legal strategies to defend himself from the abuse. "I'm still waiting for a reply from my congressman," he said, turning to look at some text on his computer screen. In a tone of voice conveying deep misery he told me of a colleague who had suffered a similar fate and had died recently at an early age as a result of a sudden heart attack brought on by the prolonged stress.

"A whistleblower feels vulnerable and under attack, and the stress can become unbearable," Zuckerman says. "In one instance, a client slapped her boss and was terminated as a result thereof. She lost control momentarily after suffering months of retaliation. This client had a perfect performance record for twenty years, but the stress caused her to act in a manner that gave the employer a sound justification to terminate her employment."

Zuckerman continues. "In other cases I have handled, chronic stress has caused my clients to have verbal outbursts that resulted in disciplinary action. But it is unfair to look at the outbursts in isolation. Those outbursts stem from my clients feeling under assault for doing the right thing. Whistleblowing often results in isolation and alienation from colleagues, and feelings that everything is on the line. For example, the whistleblower fears that he will lose his reputation, livelihood, financial stability, et cetera. This stress takes a terrible toll on my clients."

Acute stress stimulates the body, but chronic stress can be damaging or even deadly. Chronic stress impairs the immune system and taxes the cardiovascular system. Memory is stimulated by acute stress, but chronic stress weakens cognitive function and memory, because it disrupts the normal neurocircuitry in the brain; it even kills neurons. Glucocorticoids (such as corticosterone) are stress hormones that are released by the adrenal glands. These have powerful, wide-ranging effects on the body's metabolism, cardiovascular system, immune system, and cognitive function. The release of glucocorticoids is stimulated by the hypothalamus acting on the anterior pituitary in the brain, which releases hormones into the blood that stimulate the adrenal glands.

Interestingly, glucocorticoids fluctuate on a twenty-four-hour cycle, rising during the day and ebbing at night. This daily rise of stress hormones helps the brain and body operate at peak performance during the day, but it is essential that levels recover during rest. Lack of sleep causes debilitating stress in part because sleep deprivation does not allow stress hormones to return to low levels.

Neuroscientists Conor Liston, Wen-Biao Gan, and colleagues observed how neurons and synapses are affected by stress hormones by surgically installing a glass window in the skulls of mice. Then they placed the anesthetized mice under a powerful microscope that used bursts of titanium-sapphire laser illumination to penetrate into the brain for them to see what was happening. Twenty minutes after they injected corticosterone into the mouse, they saw new synapses begin to form on dendrites of neurons, and existing synapses started to disappear. Behavior tests on these mice showed that acute stress (single corticosterone injections) improved learning, and that improved performance in learning tests correlated directly with the increased rate of sprouting new synapses. Thus, the stress hormone stimulates active rewiring of synaptic connections in the cerebral cortex.

The boosted rate of forming and eliminating synapses remains in balance during acute stress (single injections of corticosterone), but chronic stress (multiple injections over a longer period) upsets the balance and causes a net loss of synaptic connections over time. The reason for this is that once new synapses form, stress hormone must return to normal levels for those new synapses to survive. Interestingly, the researchers discovered that sleep was necessary for synapse survival: The drop in stress hormone levels at night is essential to prevent them from withering away.

"Jail is one of the most incredibly stressful environments," says David Connell, who has taught a stress-management class to inmates at the Arlington County Jail in Virginia for the past fifteen years. "Because they [inmates] don't have control over hardly anything in jail."

One of the big stressors in jail is some of the guards. Some of them have issues—some of them are great, some of them aren't—and they just want to sort of poke at the inmates.

Or thinking about their court case is a huge stressor. Most of them have really stressful stories and lives. A lot of it comes from childhood, and addiction is a big one.

There are a lot of people in there because of this [snapping] response. You know, I recall there was one gentlemen; he was six foot eight, three hundred and fifty pounds. He said that when he would get angry he would almost black out, and he would wake up and realize what he'd done. I mean, someone that big who didn't have control over his physical reactions and stress—he really would hurt a few people.

"What kind of things would set him off?" I ask.

"Someone saying something—disrespecting him. Or emotionally charged family relationships, a girlfriend or that kind of thing."

Connell says he teaches several techniques in his anger-management classes to lower stress. "Physical exercise, diet, mindfulness, meditation, and visualization," he says, listing them. "Breathing, a deep abdominal breathing technique," he says, helps reduce stress and is especially useful in helping inmates go to sleep. "What you put in your body affects your mind as well," he notes, for example, caffeine, alcohol, sugar, or drugs.

"Long-term stresses will kill you, literally," he says. That is a fact, but it is also a fact that stress (both acute and long-term) greatly increases a person's likelihood of snapping verbally or physically in response to seemingly trivial incidents.

Chronic stress places the LIFEMORTS triggers on high alert because the stress indicates that a heightened state of threat exists in one's environment and this requires greater vigilance. Stress, like anger, is a bodily sensation conveyed through emotional feelings that alerts us that we are in danger, and this in turn motivates aggressive physical behaviors. Trying to understand the reasons for the underlying stress is not going to be as helpful as simply understanding that, as a biological necessity, stress will make a person more likely to snap. New research is increasing our understanding of the neurocircuitry and hormonal factors that accomplish this mental state we call stress, and identifying how these stress responses act on the triggers. Some stresses in life are avoidable, such as being late and, as a consequence, rushing to work on the highway. In this stressful situation you are more apt to experience a road rage incident because of the biology of your brain, not because of your weak morals. You may be powerless to lower the stress, but recognizing that being under stress is going to cock the LIFEMORTS triggers, so to speak, can make you more careful to avoid a misfire. Most serious life stresses are not under one's control, such as poor health, a death in the family, financial trouble, moving house, and many others. Even though such stresses cannot be controlled, it must be understood that in a state of chronic stress you are at risk of the LIFEMORTS triggers misfiring. Likewise it is important to recognize that the risk of another person snapping increases when you are dealing with someone, even a stranger, who is apparently under stress, and you are in a situation where the LIFEMORTS triggers are present.

Everyone knows what stress feels like, even if they have not fully realized all of the underlying causes of it. While trying to identify and correct these causes of stress, a person needs to recognize the fact that, for whatever reason, they *are* feeling under stress. Knowing that stress will put their brain circuitry for snapping on a hair trigger, extra vigilance must be taken. This recognition will engage the prefrontal cortex, as will be discussed in chapters 8 and 12, to proactively put the brakes on the rage response.

To make use of this newfound knowledge about the neuroscience of rage, it is necessary to develop the skill of identifying the specific LIFE-MORTS trigger at work in any threatening situation. Not only will this make what seem to be incomprehensible rage attacks in the news and in our daily lives more comprehensible, developing this skill will enable you to better control this neurocircuitry of sudden violence and aggression that resides in us all.

# 4

# Reaching a Verdict

An injury is much sooner forgotten than an insult.

**Philip Stanhope, 4th Earl of Chesterfield, in a letter to his son**

I never imagined that someone I know personally would have their life end in what was some kind of violent snap, but in the course of writing this book, the unthinkable happened.

This story does not begin, as the news stories did, with the horrible event itself; it begins two years earlier in West Virginia, at a spectacular place called Seneca Rocks. Seneca Rocks is a thin fin of jagged rock splotched with gray and white, jutting up almost a thousand feet in the air like an enormous concrete wall still under construction. In fact, the fin of rock is the hardened core of the green mountain ridge that runs for miles north to south. Over eons the soil at this spot eroded, leaving this hardened skeleton exposed. Seneca Rocks is a popular place for rock climbing, because there is no other way to reach the summit than by technical ascent up the sheer vertical wall. To stand on the summit of Seneca Rocks, at places no wider than a sidewalk but dropping down on both sides for hundreds of feet, the sensation is a spectacular, dizzying mix of exposure, beauty, triumph, and awe. The remains of what must have been a wild torrent in an early geological period now flows as knee-deep Roy Gap Run, trickling in the shade between a cleavage that breaks the wall into two parts. Roy Gap Run flows into the headwaters of the Potomac River, which runs for hundreds of miles to empty into the Atlantic Ocean. Most of the wall extends to the north of Roy Gap; the southern fragment stands like a partial Roman ruin, and is called the Southern Pillar. This is where two rock climbers, David and Kaine, were attempting their ascent. Suddenly something went terribly wrong.

"I heard the guy hit the tree as he flew through the air," recalled Brendan, a rock-climbing guide. "I looked right over and saw him fall the rest of the way." Brendan was teaching a climbing course at the Southern Pillar. The climber, Kaine, had just reached the top of a difficult climb called Judgment Seat before falling forty feet to the ground.

The sickening boom reflected off stone by the strange acoustics of mountains and reached climbers far across Roy Gap Run climbing on the northern face. "Is everybody all right?" someone called.

"We're gonna have to evac! We need a litter!" Brendan yelled.

"It's terrible," he said recalling the sound and sight of a climber falling to the ground. "I thought the worst."

Brendan ran to the victim. When he reached the fallen climber, the man was struggling to stand up. He was badly injured about the face and eyes; his fractured wrist bone was nearly puncturing the skin.

"We were worried about there being a spinal injury, but you know he had so much adrenaline going he believed he could walk out of there."

Within minutes climbers from Seneca Rocks Mountain Guides in town, only a short jog down the mountain, grabbed a metal stretcher (or litter) designed for extracting injured climbers from the mountain. The litter is kept out in front of the shop at all times for anyone to use in an emergency. Climbing instructors spontaneously marshaled students from Garrett Community College who were taking a class at the shop to form a rescue team.

"We do a lot of simulations with them, but this was full-on," Brendan said of calling on the students to help.

"By the time I got there they already had Kaine on the litter," said Jennie, another climbing guide responding to the accident.

"Dave was sitting on Roy Gap Road just looking horrible—just looking at the ground," she said. Disheveled-looking twenty-nine-year-old David DiPaolo was the partner of the climber who had just fallen.

As Jennie scrambled quickly past Dave up the steep slope, she could see them lowering Kaine down in the litter on a rope anchored to the same tree that had broken his fall. Six climbers manned the litter, three on each side, as they negotiated it gingerly down the loose rocky slope to gravel-topped Roy Gap Road, where the ambulance would be waiting.

"It wasn't a climbing accident. It was a lowering accident," Brendan explained.

Climbers typically ascend a rock in small teams. Two climbers tie themselves together with a rope 50 to 70 meters long. The lead climber ascends the rock, trailing the rope behind, and the second climber, the "belayer," threads the rope through a friction device attached securely to his harness so that he can catch his partner should he fall. As the climber ascends, he clips his rope through temporary anchor points he wedges periodically into small cracks in the rock. These are not used to assist his ascent; these points, called "protection," will catch him should he fall. The belayer pays out his end of the rope through the friction device to help him handle the enormous forces of a falling climber. The belayer makes a solemn compact with his climbing partner never to release his hand from the safety rope, not even for a second.

Rather than climb all the way to the summit of a mountain, for practice many climbers will climb only the first part of a climb (half a rope length, or "pitch"), and return back to the ground—especially if the segment is a difficult climb. In this way climbers challenge themselves and improve their skills by climbing the hardest pitch rather than simply following a multipitch route to reach the summit. This is what Kaine and David were doing.

"He had climbed it without falling," Brendan recalled of watching Kaine on the difficult route next to him as he taught his climbing course. "Very competent climber."

Kaine had reached a narrow ledge and he clipped the rope from his harness through the eyes of stainless-steel bolts drilled into the rock to provide the anchor point to lower him back to the ground.

"Ready to lower!" Kaine yelled, giving the standard command for his belayer, Dave, to lower him back to the ground suspended from his rope as it passed through the anchor. Facing the rock, he transferred his full weight onto the rope and stepped back into the air. Dave, on the ground below, began paying out his end of the rope, watching his partner carefully as he lowered him slowly. The rope warmed his palm as it slid smoothly through his clenched fist. When Kaine was about forty feet from the ground, Dave suddenly felt the rope vanish from his grip. The end of the rope had reached his friction device and snapped through it in a flash. Dave watched helplessly as his end of the rope ripped away into the air like a cracked whip and Kaine's arms and legs flailed in the air as he plummeted backward to the ground.

The rope was not long enough to lower Kaine all the way back down. The distance from the ground to the belay anchor on Judgment Seat was farther than half its length. Even so, this was an accident that should never have happened.

"You always want to have a closed system [when] climbing. So you would have each climber tied into each end of the rope, or you would at least put a knot in each end so the knot would jam up against the belay device and not go through it," Brendan explained.

David had not tied into his end of the rope.

Jennie remembers seeing Dave sitting on Roy Gap Road as she and the others charged up the slope with the litter. "When I got to the scene Dave was sitting looking very depressed," she said. She offered to take Dave back to town. "It was one of the most awkward car rides I've ever had," she said. "You know, we were trying to stay cheerful and stuff, but Dave definitely wasn't trying to be cheerful. He felt horrible and I think he was silent the whole time."

"You know, accidents happen. That's how you learn," she recalls telling him, trying to relieve his guilt as he sat silently in the backseat on the drive back to town. "But he felt horrible."

Dave didn't accompany his partner to the hospital either. The two didn't really know each other. In fact, Dave, who is a strong climber, had no regular partner at Seneca Rocks. He roamed the two climbing shops in town looking for anyone to climb with him. That day, it had been Kaine, someone he'd apparently just met. The consequences of that meeting would utterly consume Dave and dramatically change the course of his life.

It was not until a year later that the guide, Jennie, would see Dave at Seneca again. "He finally came up to me and he stuck out his hand and said 'Do you remember me, I'm that guy who dropped that guy at the Southern Pillar?'"

"I just said, 'Yeah I remember,' and I tried to make light of the situation, but that's definitely how he identified himself: as the guy who dropped his partner."

"From then on he was known as Carderock Dave—he dropped his partner. He's so incompetent. How could he do such a horrible thing? What an idiot."

Carderock is the name of a climbing spot in Maryland, where Dave had learned to climb from his mentor, Geoff Farrar.

# Carderock

David had been introduced to climbing as a kid at the popular climbing area near Washington, DC, called Carderock, an outcropping of rugged cliffs carved away by the Potomac River. Geoffrey Farrar, a lanky senior, his hair the color of snow on slate and wearing glasses with large lenses long out of style, had spent decades climbing at the popular suburban climbing spot and he made sure everyone knew it. He knew every route, every feature of the rock, and the correct sequence of moves to execute each climb.

"If you spent much of any time at Carderock, you were bound to at least *hear* Geoff's booming voice, typically giving advice," a climber said tactfully of him.

Among the climbing community he was known as Carderock Geoff, because he seemed to always be there dispensing unsolicited advice and political opinions. He struck up fervent monologue-loud conversations with everyone he encountered.

"Talked until blood shot out of your eyes," another climber said.

Nearly two decades ago, Geoff had taken Dave under his wing at the age of eleven and taught him to climb. Over the years Dave came to be known as Carderock Dave. Geoff called him Little Dave, as did others. Many, however, had less favorable nicknames for him, such as Stoner Dave.

"He always seemed more than a bit unusual—not violent per se, but definitely somewhat manic and certainly reckless. It was difficult to carry on a coherent conversation with him, and I thought he was going to kill himself the way he placed pro[tection] and climbed. I always felt the need to watch my gear closely when he was around," a climber recalled.

Geoff favored climbing without a rope, and he often did so right through other parties of climbers ascending under belay. When done a safe distance from the ground, ropeless climbing, called bouldering, is a challenging workout that quickly builds climbing muscle and skill. Those who practice the sport set up intricate traverses, moving laterally along the face of the rock just above the ground, that require awkward body positions, delicate balance, and a very strong grip. Some, however, begin to climb higher than is safe to do without a rope. Geoff would do this habitually. For many, it was difficult to see it as anything other than Geoff showing off. Such climbing without a rope is called free soloing, and it is extremely dangerous. A fall means certain injury or death. Geoff challenged David to do as he did, and soon Dave was free soloing some of the most difficult routes at Carderock that very few climbers could ascend without falling.

At age thirty-one, two years after dropping Kaine at Seneca Rocks, Dave went with Geoff for a climb at Carderock. It was one of those glorious breaks in winter when the frigid temperatures spike briefly up into the 50s. The next day, Sunday, there would be cold rain. The brief window of opportunity drew a few die-hard climbers to escape cabin fever three days after Christmas 2013. The air temperature on this cloudy day was reasonable, but the stone was cold, sucking warmth from bare fingers.

Another member of the Carderock family, John Gregory, arrived to climb that day too. Known as the Mayor of Carderock after thirty-five years of climbing at the site, he was often around. Fellow climbers found him outgoing while at the same time soft-spoken. John was never obtrusive, as many found Geoff to be. John wouldn't criticize or boast, but he was always available to help anyone who had any questions. John knew the rocks and all the history and gossip surrounding them. He wrote the official climbing guide to the region.

Geoff and David met in the parking lot about noon on that glorious winter day, but an argument ensued. John found Geoff distraught when he encountered him on the trail between the parking lot and the rock. "Well, I've been a friend to Dave for twenty years, but I guess it's over now," Geoff told John. The two parted to climb different segments of the cliff, John setting up downriver. John clipped the midpoint of his rope to an anchor at the edge of the cliff and tossed the two free stands over the precipice. Then he hiked down the same path Geoff would have taken to the base of the cliff to tie in and begin his climb. Rounding the bend he found his friend crumpled unconscious on the ground, bleeding profusely from his head.

John Gregory wrote the first account of the fatal accident, which was posted on the Potomac Mountain Club forum on December 29, 2013.

Geoff Farrar, AKA Carderock Geoff, was seriously injured falling from the traverse near Cripple's crack at Carderock this afternoon. He hit his head on a sharp rock and the edge of a railroad tie, sustaining head injuries including a fractured skull and jaw. NPS

[National Park Service] airlifted him out from the base of the cliff to suburban hospital. He is in the ICU, his wife is coming from WVA.

Geoff died later that day at the hospital. News of the accident swept through the climbing community via Facebook and online climbing forums. Many climbers, although stunned and saddened, were not surprised to hear the tragic news. Free solo climbing is a daredevil death-defying act. A local climbing gym posted an online note about the accident, which it twisted into a crass promotional pitch to climb in its gym with padded floors because that was far safer than risking one's life climbing outside on rock.

But the story that spread like wildfire over the Internet never made the newspapers. This was odd, considering that the popular recreational park was well known to most residents of the Washington, DC, area, and that this was a sensational accident involving the death of a local climbing legend.

The reason for the silence was a bloody claw hammer reportedly found next to the body. The autopsy showed that the injuries were inconsistent with those from a fall. Geoff's skull and jawbone had been crushed by the hammer.

From *The Washington Post*: "Pictures that have surfaced of DiPaolo show long scraggly hair and rough beard. His dress style could be described as grunge, and he is known for wearing mismatched shoes."

Several days later, police traced a call that David had placed from a pay phone at a gas station in upstate New York near his father's home. A New York state trooper pulled David over near Glens Falls, and took him to the police station.

"I'm sorry this happened. I didn't want it to happen. I didn't know it was going to happen," David DiPaolo told police in a written statement.

My last memory of Geoff was about a month before his death.

As I coiled my rope at the end of the day, he was shouting unsolicited beta (climbing advice) to a climber as she tackled a difficult boulder problem, telling her loudly that she was doing it all wrong. Geoff yapped away like a neighbor's annoying barking dog. His beta was solid, but moves ideal for a six-foot man were useless to a five-foot woman. When she triumphed, he congratulated her genuinely. "I've never seen anyone do it that way."

That was Geoff. He meant well. He did enliven the place and help many climbers new to the rock. He was a character. A nice guy. David was too, and although I knew him less well, I am familiar with his difficult backstory.

Tom Cecil, rock climber and owner of Seneca Rocks Mountain Guides, says about Dave: "I think he kinda looked up to me, as I think he did all climbers who were known." Cecil had established several of the popular climbs at Seneca Rocks, making numerous first ascents there and around the world. "Kind of a hero-worship thing—I'll bet he really admired and respected Geoff.

"I've always had a soft spot for that type—to a point," he said, meaning people who'd fallen on hard times. "The last time I saw him he looked very bad. I actually considered offering him some odd jobs to try to help him out, but he was just too fucked-up-looking and -acting," Cecil said, referring to David's appearance and druggie manner. While shabby dress and idiosyncratic personalities are a part of the climbing scene, they are not acceptable in a professional climbing business.

Tom Cecil recalled the accident two years earlier, when Dave had dropped his partner at Seneca Rocks.

"The accident was only partially his fault. The guy he was climbing with had cut his own rope the day before [to remove a dangerously frayed portion], but he did not tell Dave. They should have tied knots, but we all make mistakes," Cecil observed.

Another climber said of Dave, "I just thought he was sort of a burnout, that he had done too many drugs and possibly still did."

The climber continued, "I certainly never saw him as someone who would hurt anybody else. We always had a standoffish or eerie feel about him, but only in the sense that we were judging him because of his appearance and his inability to complete a sentence."

David DiPaolo was charged with voluntary manslaughter committed in self-defense.

David admitted to police that he and Geoff had had an argument at Carderock, and the next thing he knew Geoff was choking him. He and Geoff fell to the ground struggling. As David began to lose consciousness, he found a claw hammer on the ground and he clubbed Geoff with it until he let go. David provided a written statement insisting that Geoff had his hands around his neck during the entire struggle.

"I'm sorry this happened. I didn't want it to happen. I didn't know it was going to happen."

The Saturday morning right after the news of Geoff's death was published I went to Carderock. A deep sadness permeated the place. No one was there. Every crevice in that lizard-green-gray rock flashes a brief memory clip. Birthday parties with kids overcoming their fears on the rope, persevering on slick rock joyfully despite the rain. Sacred spots on the mud path at the base of climbs where friends once lay injured after falls. The psychic scars now healed like the bones that were broken.

When I first heard Geoff had died from a fall I was one of those who was not surprised. His free soloing was something I could not abide. For most people, climbing is about managing risk, not inviting it. But when we found out that his death was not from a fall, it was a different kind of emotional weight. A pall hung over the rocks on that cold silent morning in January. Whether manslaughter or murder, these two had been friends. It is a message that sinks in slowly.

A day hiker walked past and I thought how much harder this is for climbers. In all the cracks and bumps we see something others do not—hands gripping the stone now gone.

Newspapers reported that the police were investigating the possibility that Geoff suddenly attacked David during an argument and tried to strangle him to death. It is true that Geoff berated David for dropping his partner at Seneca Rocks. Geoff's incessant nagging and badmouthing of David to other climbers and even to strangers went on unrelentingly. "You know David? He dropped his partner at Seneca!"

*David dropped his partner!* That critical accusation spilled out loudly and automatically as if it were a hot news bulletin. Even two years after the accident, Geoff's boisterous monologues would boil up and scold David with disdain.

"Geoff was obnoxious," a climbing instructor admitted soon after Geoff's death, trying to puzzle out how this killing could have occurred.

The climber recalled seeing Geoff about a month before his death. "We had some students at the cliff [Carderock] and he came up and was just bouldering and soloing.

"He wasn't there with anybody and he was just talking. He was talking shit. You know, he was just trash-talking. He trash-talked."

"Do you know Dave?" Geoff had asked her, and he described Dave's appearance.

"Yeah I know Dave," she'd told him.

He said, "Yeah, do you know he dropped somebody?"

This, again, was nearly two years after the accident at Seneca Rocks.

She said, "And I was like, *Yeah, I know he dropped somebody.* Then he told me that a week before he did the same thing at Great Falls. I have no idea whether or not that's true."

She continued, "I just saw him as background noise—just clutter. I call it background noise because I could tune it out, but he was very much in your face."

Did she see Geoff as a strangler? I asked her.

"No. I don't believe that that happened. I do believe they were in a heated argument, in a yelling battle. I could see that they could get in a pretty heated verbal argument, but I don't see Geoff strangling him, and I also don't see this young strong Dave being helpless against Geoff."

On the other hand, it is difficult to understand how David could bludgeon his mentor to death.

"I was very surprised to hear that he hurt somebody or that he killed somebody," Jennie from the Kaine evac team said of Dave:

He was certainly never mean to anybody. You know, people just saw him as sort of a loser. I didn't think he was harmful. I just thought he was a burnout, a deadbeat; he just did too many drugs. But I really never felt threatened by him or anything. He was actually very soft-spoken. He was always just looking to have a conversation. He wasn't in your face like Geoff could be in your face. He was just into rock climbing and that was always what he would talk about—rock climbing.

I never felt threatened by him but I never sought him out either. I would see him in town and walk the other way. We never thought he would steal from us or hurt anybody, but at the same time we didn't want to be associated with him.

I'm not surprised to hear that he ran. You know, I think he's . . . I hope it wasn't premeditated in any way, and I do genuinely believe that he didn't go to Carderock thinking he was going to kill Geoff, but I could easily see him just freaking out not knowing what to do.

He really does seem like the perfect murderer just by his appearance. It is going to be hard for anybody to forgive him in any way because of his appearance and stuff [if it was a case of self-defense].

But Geoff was not violent either. "Never saw him get in [physical] disputes. I certainly never had any altercations with him," Jennie said.

Another climber related how he was devastated after a climbing accident in which his best friend fell thirty-five feet to the ground. He was seriously injured and had to be evacuated by helicopter to the hospital. "It took me years to get over it," he said. Although he was not at fault—the rock had fractured under his partner's weight—the climber blamed himself. "I felt responsible," he said. "If someone had harangued me like [Geoff] did David, I never would have gotten over it. It would have destroyed me.

"Geoff was like a dog with a bone," he explained. "Relentless. He'd never let go. He pushed his opinions on people."

On December 28, 2013, a smoldering argument between friends seems to have reached a flash point and exploded. Whatever happened that day at Carderock, this tragedy was clearly the result of someone snapping in rage. Different scenarios for what triggered the violence are consistent with what is reported to have transpired. David may have acted in self-defense as he claims and consistent with the police charge of manslaughter rather than murder. A manslaughter charge arises in this case because David's act of self-defense did result in a person's death, and the situation is compounded by questions as to whether such deadly violence was necessary for David to defend himself against Geoff. If this is what happened, then Geoff would have had to suddenly snap in rage and find himself with his hands clenched around his young protégé's throat, strangling him. By all accounts Geoff was not considered violent, but he was known to provoke angry arguments routinely that could have escalated to a point where he could have conceivably snapped violently. But if that is what happened, what could have triggered such rage in Geoff?

Another scenario is that David became enraged and brutally attacked his mentor, killing him. This would be murder, not manslaughter. Some have speculated that authorities may have brought manslaughter charges against David because the high standard of proof for a murder conviction may have been difficult to meet with the evidence at hand. Remembering the brutal killings in the OJ Simpson trial, one can perhaps understand a prosecutor opting to settle for a plea bargain and having the attacker jailed for fifteen years for voluntary manslaughter rather than risk a murderer going free. OJ Simpson was found not guilty in a criminal trial where the standard of proof is high, despite incriminating DNA and mounds of physical evidence implicating him, but he *was* convicted of the killings in a civil trial. If the Carderock killing was a case of murder committed in rage, what could have caused David to snap and embark on a savage homicidal attack against Geoff, his lifelong mentor, using a claw hammer to kill him?

The trial has not yet taken place, so we have no verdict. It is crucial to maintain the presumption of innocence for all parties until the legal process completes its course. While we wait, let's consider how neuroscience and the LIFEMORTS triggers might provide some understanding of this horrible and perplexing death resulting from one or two friends snapping in rage. If, as David states, he was being choked by Geoff, his violent reaction makes sense. Being throttled would trip the Life-or-limb trigger to fight for your survival. Without air filling David's lungs or blood flowing to his brain, his own death was only minutes away.

If this is what occurred, David could be viewed as the unfortunate Lennie in Steinbeck's *Of Mice and Men*, a tale that was partly inspired by a pitchfork rage murder the author had witnessed. Lennie unintentionally killed someone and then faced a lynch mob. But to make the analogy fit, it would have been Lennie's protector and mentor, George, who would have become the inadvertent victim, not a flirtatious woman he'd just met. If Geoff died at the hands of Dave in self-defense, what happened at Carderock is even more tragic than Steinbeck's tale.

Even in self-defense, though, was such savage lethal force—crushing Geoff's skull repeatedly with a hammer—necessary? It seems difficult to accept that a very fit thirty-one-year-old man could not break away from a sixty-nine-year-old without resorting to such a brutal killing. The thing to understand is how this all plays out rapidly in the brain during a violent encounter, rather than judging afterward what might have been the ideal reaction. One does not have the luxury of hindsight where precious fractions of seconds are spent in deliberation. A response in a life-and-death battle has to be instantaneous.

If the situation had reached what was indeed a life-or-death struggle, *unlimited* violence to kill the attacker—no matter how savagely—takes over. Survival instinct kicks in. Recalling my physiological response in Barcelona, I can vividly recall the superhuman surge of power that infused my body in a flash with adrenaline-fueled strength as I consciously fought with all my will not to pick up the hoodlum squaring off against me, hoist him over my shoulders, and hurl him into his comrades like a battering ram to send them all tumbling down the subway steps. That internal struggle was my frontal lobes calculating furiously and suppressing the deadly violence that was on the verge of being released by defense circuits deep in my brain. Once a situation explodes into a fight for survival, the body is instantly and fully committed to fight to the death if necessary to kill the attacker with any means available. This potentially deadly action is a biological imperative. It is the consequence of the harsh evolutionary logic of survival hardwired into our being. There is little or no choice when a violent battle reaches this point, because your opponent's brain has also engaged these same automated circuits of violence with the intention to kill you to preserve his own life. Your brain "knows" that. At a level beneath the realm of conscious deliberation—this is the last-ditch equivalent of the unthinkable intercontinental "nuclear response" to threats our nation has faced, or imagined.

But this brain wiring means that nearly all violent struggles will trip the Life-or-limb trigger at some point in a violent altercation, just as Ray Young, the sixty-seven-year-old man who was convicted of knifing a stranger who he thought had cut in line at the post office, perceived. "I was defending myself . . ." Our legal system recognizes the biological imperative for self-defense by absolving a person from fault if the homicide is *truly* the result of necessary self-defense in an unprovoked attack. But self-defense means that the Life-or-limb trigger must be what *unleashed* the deadly violence, rather than becoming tripped at some point during the altercation, because the Life-or-limb trigger will always get tripped in both parties in a violent struggle. If the hammer hadn't been lying there, as reported, within Dave's easy reach during his struggle with Geoff, the episode might have had a very different outcome.

"What in blazes was the purpose of a claw hammer at Carderock?" one climber wondered.

Climbers once used hammers to drive pitons into cracks for protection, but pitons are almost never used anymore. A climbing hammer is much different from a hammer you would find at a hardware store, and pitons are never used at Carderock. A claw hammer might have been lying there being used "to replace wood at the base of something," another climber speculated, discussing the homicide online.

Facts point to the Insult trigger regardless of which person instigated the deadly violence. David's failures in life—in dropping his climbing partner and his recklessness climbing in general, in his slovenly personal appearance and dysfunctional social manner, in not achieving independence and success in life by the age of thirty-one—could all have been taken as an insult by Geoff after twenty years of mentoring him like a son. If David was a failure, then so was Geoff. Did the strain of the persistent and growing insults, of his protégé David becoming the object of his derision, a kind of pariah of the climbing community, push Geoff to rage? On that Saturday afternoon, sparked by some heated argument, did a breaking point trigger Geoff's rage attack against Dave?

We don't know for sure, and it just as easily could have gone the other way. There is no question that David was the object of repeated and prolonged public insults by Geoff, especially in the last two years. With the mounting tensions and apparent stresses in David's life, it is easy to see how if the argument in the parking lot had involved a direct insult to Dave, the mounting pressure could have tripped a violent, even murderous, reaction against Geoff.

This case study of violently snapping typifies so many rage attacks, and it illuminates why they always strike us as shocking and often incomprehensible. Human beings are complex, but the essential complexities underlying most rage attacks that we read or hear about in the media are rarely if ever made public. News journalists face impossible challenges. They must gather and synthesize information into an article within hours of the event—finding original sources to interview and sorting through all the information to compose an accurate report of what occurred and why. The feverish broadcasting of news and gossip through blogs and Twitter have only exacerbated the problem, increasing pressure on reporters to work faster while increasing the clutter of false leads and rumors they must sift through that come from amateur reporting over the Internet. To this, add the severe constraints in criminal investigations where sensitive information released to the public about a suspect could sabotage an investigation or undermine a conviction. The Carderock homicide is no different. Except that in this case I unfortunately happen to have some insight to what happened from knowing the people and their background, which is never possible when hearing a news story secondhand. My hope is that by forcing us to look in detail at this tragedy, something of value can come of such senseless loss of life.

The few days after David DiPaolo was indicted on federal charges for voluntary manslaughter, I went to Carderock to meet with John Gregory. About five weeks had passed since Geoff's killing, but there had been no new information in the news since David's arrest on January 8. As the "mayor" of Carderock, John was eager to speak with me about my research into the triggers of rage. He had much to tell me that would be pertinent, he promised.

A frigid cold snap had left sheets of ice lining both banks of the green Potomac River. Snow still collected in hollows and shadows, but the temperature had spiked and climbers were out this afternoon with the winter sun. I met John on the footpath at the base of the crag. I spotted him from a distance by his neatly cropped full gray beard and mustache, his trademark cap, and his daypack slung over his shoulders.

He took me to where he'd found Geoff, head resting on a railroad tie bordering the outer edge of the trail, his body lying diagonally across the path. He molded the shape of the body with his hands in the air as he evoked the scene. John had been the first to find Geoff. He showed me a shallow hole next to the wooden tie about the size of a volleyball where there had been a rock covered in blood. The police had dug it out and taken it as evidence.

We left the chilling spot and walked back along the trail as John recounted what he had seen that Saturday afternoon.

Describing how he was first on the scene he said, "I couldn't recognize him. I couldn't recognize it as being human at first. It had no head, just a purple bloody mass." Geoff's head had swollen up massively and John didn't realize it was him until he recognized his climbing shoes. "I recognize you as human because you have a head and a face. Geoff did not." Geoff's skull was crushed; his jaw broken; his eye out of its socket.

John said he realized immediately that it wasn't a climbing accident, from where the body lay with respect to the climbing route and because the victim had no scrapes or other injuries on his legs and arms that a climber would have suffered in a fall.

"The single most important thing to know is that for both the victim and the perpetrator, the argument in the parking lot was the end of a twenty-year relationship. That changed everything. Geoff said so," John explained.

"Saturday the twenty-eighth, about one o'clock, I'm meeting my partner here and I'm unloading my gear from the back of my Jeep in the center of the lot. Geoff pulls in and says, 'Have you seen Dave?' And he says, 'Well, he practically got himself punched out yesterday,'" referring to an incident in which Dave got into an altercation with other climbers and made comments so offensive the others nearly came to blows.

> Geoff had what a gambler would call a "tell." He had little tics that he did whenever he was excited or agitated or arguing. He would start to swing his arms like this and rock back and forth on his feet. It's the damnedest thing.
> And he was starting to do that while I'm getting my pack out and telling me, "You know, Dave tried to steal a jacket, and Dave got in an argument," and on and on and on and it didn't seem to make much sense.
> This goes on for a while and then right across the parking lot there is a silver minivan, and I didn't realize that that was Dave.
> He gets out of the car and he comes over. You know, I didn't know he was there. He's alone and as disheveled-looking as usual, if not more; wearing one white shoe and one black shoe; clothes that look like they have come out of a Dumpster.

An argument immediately erupted between Geoff and David. Angry insults were hurled back and forth. One man reportedly challenged the other to a fistfight, but the argument in the parking lot never boiled into anything physical, only yelling insults at each other.

Until the trial concludes we will not know who, if anyone, is criminally responsible for this homicide, but in the meantime we can try to comprehend from a neuroscience perspective alternative scenarios that could have unleashed such unthinkable violence between longtime close friends.

"I mentioned he could be antagonizing and he *was*. . . . *God* yes," John replied to my question of whether Geoff's penchant for arguing could have fueled the dispute. "If he'd kept his fucking mouth shut we wouldn't *be* here! If he'd just kept his fucking mouth shut!" John whines with a deep emotional mixture of anger, regret, and fatalism. "Oh my God. *Oh* my God! You know? Did you *have* to do that? Did you *have* to do that [engage in a yelling verbal argument in the parking lot with Dave]?

"But he would confront people and sometimes provoke people," he says, releasing his exasperation.

I mentioned to John that every time I saw Geoff he would rag on David, out of nowhere.

"Yes. Out of nowhere!" John agreed.

I asked John if Dave dropping his climbing partner had changed their relationship.

"I think so. I think so. Geoff took it personally, because he was the one who had taught David to climb.

"Geoff did mention [dropping his partner] a number of times. Other people heard him mention it. Geoff was harping on that, and that Dave did not belay very well when he was here. People were getting to the point where they just stopped letting him belay."

On the other hand, David could provoke altercations, as apparently he did only the day before if Geoff's accusations about David attempting to steal a jacket were accurate. Many saw Dave as a drug user, reckless climber, and someone to be avoided.

David expressed remorse in his statement to police when he was arrested. The altercation that resulted in homicide is a deep tragedy and a profound loss for everyone involved. Geoff is gone. David is in custody. Friends and family are grappling with the sudden loss and are grief-stricken. It is horrible.

"No one would have thought that what happened there [the argument in the parking lot] would trigger lethal results," John says in disbelief.

"The first thing I did is look up [at the route from where Geoff had apparently fallen]," John said, coming back to recount his description of the tragic scene.

"When I first see it, I didn't know it was human," he reiterates from before. "Then I realize he is breathing very heavily, blood gurgling. I heard my partner behind me say, 'I'll go find a phone.'"

John went forward. "His breathing was OK—not great, but OK. He wasn't going to expire right there. He had a lot of blood, but he was not bleeding to the point that he was going to bleed to death here. At that point I see there are legs—OK, this is a person. Then I looked down and saw his old Mariacher climbing shoes, and I said, 'Shit! It's Geoff.'"

John maintains his professional composure, but as he relives the trauma of finding Geoff lying there dying his nose begins to run. His metered sentences are precise, objective, and clinical, but they are punctuated by sharp sniffs to check his running nose.

"At that point I knew who it was." He called 911 and then fire and rescue dispatchers. "This kid Jimmy comes up and says, 'I'll go get his vitals.' Gives the dispatchers the pulse and respiration. They ask me, 'Do you need the boats?' and I said, 'Yeah give us the boats. You are going to need a Stokes [litter], a backboard, a C collar, and oxygen.'"

The first fireman showed up with all of those things. No paramedic. Everybody looked at that injury and stepped back. They didn't have a dressing big enough to put around it. A park policeman followed the sirens and got a helicopter. Did the most amazing extraction I've ever seen. They swung the cable in, attached it right over the river. That guy was hovering right above the trees. I didn't get to see too much of it because you had to get down on the ground and literally hold on to everything you had and you were getting pummeled by all the dead tree limbs, and all of that mulch [from the trail] now flying at you at about a hundred miles per hour. I was hiding behind that big rock to get out of the wind. He pulled that thing in there and they were out of there and gone to Suburban [Hospital].

It was a very quick extraction and far as the "golden hour" goes, Geoff had everything going for him. But I remember, you know, very early on, in fact when I was dialing the phone, saying, "Shit. I hope he doesn't live." [He sniffs again sharply.] Because this is *terrible*. This is just terrible. You know, *surviving* this is not good. It is not the best outcome [sniffs again].

His wife calls on her way to the hospital. They're asking her, "What are his wishes?" And they say, "Look, his brains have come out of the cracks in the sides of his skull. Things are not going to work out here."

She said, "Well, let him go."

[John sniffs sharply again but maintains his composure.] He died at eight o'clock that night. She arrived around nine thirty.

So, you know, that's uh . . . [long sigh] . . . That's about the size of it.

It's a . . .

He looked down, digging his toes into the path. Silence completed the sentence.

At the time of this writing, the public has not yet heard this account. A trial or a conclusion to the legal proceedings through a plea bargain awaits at some future date. Even if John's witness account comes out among climbers communicating online, it will not reach the papers or news stations. There have been half a dozen murders in the community every day since Geoff's death. This is old news.

A few months after I talked with John Gregory about the violent killing, a remembrance was held for Geoff Farrar at Carderock. Billowing cumulus clouds skated on a stiff warm breeze through a blue sky as spring blossomed at Carderock. The remembrance was held in the parking lot—the same parking lot that had been the scene of the deadly dispute. Geoff's friends and family came along with many Carderock regulars. Nothing formal had been planned. Introductions and small talk filled time as people gathered.

"Chris went to the hospital with Geoff," his wife said, pointing out Chris and introducing different people in the small assembly of about twenty-five. Det. Glenn Luppino and his partners from the US Park Police were there to ask "anyone with any information to please come forward." He passed around his business card.

"He worked twenty-four hours a day to track down Dave," John Gregory said to the crowd, and gestured toward Detective Luppino.

People took turns recalling snippets of Geoff's life and then everyone lunched on hot dogs, pork and beans, and potato salad set out on folding tables at curbside under a tree.

Whatever the reason for this homicide within our climbing family, whether legally defensible or criminal, anger and vitriol toward David wells among many of those mourning Geoff's death.

"If I saw him now, I would kill him if I could," someone who had climbed with David said.

"That would solve nothing," someone quietly rebuked.

"I know."

The case would be submitted to the grand jury that week, and many of those present were going to testify, "To seek justice."

Suspended in the atmosphere was a heavy feeling of senselessness, emptiness, anger, and sorrow. Mostly it seemed unreal. In the end, with everyone standing around in a circle—all the characters in the drama, including the detective—it felt like a curtain call at the end of a play when the actors stand onstage and the illusion shatters. Villains and victims, they are just people.

After rejecting the manslaughter plea that was offered him, David's public defender entered a plea of Not Guilty by Reason of Insanity. The plea incited intense disgust among many who knew David and Geoff.

Insult is a trigger for violence that extends through all cultures, passing back in time to our earliest ancestors. It originates from violent encounters between social animals, especially males, as they sought to establish dominance. Seen in contrast to all civilized morality and sense of decency today, snapping in violence in response to insult is not tolerated. However, the brain wiring that ignites this intense anger and the commitment to rapidly engage in violence to achieve social dominance is understandable from the perspective of where the human brain originated—in the harsh animal struggle for survival. Human beings, like other highly social creatures, cannot survive outside a social organization. One's access to resources, which translates into survival, is determined by social dominance. So it was for humans in prehistoric times, battling for dominance over access to the basic necessities of life (food, shelter, reproduction), and so it is today with regard to the same necessities of life, although they are represented more abstractly in our highly evolved and complex society. One's rank in society ultimately translates into resources and comfort. People will fight for social status within their group. Language is the mechanism of social interaction between humans and used in achieving social rank, making insult a personal challenge to an individual's dominance and thus a real threat. Why else would you feel a rush of intense anger when another motorist elevates his middle finger toward you?

Reportedly the argument between Geoff and Dave in the parking lot, whatever sparked it, rapidly escalated to challenges between the two men to resolve the dispute in a fistfight. If only the parties involved, or even any bystanders, had been aware of the neuroscience of snapping and the potentially deadly implications of insult, especially when delivered on top of the weight of prolonged stresses pressing on that same LIFEMORTS trigger, Geoff and Dave might be climbing together at Carderock today. Or their twenty-year-old close friendship could have still come to an abrupt end in the parking lot but this horrible killing might never have occurred. Insights into the neuroscience of snapping are only now emerging from new research, so this new perspective is not generally known. My hope is that this will change. I hope that people will come to recognize that we all have the potential for explosive violence as a necessary brain function and that children, especially teens, will be taught in school about this dangerous aspect of our biology. Understanding the biology of snapping in rage is the first step toward controlling it and to preventing such horrors.

## Intolerable Insult

Until fairly recently it was considered acceptable to respond violently to insult, to defend one's honor in a deadly contest. The use of deadly violence in this way was even celebrated, provided that strict rules for the violent resolution were followed, just as it is with animals that butt heads for dominance.

Aaron Burr took deliberate aim and fired first. Alexander Hamilton fell mortally wounded. A dispute between political rivals—US secretary of the Treasury Hamilton, and the sitting vice president Burr—was settled by a duel early in the morning in 1804. Burr walked over to the wounded man and feigned an expression of regret. Then he left the field where the two men had agreed to settle their dispute with pistols. The rest of the day Burr conducted business as usual, tending to real-estate matters. The bullet had penetrated Hamilton's rib, passed through his liver and diaphragm, and embedded in his second lumbar vertebra.

Mortally wounded, Hamilton was carried to the residence of a friend, William Bayard. "Let her be sent for," one of the men present said, referring to Hamilton's wife, "but break the news gently to her and give her hope."

Mrs. Hamilton and six of their children arrived and were brought into the room where the dying man consoled his wife with his dying breath: "Remember, Eliza, you are a Christian." Hamilton suffered in great agony, lingering until the next day when he died at two in the afternoon.

Throughout history, dueling has been accepted across cultures and around the world as a manner for settling disputes. The practice extends throughout Europe, South America, North America, and Asia. Mark Twain narrowly avoided a dueling contest with a rival newspaper editor. The seventh president of the United States, Andrew Jackson, fought at least two duels. President Abraham Lincoln, when he was an Illinois state legislator, met to duel with state auditor James Shields, but the two backed down at the last minute. Gunfight duels of the Wild West are legendary: Wild Bill Hickok and Davis Tutt, Doc Holliday and Mike Gordon, along with thousands of others settled their disputes and defended their honor in a formalized deadly contest intended to leave only one man standing. Duels in the seventeenth and eighteenth century in Europe were fought with swords. Regardless of the deadly weapon used, the victor in a duel was not regarded as a murderer but rather as a brave hero, respected for standing up for his honor. The winner of a duel frequently enjoyed increased status; conversely, losing or walking away from a duel decreased it.

It is difficult to accept that we humans, and especially males, are wired to use quick violence to establish dominance, but this is our biological legacy. Our brain developed the reflexes to generate an intolerable sense of anger and rage in response to insult because our brain evolved in an environment where this was vital. Our environment changed radically and very quickly, but the brain today is the same organ it was when our ancestors lived in caves.

Modern society is so unimaginably different from the world our ancestors lived in during the period when the human brain evolved that the brain's innate capabilities for violence seem incomprehensible. To recognize the situation more clearly, imagine yourself in the wild environment our ancestors inhabited. Imagine that you awake alone in the wilderness. How are you going to find food and water, shelter, and protection from dangers? So removed are we today from the natural environment that the human brain and body evolved to inhabit, being alone in the wilderness now is cause for launching frantic search parties to rescue the individual before they die.

So, imagine you awake and now have a few hours to find something to eat. Humans don't even have natural "clothing" to survive outdoors. How will you survive alone? The answer will differ somewhat depending on whether you are male or female. Let's say you manage to catch a squirrel— very unlikely, but for the sake of argument we'll say that you caught one. Forget about cooking it; you must eat it raw. Just as with other animals, nature provided you with nothing to make fire and no cooking vessels. You have no knife, fork, spoon—nothing. You must skin and gut it with your teeth and fingers. Now the wide variety of specially shaped teeth we have evolved will become evident. They are nature's equivalent to a Swiss Army knife for survival. Your canines are for grabbing meat to tear flesh, incisors for cutting, molars for grinding and mashing. You must succeed at this every few hours to obtain enough food and water every day to survive.

In place of tooth and claw, nature has given humans incomparable intelligence to fashion tools for survival, but individual intelligence is not sufficient. It is the ability of the human brain to share individual knowledge with other members of the species through language that enables our species to flourish. Would you alone be able to invent how to make fire, to tan leather, or to make pottery? And in the centuries ahead, could every human independently invent algebra or make an iPhone? Human knowledge—as a society—grows exponentially like compound interest, building upon all the accumulated intellectual wealth of other minds in times past. Isolated from society, a big brain is like a microprocessor without a circuit board.

Now imagine in prehistoric times if you are seven months pregnant; how would you survive in nature alone? Imagine you are a male and your wife is nursing a newborn and tending a toddler; how would your roles have to differ by sex? If you were a man, your job would be to kill a woolly mammoth with a sharp stick. That takes a certain amount of fortitude, bravery, bodily strength, reflexes, and brain circuitry to enable you to approach dangers and to work with other men to bring down such dangerous prey for food. You, man or woman, need to become a part of a mutually interdependent society to survive.

These imperatives are what directed the specialization of the human brain and body. This is what made our brains what they are today. Just as we may not appreciate our Swiss Army knife dentition in a world of tender and nicely prepared cooked food, it is easy to overlook that large portions of our brain are devoted to threat detection, social interaction, and rapid physical response to danger. The imperative of cooperation for survival is what made male and female roles different and modified the brains and bodies of men and women somewhat differently for maximal success for their distinct roles in the wilderness. Human survival in the wild meant the formation and protection of family and tribe and an adherence to strict rules of conduct and social interaction. None of these imperatives have changed. Viewed from this perspective, the automated LIFEMORTS triggers of rage, which seem perplexing and disconcerting at first, are understandable as vital neural circuitry.

# Insane

Do you believe in the insanity defense? It sounds so simple, even self-explanatory. But it is most definitely not.

A man on trial for murder testified that he heard voices. He imagined his girlfriend sucking his blood. He thought she was a devil. "If this devil is not dying, I would be dying," he told the court, explaining his mind-set at the time of the grisly murder.

"If a person was not of right mind," the man's defense attorney, David Martella, argued in court, "they should not be held criminally responsible." (The legal term "not criminally responsible" is commonly known as the insanity defense.)

The killer testified that he must have blacked out. All he remembers is finding himself sitting on his girlfriend's body with a bloody knife in his hand. He does not remember stabbing her at all.

No one questions the fact that he did slay his girlfriend with a knife in her own bedroom. Barry Kin Lui says that he was sleeping on the floor in a storage room in his home. To be precise, Lui was no longer the legal owner of the property. He had transferred the deed of his house to his girlfriend a year after their relationship began. His girlfriend, Lan Mu Do, kept barging into the storage room that fateful night, flipping on the lights every half hour to argue with him. The lack of sleep and stress from constant arguing caused the sixty-three-year-old man to hallucinate, Attorney Martella explained.

He must have hallucinated that he was stabbing "a Dracula," the defense contended in arguing that Barry Kin Lui should not be found criminally responsible for the savage killing.

No one claims the killer had a history of mental illness. Such a history, however, is not necessarily required for the insanity defense. "What matters is the person's state of mind at the time—temporary insanity," says University of Alaska psychologist Bruno Kappes, an expert on the insanity defense.

County prosecutors, however, said that Lui did not appear to be emerging from a blackout at the time of the murder. Hearing his mother's blood-curdling screams her son rushed into the bedroom and found Lui sitting on his mother.

"Call 911," Lui told Do's son. "I've killed your mother."

This seems to prosecutors to be a rather lucid assessment of what had just transpired, and to whom.

Lui contends that he began to suffer a mental breakdown after Do black-mailed him into signing over the deed for his house to her. Lui told the jury that his girlfriend had threatened to tell authorities that in exchange for $35,000, he had entered into a fake marriage with another woman who needed a green card. The trial will determine whether Lui should be confined to a mental institution or sent to prison for murder.

Bruno Kappes has served as an expert forensic psychology witness in a wide range of court cases. According to Kappes, the insanity defense is highly ambiguous and confusing. In fact, there is no uniform insanity defense across the United States; it varies greatly among different states. The following range of insanity-defense verdicts is available in different states:

Guilty but Insane

Guilty but Mentally Ill

Acquitted by Reason of Insanity

Not Guilty by Reason of Insanity

Not Guilty by Reason of Mental Disease

Each of these verdicts is quite different. When you answered the question at the beginning of this section, which of these were you thinking of? Were you thinking that a person committing a homicide was insane if he could not distinguish right from wrong? This is the McNaughton definition of insanity, but that is not the only way the insanity defense is viewed. According to the Durham rule used in other states, the insanity defense can only be applied in cases of mental illness. Other states accept "irresistible impulse" as an insanity defense. (Would that define snapping in some instances?) The American Law Institute requires both the conditions of the McNaughton rule (knowing right from wrong) and that irresistible impulse occurs at the time of the crime.

After the Insanity Defense Reform Act of 1984, the burden of proof moved to the defense to provide clear and convincing evidence supporting an insanity plea. This act was put into place in response to public outrage after John Hinckley Jr. was acquitted for the attempted assassination of President Ronald Reagan with an insanity defense. This change in law makes it significantly more difficult to obtain a verdict of Not Guilty by Reason of

Insanity, because the government is no longer required to prove that the defendant was sane at the time of a murder beyond a reasonable doubt. Furthermore, expert witnesses for either side were now prohibited from testifying directly as to whether the defendant was legally sane or not. "Meaning, you cannot in court say that, in my opinion, the person is sane or insane," Kappes explains from his perspective as an expert witness. "Because that's not our language; we talk mental illness. We don't talk about insanity—that's a legal term."

And because the legal definition of insanity is not fixed or uniform, mental experts are placed on the stand to speak in favor of the defense or the prosecution, but neither testimony has any ultimate legal certainty. The jurors simply listen to the opposing opinions and do what they think is best.

Four states have no insanity defense at all: Montana, Idaho, Utah, and Kansas. These same four states support the death penalty. Twenty states use the "Guilty but Mentally Ill" defense.

The insanity defense as presently defined and practiced, says Kappes, is a tragedy. No verdict or definition seems acceptable to everyone, because the insanity defense stretches between two opposing objectives: to provide justice for criminal acts, which is essential to maintain social order, but also to show compassion for the ill. The interface between psychology and the legal system has little overlap to help unite these competing objectives of justice and compassion. Psychology is motivated to provide treatment, understanding cause and effect, assisting the patient in recovery. The role of the legal system is to assign responsibility and to assist the court in making decisions regarding punishment to maintain social order.

Regardless of the various definitions for the "insanity defense," Kappes believes that jurors, who are the ones required to make this difficult judgment, are ill equipped to do so. According to Kappes, 67 percent of jurors believe insanity is a medical term. That is not the case. Insanity is a legal definition without clinical meaning. Many argue that it makes little sense to give people who have never grappled with mental-health issues the authority to make these difficult and often life-or-death determinations.

Some have speculated that one of the Boston bombers, for example, was suffering from schizophrenia—namely the older brother, Tamerlan Tsarnaev. This assessment is supported by little more than the public statements of an infirm old man, Don Larking, who himself suffers from traumatic brain injury, but many people who are struggling to comprehend the heartless and horrific act committed during the Boston Marathon on April 15, 2013, turn to insanity as the only way to reconcile the cruel crime. How could anyone commit such a vicious crime that wounded 264 innocent people, many of them grievously, and killed three, including an eight-year-old boy who was watching the race with his family when the bombers planted their explosive backpack next to him? Another person was killed later, a policeman who was ambushed and executed by the bombers during a failed escape attempt. With Tamerlan killed in the shootout with police, speculation about his possible insanity is now moot.

Speaking in his office at the University of Alaska about the younger bomber, Dzhokhar Tsarnaev, a year before the trial was held, Dr. Kappes asks, "What are we going to do about this guy? Right now they are not going to go for Not Guilty by Reason of Insanity; they are going to go for Diminished Capacity, because they are going to say that his older brother influenced his behavior."

Dr. Kappes explained that a common example of the Diminished Capacity defense is when a person commits a crime while drunk. They are not mentally ill, but they do not have full control of their actions either. This will distinguish murder from manslaughter. "Alcohol or drugs diminished their capacity to make a reasonable judgment," he explains by example. "In this case, they are going to say that his older brother influenced his behavior, because he was a nineteen-year-old at the time. This would have diminished his ability to have planned and plotted to do this whole thing, and I think they are going to succeed on that; however, the Boston people are going to be angry. They want death."

(A year later, when the trial was held in Boston, Dr. Kappes's predictions would prove to be remarkably insightful. The defense did adopt the strategy of blaming the older brother for leading his younger brother astray, and the jury rejected the argument, finding Dzhokhar Tsarnaev guilty of murder on April 8, 2015. After the verdict, Michael D. Kendall, a former federal prosecutor, predicted how the jury would decide on Tsarnaev's sentence: "The big choice for the jury is going to be which is more cruel, life without parole for a young man or the death penalty. They'll pick whichever they think is worse." On May 15, 2015, the jury sentenced Dzhokhar Tsarnaev to death.)

The movie-theater shooter from Aurora, Colorado, James Holmes, is another prime example. Here's the quandary: Looking at Holmes's bizarre appearance and demeanor at the time of his arrest, any jury would likely take it as corroborating evidence that he was insane at the time of the shooting. Thus, the death penalty would not be permitted. In pre-trial negotiations Holmes's defense offered to plead guilty provided that the death penalty was not imposed, but prosecutors, appalled by the horrific crime and armed with overwhelming evidence, refused. Now prosecutors will have to hear the insanity defense that will preclude the penalty of death they desire, and instead could result in the defendant being committed to a mental illness facility rather than sent to prison.

The fact that Holmes had been treated for mental illness prior to the crime does not necessarily prove the insanity defense. "It doesn't matter what your prior mental issues were," Kappes says. "What are they at the time of the crime? You may have a mental illness, but it may have nothing to do with how you behave at the time of the crime." This is a very murky and difficult judgment for twelve untrained average citizens to make by listening to arguments in court, he says.

Both the conscious and emotional brain that we have been exploring in understanding why we snap enter into an insanity defense. In some states the belief is that the McNaughton defense of "knowing right from wrong" refers only to the cognitive level, not to the emotional level. "For example, Andrea Yates, who drowned her five kids in Texas . . . she knew cognitively that it was wrong, but emotionally she didn't. Emotionally she believed that she was a failure and that she was hurting her kids by being an unfit mother." Texas did not have the "volitional standard" in the law, Kappes says: "They prose- cuted her on the cognitive standard that she *knew* it was wrong because she called the police." A different state would have had a different set of criteria for the insanity defense applied in Yates's crime.

Guilty but Mentally Ill, "That's an oxymoron," Kappes says. "If you are mentally ill, you *cannot* be guilty," he cries. "You *cannot!*" The distinction among these different "insanity" defenses matters a great deal. "Either you are treated by the criminal-justice system or you are treated by the mental-health system."

In his research, Kappes studied 337 people who were asked to read the facts of the Holmes murder case in Aurora, Colorado, and select the verdict and punishment.

Thirty-four people voted Not Guilty by Reason of Insanity for Holmes. The vast majority said that he was guilty. By profiling the people in the study, Kappes found that the people in these two groups were quite different in several ways. Those voting guilty already had very negative views of the Not Guilty by Reason of Insanity (NGRI) defense. They believe the insanity defense is overused. They believe that a person convicted under NGRI will spend less time in jail. They believe that the person is usually faking. They believe that the defense attorneys are hired guns only looking for acquittal.

How often is the insanity defense used in murder trials? Kappes says 76 percent of people believe that 5 percent of all trials use it. "In fact, the insanity defense is only used in one percent of murder trials."

He also found that 85 percent of people in the study believe that "competency" in the legal sense pertains to the killer's state of mind at the time of the crime. In reality the term relates to the individual's present state of mind *during trial*. Competency is whether the individual is able to aid in their own defense in the court proceedings. "So a lot of the terms [jurors] are being asked, they don't know—don't even have a clue about it."

Kappes then found that if these same people were instead given three possible verdicts, which included Guilty but Mentally Ill, there was a different outcome. When given two choices, 88 percent said Guilty and 12 percent said Not Guilty by Reason of Insanity. However, half of the people in the study selected the third choice when it was available. They did so because they mistakenly believed that individuals convicted as Guilty but Mentally Ill would get treatment. On the contrary, Kappes says, this group receives dual punishment—subject to both the criminal-justice system and the mental-health system.

"It is a misnomer. It is an attempt to relieve our frustration of what to do with these people. By giving jurors this choice, they think that they are doing something on behalf of the client, but the judge is not allowed to tell the jury what are the consequences of these various verdicts. Because if they do so before jurors make the verdict, there is the fear that that would influence the verdict that they would use. So you are being asked to give a verdict without any knowledge of the consequences of that verdict, except in the death-penalty states."

Whether individuals in the study had a history of mental illness, currently suffered mental illness, had been convicted of a crime, or whether they were a psychology major or a criminal-justice major had no influence on whether they favored the Not Guilty by Reason of Insanity verdict. People who favored the insanity verdict did, however, share an opposition to the death penalty.

Four states have no insanity defense but do have the death penalty for murder. "If you are caught in one of these states, you don't stand a chance if there is a psychological reason as the basis for your criminal conduct," Kappes says. In arguments put to the Supreme Court that the death penalty in cases of mental illness is unconstitutional, the Supreme Court has determined that "The state's right to put someone to death supersedes the individual's right to a fair trial. The state has a right to carry out its business," Kappes says.

"In Texas, they just put this guy to death who had an IQ of sixty-one. Yet we believe that children, minors, animals who are not cognitively capable of making rational decisions, do deserve some kind of compassion on our part.

"It has been going on like this for hundreds of years!" he exclaims. "I thought I could bring some sanity to insanity [through academic research], but a lot of this has to do with reconciling psychology and the law—mental illness versus insanity. Psychology is interested in the laws of nature. The legal system is interested in the laws of man."

Kappes is still collecting data for this research. He brings up the most recent data on his office computer screen. "We should have five or six hundred [respondents] before long.

"We are interested in that twelve percent [who chose Not Guilty by Reason of Insanity in the Holmes case]. What makes them different?"

I argue in favor of that 12 percent viewpoint in the case of Holmes. "I just can't see how anyone in his right mind could do what he did, walk into a movie theater and throw a bomb and start shooting, unless he was mentally ill," I say.

"I can. When you take someone who was supposedly brilliant, doing well, a four-point-oh grade average and so on, and then I think in this case he missed a very important exam. He had a break from what was his lifeline. What was his identity. If you are a neuroscientist [as Holmes was studying to be] and that is your identity and you are planning on being successful in this area and you have been rewarded and recognized for your achievements, and then all of a sudden you are failing. And you may be failing for reasons having to do with something biological in your own being that is preventing you from cognitively functioning as you did before [but are not mentally ill]. And you can't understand why this is happening to you because there is a cognitive deficit going on. And then you lose everything. And you may be going to a psychiatrist, as he was, and talking about his aggressive impulses and so on."

(I later interviewed faculty members, neuroscientists, at UC Irvine and the University of Colorado who had taught Holmes, and both said that Holmes was a bright student and aloof, but they never imagined that he could commit mass murder.)

Kappes continues his line of inquiry. "We had the same thing happen with the Texas Tower Massacre—the guy who shot all those people at the University of Texas who had a brain tumor and so forth [but had no mental illness]. So there *is* some biological basis for aggression and violence. We know that," Kappes says, explaining how someone who is not insane could nevertheless deliberately murder innocent people.

I push him on this. The scenario he just proposed sounds exactly like what would be caused by the Insult trigger of rage, but Holmes looked and acted so bizarre immediately after his arrest, and what he did was so senseless. Becoming enraged because of personal losses or threats to one's identity could cause someone to snap in a murderous rage, but it is difficult to imagine that it could cause a normal person to do what he did—walk into a movie theater and start to shoot everyone. Possibly, I suggest, the same violent actions could be committed by a sane person in some other circumstance. Admittedly, this is difficult to imagine, but to explore the point, I concoct a scenario of taking revenge on an enemy watching a movie during a war. But could a sane person ever justify what Holmes did?

"Now, you are thinking that *justification* would be the key to insanity or not," Kappes counters.

"Sergeant Bales, who snapped—is he insane?" I ask, pointing out that he did not enter a plea of insanity or diminished capacity due to PTSD or drugs, for example, in his defense. His attorney argued that Bales had simply "snapped." But that explains nothing. "From a psychological perspective, when someone snaps, what does that mean?" I ask.

"You lose contact with reality. This happens a lot with PTSD; it's called the shattered self," he explains. His point is that this response is different from being insane under any of the legal definitions.

"I'm having a hard time with this," I say, taking the role of devil's advocate. "Anytime you take a gun and shoot your girlfriend through a bathroom door, that has to be a break with reality."

"Yeah, but it is always a question of his mental state at the time of the crime," Kappes says. "It is called a psychological autopsy, meaning that you have to go back in time to see what was happening. What were the circumstances? You have to try to map out all of the conditions that were happening at the time to put yourself in that person's shoes. It is not always as simple as it seems." He likens the process to the forensic analysis of a crime scene, tracing the angle of bullets shot through the door and sifting through other evidence to reconstruct what occurred. "What were the elements that led up to the person's state of mind at the time of the crime?

"Admittedly, it is very difficult, because you are making assumptions. Can you *ever* be in somebody else's brain? Did you *know* what they were thinking? And more importantly, knowing whether it is right or wrong is *not* sufficient! It is an important element, but it is [not] or should not be the only variable that determines culpability. Culpability has to also be decided by motivation—by emotional factors, not just cognitive factors and behavioral factors."

I ask about the Norway shooter, Anders Behring Breivik. He was not insane in the sense that he knew what he was doing. He planned to do what he did for a very long time and he carried out his plan, and his worst fear was that the court would judge him insane, but such thinking is delusional.

Kappes cites another similar example—the Unabomber. "He also did not want to be judged insane. I would say the Unabomber *was* insane, because of his state of mind and consciousness. He was schizophrenic anyway," he says. "He already had a history [of mental illness].

"Most of these people do not, as you say, just snap. Now, I can understand the snapping occurring for PTSD and often tied to brain injury. Often [people who snap and serial murderers, for example,] have that element of child[hood] sexual abuse or some early aggression. That manifests itself in a sense of anger. When the prefrontal cortex is damaged by injury, then it releases one's volitional control. People who have been abused don't necessarily become violent or serial murderers, but when you combine the two [past abuse and a sudden traumatic event or injury to frontal-lobe function], it makes it more likely that they are able to engage in serial murders. Brain injury can occur from a lot of different things—drug use, among others.

"So it is complex. And you have to go state by state [to apply the specific law to the psychological factors]. So, even if you and I were to decide, 'Oh! Here's what it *is* once and for all,' each state is going to have their own definition of what insanity is.

"I get so frustrated. I can't believe so many times in court that they are not willing to listen to science." For example, he cites the science showing how unreliable human memory, and thus eyewitness testimony, is. "But then, I can understand them. They look at us and say, 'You guys can't even define insanity for us.'

"In psychology you have to have a high tolerance for ambiguity and appreciate that there are different ways of looking at things; there's not just one way—it is not like gravity."

I asked him about Carderock: "Would you say Dave was insane?"

He told me that he would have to see evidence of diminished capacity, but on the contrary, "I think anyone is capable of it [doing what Dave is accused of doing].

"You have to look at the person's psyche for a longer period than that one act. You usually can find the circumstances that led up to it. I think we all believe in cause and effect. That things just don't happen out of the blue. If you look at them, you will start to see early examples of anger issues that he had with a girlfriend, or other things. Unfortunately, we don't act on those issues. There is evidence early on of certain behaviors that don't rise to the level of attention that would say, 'We need to do something.'"

It should be obvious, but to make it explicitly clear, Kappes is not rendering a professional or personal judgment on the Carderock crime or any other crime. He has not been presented with all the facts, and rendering judgment is not the purpose of our discussion. In talking with an expert about the complex issue involved in human psychology and the insanity defense, snippets of various crimes are brought up as mere examples to help illuminate critical aspects of the various definitions of insanity.

(To update events a year after my interview with Dr. Kappes, the trial of James Holmes for the Aurora, Colorado, mass murder has now begun. The testimony thus far of many psychiatrists who evaluated Holmes's psychiatric condition is that Holmes was not insane at the time of the shooting. Holmes's notebooks showing detailed planning of the massacre are cited as damning evidence that Holmes was fully aware of what he was doing. These psychiatrists argue that Holmes may have suffered psychiatric illness but that he was not criminally insane at the time of the murders. As this trial is still under way, a final verdict has not been reached. That Holmes committed the murders is not in question. He has admitted to the Aurora theater massacre, but presumption of innocence must be maintained until the trial is completed. Arguments in court thus far, however, conclude that the horrendous mayhem and cold-blooded slaughter of innocents in the Aurora movie theater was not an act of insanity—it was an act of rage.)

For a week the jury heard testimony from psychologists to decide whether Barry Kin Lui suffered from a mental disorder the night that he brutally murdered his girlfriend with a knife. Their decision would determine whether he would serve his sentence in a state mental institution or in prison. In May 2014, the jury drove a stake through the heart of Lui's "Dracula" defense, and sent him to prison for thirty years.

One of Kappes's last comments as our conversation in his office came to an end still rings in my head: "*I think anyone is capable of it.*"

## A Crying Baby

Squeezing her hand over the toddler's nose and mouth, she smothered him to death because he would not stop crying. Twenty-two-year-old Jessica Fraraccio pleaded guilty in a Virginia courtroom to felony murder of twenty-three-month-old Elijah Nealey in the summer of 2012. No one in their right mind could conceive of committing such a horrible act, but babies *are* tragically killed or left severely brain damaged by shaken-baby syndrome inflicted by a parent, family member, or caretaker frustrated by a child's incessant crying. Dismissing those who act with depraved minds, how can we comprehend such sad stories as this one?

Jessica Fraraccio was well known to the Nealeys when they hired the young woman as a babysitter for their son, Elijah, and his two sisters in their Northern Virginia middle-class home. The Nealeys knew Jessica's parents well. Jessica was a devout Catholic who was studying child development in the hope of owning her own day-care business someday.

Friends and family testified in court that Fraraccio was a kind and giving person. According to a newspaper article in *The Washington Post*, "She is incapable of wishing evil; in high school she wouldn't even gossip about the girls everyone 'hated,' one friend wrote in a letter to the court."

"We never would have dreamt of this," Mike Nealey said of the young woman they thought they knew. "I don't know how to process it," he said after hearing her shocking and remorseful admission.

The young woman faces the prospect of fifty years in prison, and the

Nealeys are living with horrendous grief over the murder of their son, who would have had his second birthday the following month, September 2012. Nothing can excuse such a horrible crime, but to reduce the chances of another child suffering a similar fate, it is necessary to seek an understanding of what went wrong.

Babies and toddlers are especially vulnerable to brain damage caused by fierce shaking because of their comparatively large heads and weak supporting neck muscles. The violent whiplash caused by shaking a baby smashes the infant's brain against the internal walls of their skull, inflicting severe trauma. According to a recent study, 18 to 25 percent of babies who are hospitalized after being shaken in frustration will die. And 80 percent of children who survive are left with significant lifelong brain injuries.

The trigger is crying. It seems paradoxical that crying could trigger someone to murder a child on impulse. How could evolution give us that?

An infant's first act in life is to cry. This stimulates concern and provokes an urgent caregiving response in those who hear it. Neuroimaging shows that infant crying stimulates brain activity in areas involved in parenting behavior, empathy, attention, and stress. Mothers are more sensitive to the cries of their own infants than to the crying of unfamiliar infants, and neuroimaging illuminates this behavioral preference in the level of activity evoked in the mother's brain.

The impulsive murders of crying infants are often as bewildering to the perpetrators, who are frequently otherwise devoted parents, as it is baffling to others who struggle to comprehend the horror. The perpetrators of shaken-baby syndrome are most likely to be males related to the child. This is followed in frequency by boyfriends or stepfathers; then mothers; and at a lesser frequency by temporary caregivers. The helpless victims are much more likely to be male than female. These statistics must provide clues to understanding how the unthinkable can happen.

The leading hypothesis for the greater number of victims who are boys is cultural. Males should not cry, but females are allowed to. Thus incessant crying in a female child is more tolerated.

Adult men are more aggressive and stronger than women, so their violent actions are more powerful and deadly. Crying does evoke different responses in the male and female brain, though. Neuroimaging shows that infant hunger cries strongly interrupt "mind wandering circuits" (the dorsal medial prefrontal and posterior cingulate) in women, whereas men tend to carry on with less interruption in thought. Women in general show greater response to crying and laughter in the brain regions that process empathy than do men. Cognitive control by the left auditory cerebral cortex actively inhibits activity in the right amygdala in response to laughing or crying, and this reduces stress and anxiety.

Rather than dismissing a violent response to a baby's cries as psychotic behavior, Ronald Barr, professor of Pediatrics at the University of British Columbia, concludes that shaken-baby syndrome is a tragic failure of an otherwise normal, common interaction between infants and caregivers. According to this analysis, crying is an ambiguous signal. It can provoke positive, supportive, survival-promoting caregiver responses as well as negative, destructive, survival-endangering caregiver responses.

The onset of crying provokes a caregiving response, but if the caregiver is capable and provides "good care," the baby should cease crying. If instead the crying cannot be stopped, the caregiver may unconsciously interpret the crying as an indicator that they are not capable caregivers. Thus, rather than triggering a biological response to satisfy the needs of the infant—hunger or the need to change a diaper, for example—crying now signals personal criticism. These opposite responses to crying are not deliberate, conscious thoughts generated in the cerebral cortex; they arise in the unconscious emotional brain.

This explanation is only a hypothesis, but the statistics indicating that family members close to the child have a greater probability of violently shaking their baby than do outside caregivers seem consistent with this interpretation. The cries that will not stop are a hidden and unconscious personal insult in the mind of someone who cares. If this suggested explanation is true, such tragedies are cruel ironies. This analysis might account for the paradox of how someone who is apparently devoted and caring could respond to the incessant cries of an infant with rage instead of love.

## Our Environment

A small gray bird lay on its side with feet curled, looking like a stuffed museum specimen but resting on dried leaves blanketing the ground. I felt the urge to stroke its satin feathers, but resisted, knowing they would be cold.

It was resting on the ground beneath a plate-glass window, which instantly explained my initial shock and answered my urgent question about what had killed it.

It is astonishing how birds can fly with ease through thick forests at several times the speed we can move about in our world. Soaring on wing, how do they effortlessly avoid collisions with a deadly thicket of branches and twigs in their path? The sharp eyesight and quick reflexive maneuvering of birds in flight are astonishing. Humans can scarcely tromp through a thicket without pokes and scratches from twigs and branches, but birds sail through the canopy of forest trees effortlessly and at such speeds they are but a streak to our eyes as they thread through their aerial obstacles.

But the quick instinctive reflex to dart through the center of a clear path between branches fails them in an environment that has suddenly changed. Their lifesaving intricate behaviors, honed over the tens of thousands of years of evolution that created intricate neural networks in their brain to accommodate such natural dangers, are rendered not only useless but deadly in a world that did not exist when their brain evolved.

And so it is with humans and their LIFEMORTS triggers. This is the story of evolution. It is the same theme in the score of life repeated endlessly in infinite variety. The dynamics that drive evolution in a constantly changing world affect all creatures—even the mightiest, as dinosaur bones encased in rock now attest. An entire branch of the evolutionary tree of life perished in a sudden environmental change, the Ice Age. The sparrow and the plate-glass window and the human being and the automobile, both are a clash between the environment of the past that sculpted brain and body and the modern world that confronts individuals with strange new conditions and threats.

The vast majority of us will never commit a sudden act of violence; nevertheless, the neural circuitry of snapping in rage is very relevant to everyone. Nearly everyone snaps in anger and does something destructive or says hurtful things while enraged that they immediately regret. These familiar bursts of anger often have regrettable and long-lasting consequences. Snapping can cause strife within families, among coworkers, and in everyday interactions with other people. Everyone desires to be in control of their emotions and to become a better person by understanding and controlling the rage circuit we share biologically with beasts. Understanding this circuitry will equip anyone to better control the preprogrammed snapping response in themselves, and just might save a prized golf club from getting wrapped around a tree!

When you suddenly feel an explosive rush of anger rise inside you, ask yourself why. Rather than trying to "put a lid on it," ask yourself *why* you are suddenly provoked to rage by this situation. Many things can (and should) make a person angry, but is the sudden situation pressing on you one of the LIFEMORTS triggers? If it is, then you will understand that the sudden release of the emotion we call "anger" and the powerful physiological response it launches in your body is the ancient survival instinct that prepares you to fight. Is this a situation in which a physical fight with someone is appropriate? Humans no longer live in the natural world. They have created their own environment, and they have the ability to comprehend it, to change it, and to adjust to it.

The modern world is still filled with dangers and threats. Sometimes these triggers of sudden violence backfire with tragic consequences, but most often it is the opposite. These automated defensive brain circuits have a positive, lifesaving role in everyday life.

PART TWO

# ALERT
## AND **AGILE**

# 5

# To Do the Right Thing Fast

When I am angry I can pray well and preach well.

Martin Luther, *Tischreden* (*Table Talk*)

L ucky, what are you doing? Get your butt up here and let's go!"
Heather "Lucky" Penney was a twenty-six-year-old DC Air National Guard lieutenant stationed at Andrews Air Force Base in Maryland. Jolted by urgency, she aborted the methodical checklist, climbed in, ignited the engines, and screamed for her ground crew to pull the wheel chocks. Within minutes the petite blonde was piloting the F-16 fighter jet screeching at top speed through the crystal-blue sky with her commander, Col. Marc Sasseville, piloting his own jet on her wing. The date was September 11, 2001.

Both of the Twin Towers and the Pentagon had just been hit, but there was a fourth passenger jet commandeered by Islamic terrorists targeted at the United States Capitol or the White House.

Obviously Heather knew there was no chance that any of the men, women, and children on board the United Airlines 757 passenger jet she was intent on intercepting would survive. In an improbable twist of fate, there was a good chance that Heather Penney's own father could be the captain on United Flight 93, which she was determined to destroy.

"This sounds coldhearted; I mean, that was my daddy," Penney said in an interview afterward. "I couldn't think about it. I had a job to do," she said later to her mom.

"We don't train to bring down airliners," Colonel Sasseville said, describing the gut-wrenching, unthinkable act he and Heather Penney were about to commit.

Training would have been useless in any case. Neither of their jets was armed. No one had anticipated the need to have fully armed fighter jets at the ready to protect against an aerial attack originating from within the United States.

"I'm going to go for the cockpit," Sasseville said.

"I'll take the tail," Penney replied without hesitation.

"I genuinely believed that this was going to be the last time I took off," she said, now a single mom with two girls. "I had already given myself up, knowing what my duty was." She was fully committed to ramming her jet into the passenger plane to bring it down.

As fate would have it, her father, Col. John Penney, was not piloting Flight UA 93 that morning; it was his good friend Jason Dahl. "With Jason on the plane, it would have been an additional level of grief," John Penney said later.

But Lieutenant Penney and Colonel Sasseville never had to execute their kamikaze mission.

"Let's roll!" said thirty-two-year-old Todd Beamer, a passenger on the hijacked flight. Those were the last words of the soon-to-be father ending his conversation with telephone switchboard operator Lisa Jefferson, whom he had been relaying information to for the last thirteen minutes of the hijacking. They had just finished reciting the Lord's Prayer and Psalm 23 together. "*Though I walk through the valley of the shadow of death, I will fear no evil; for Thou art with me; Thy rod and Thy staff they comfort me.*"

Beamer and other passengers stormed the hijackers. Jeremy Glick was a six-foot-one judo champion; Mark Bingham was a rugby player; Tom Burnett had been a college quarterback, Louis Nacke was a weightlifter, and William Cashman was a former paratrooper. The cockpit recorder captures the sound of food carts ramming the cockpit door and the cries of the terrorists screaming at one another to hold the door closed.

The door bursts open. "Let's get them!" is heard on the recording as the passengers overwhelm the terrorists.

"They were the true heroes," Heather Penney says. "These were average, everyday Americans who gave their lives to save countless more. The selflessness reminds us that we are part of something greater than ourselves, that there are things in this world more important than ourselves."

Today there is a memorial on the green pastures of southwestern Pennsylvania, and the White House and United States Capitol stand spared from destruction.

This is the power of the Tribe trigger to life-risking violence. Gangs, wars, and racism are its ugly dark side, but this bit of neural circuitry is very much a part of what makes us human. It can unleash the best in human nature; selfless sacrifice for others. It is the trigger of heroism, and as the monument now standing in the field in Pennsylvania reminds us, it is a vital part of every one of us.

"I believe it's a human instinct. I didn't weigh it or think about it. I just did it," Leonard Skutnik said.

On January 13, 1982, Air Florida Flight 90 took off from Washington National Airport on a miserably cold and snowy morning. Its wings laden with ice, the aircraft failed to gain altitude and the Boeing 737 passenger jet with seventy-four people on board crashed onto the Fourteenth Street Bridge, which spans the Potomac River. The aircraft crushed six cars, killing four motorists as it skidded off the bridge and crashed into the ice-covered river. The aircraft broke apart on impact and sank beneath the icy surface with everyone aboard strapped into their seats, but a piece of the tail section bobbed on the surface, slowly sinking with six passengers clinging to it for their life. The water temperature was one degree above freezing, which gave the hypothermic survivors, many of whom were badly injured, only thirty minutes at best to survive. Blocked by the fractured ice, rescue crews were unable to reach the survivors in inflatable rafts and they crowded helplessly along the shoreline. It took nineteen minutes for a rescue helicopter to arrive, leaving only about ten minutes to extract the six survivors from the Potomac.

The helicopter hovering above whipped the frigid, icy slush into a frothy gale as rescuers dangled a lifeline to the wreckage. Arland Williams Jr., a forty-six-year-old federal bank examiner, grasped the line with one hand while clinging to the aluminum tail section with his other, and he handed it to the woman next to him. She was hoisted to safety.

"He seemed sort of middle-aged and, uh, maybe balding," was all the chopper pilot could describe afterward. Williams had spent the last twenty years sitting quietly in a bank office reviewing accounts. He was father to a teenage son and daughter, and he was his parents' only child.

The life ring dropped again and once again it was Williams who grasped it. With only a few desperate minutes before all of them would perish of hypothermia, he passed the lifeline to another person.

Meanwhile, Lenny Skutnik, who had witnessed the crash while on his way home from his job in the Congressional Budget Office, had rushed to the site and was watching the desperate rescue in horror from the side of the bridge. Priscilla Tirado looped her numb arm through the life ring as rescuers tried to haul her to safety, but she was too badly injured, too cold and weak to hold on. She splashed back into the freezing water helplessly. The life ring was trailed back to within her reach again several times. Each time she made feeble attempts to hang on but failed. The last of her strength now fully sapped, she began to flounder and drown beneath the tempest driven by the helicopter blades hovering above her.

Leonard Skutnik tore off his warm winter coat, kicked off his winter boots, and plunged into the freezing water, kicking and splashing with all his strength to swim out to the stranger slipping beneath the surface in the middle of the Potomac River. Reaching her, he threw his arm around her and towed her back through the river currents and icy slush to safety on the shore, where he collapsed in exhaustion, shivering violently.

Flight attendant Kelly Duncan, who had given her own life vest to one of the passengers, and Arland Williams were the last survivors still clinging to the tail section of the wreckage. The helicopter returned and dropped the lifeline to them. Williams reached it and, for the third time, passed it on. He handed the life ring to Kelly and she was pulled to safety.

Returning immediately, the helicopter crew dropped the lifeline for the last time onto the tail section, but Williams was gone. He had slipped beneath the surface. When the bodies were recovered and examined at autopsy, Williams was the only one of the passengers to have water in his lungs. He had drowned saving the lives of three complete strangers, leaving his own family without a father.

The Fourteenth Street Bridge is now officially named the Arland Williams Bridge.

On January 25, 2011, seventeen-year-old Nicole Bean and her eighteen-year-old boyfriend, Kevin Minemier, were shopping at a Wegmans grocery in

Henrietta, New York. Sixteen-year-old high school student Christopher Patino and his friend Mustafa Said were shopping one aisle away when they heard Nicole's bloodcurdling screams. They rushed around the corner to the frozen-food aisle and saw Kevin Minemier stabbing his girlfriend in the face and eye.

Christopher instantly rushed to the assailant and slugged him in the face, battling the knife-wielding attacker with his bare hands. Mustafa joined in, grasping the knife and flinging it away, severing tendons in his hand in the process.

"Chris took the dude and put him on the ground and I took the girl to safety. I told her that no one was going to hurt her. . . . I was terrified for my own life and my hand was gushing out bleeding and her eye was bleeding also," Mustafa said afterward.

Christopher struggled with the assailant on the ground and held him down until police arrived. "I was just like, Oh wow! She's really hurt. I'm not just going to stand here and watch her possibly die. So I just ran up as quick as I could and tried to get the knife out and try to make sure she didn't die there. She was screaming. She was literally dying right there. I felt horrible."

Other shoppers, most of them adults, watched the vicious attack immobilized in fear, but the sixteen-year-old and his buddy snapped reflexively to action with selfless heroism.

"It was just like, It's now or never," Christopher said. "I had to do it just because he had a knife. I'll admit I was like, What if he stabs me? But then I was like, I don't care. I got to do this. So I went for it." Christopher was battered and bruised and Mustafa's hand required surgery, but both eventually recovered and Nicole's life was saved.

"There were people watching and, well, I don't have that kind of heart to stand there and watch while a girl gets killed," Christopher said. "He was sucking the life out of her and some guy was taking pictures and we were like, We can't just stand there. It's time to act."

Alexander Travis, seventeen, and his five-year-old sister were riding in their grandfather's car returning on a twisty rural road from an overnight visit to his home on August 15, 2013. As they rounded a bend they saw a car had skidded off the road and was sinking in the Bradley Brook Reservoir with ninety-three-year-old Stuart Deland trapped inside. His grandfather had barely pulled the car to the side of the road before Travis unbuckled his seat belt and leaped from the car. "I didn't really think. It was just like an instinct," he said.

He sprinted to the reservoir, where he stripped off his shoes, threw down his cell phone, hurled himself down the steep embankment, and dove into the water. The front end of the car was already submerged and the vehicle was sinking quickly.

"I opened the back door and it just pulled me into the car because it was sinking," Travis said. Travis stayed with the car as it sank beneath the surface, descending into the muddy reservoir. "I was so deep my ears were ringing and there was a lot of pressure."

Travis struggled in the murky water to pull the driver out of the front seat into the backseat to escape, but it was impossible. His lungs bursting, Travis shot to the surface for a quick gulp of air and dove back down to the submerged vehicle. He managed to wedge his fingers through the partly opened driver's-side window. Planting his feet against the door he pulled with all his strength and snapped the window glass into shards. Reaching in with both hands he grabbed the man and wrenched him out of the front seat through the window and brought him to the surface.

He did not notice the cuts on his hands and arms until it was all over.

He admitted that he was afraid, but said, "I would rather die trying to save someone's life than live with the guilt of watching someone die when you could have done something about it."

Don Holler, sixty-five, could fix anything. He enjoyed the outdoors: gardening, hunting, and fishing. A friendly person, Holler was one of the regulars at Fat Daddy's Place in Pennsylvania, arriving to socialize every afternoon at four thirty.

"This is a neighborhood watering hole. He knew a lot of people and a lot of people knew him," a patron said of Don.

On July 11, 2011, golfing buddies Kirk Haldeman, a fifty-one-year-old insurance agent, and Michael Ledgard, fifty-three, were rained out, so they headed to the pub for a drink instead. Don Holler was sitting at the bar watching the TV screen.

Stephen Fromholz, a forty-three-year-old Army veteran from San Antonio, entered the bar. The six-foot man stood out from the local crowd in his black-fringed leather vest, knee-high moccasins, and big black cowboy hat.

Fromholz asked the bartender to change the television channel, which was showing scenes from the war in Afghanistan. Holler said he was watching the news, and that if he didn't like what was on TV, there were a lot of other bars in town.

Fromholz glared back.

"I don't want any trouble," Holler said.

Fromholz put down his half-finished beer and walked out.

"Within thirty seconds, I could see him coming back in the dining room door," Haldeman said. "He had a semiautomatic rifle on his side [an AR-15]. He took three steps inside the bar and said, 'I'll show you what war feels like.'" Fromholz raised the rifle and shot Don Holler.

There were ten people in the bar when the shot was fired. The exit door was closer to Haldeman and Ledgard than the gunman, but rather than run for cover or escape, Haldeman attacked the man firing the assault rifle. "I jumped out of my seat, grabbed him, and drove him back into the corner of the bar," Haldeman said.

"I took a deep breath and tackled them both," his golfing buddy said, explaining how he joined in to help Haldeman, who was struggling with the rifleman. "I was really afraid what would happen next."

Another earsplitting shot rang out but missed. The golfing buddies who had only moments earlier been consoling themselves with a cold beer now held the gunman pinned to the floor until the police arrived. In an instantaneous reflex that beat out a rapid-fire semiautomatic rifle, the golfers had rushed to the aid of a stranger neither of them knew, risking their own lives. They probably had saved the lives of many others in the bar, but by the time the police arrived, Don Holler was dead.

Such acts of heroism happen every day. This selfless reflexive response is never referred to as snapping, but from a neuroscience perspective, both heroic behaviors and rage behaviors are driven by exactly the same brain circuits. We would not have these circuits and LIFEMORTS triggers that set them off if they were not of benefit to our species. As peculiar as it seems that the human brain is wired to set aside reason and self-interest and instantly engage in violence at the risk of death for another human being, this nobility is at the core of humanity.

"The shark was around me and she's bleeding," Richard Irvin Moore, age fifty-seven, said after jumping into the ocean to save a stranger from a shark attack. "We look out and there was blood everywhere in the white water around her." The shark had bitten off the twenty-year-old woman's arm.

After swimming out 100 yards from shore Moore reached the woman and began towing her back to the beach, backstroking through the strong ocean currents. "It dawned on me—I was in danger now," he said. "I start praying out loud. God, God protect us. . . ."

"She said 'I'm dying. I know I'm going to die.'"

Despite Moore's heroic actions in rescuing her from the shark and getting her into an ambulance, the young woman did not survive.

The Family trigger: Family members escape in the middle of the night from a burning home in Gloucester, Virginia, on January 16, 2013, only to realize that Gabriel, six months old; Michael, age two; and Thomas, age seven, are still inside the blazing home that is fully engulfed in flames. The children's grandmother, fifty-four-year-old Virginia Grogan, who had just escaped with the others, realizing the children were still inside, rushed back into the inferno to rescue them, only to die in the flames alongside them.

On an outing with his family to Yosemite Park, sixteen-year-old student Alec Smith was hiking along the Mist Trail above the 317-foot Vernal Fall when he heard a mother's screams.

"Save my baby! Someone save my baby!"

A nine-year-old boy had fallen into the rapidly flowing torrent just before it spilled over the precipice of Yosemite's roaring falls.

"When I heard the scream and saw the boy, I thought, 'Oh, no! Oh my God, no,'" Alec's mother said. "Then I blinked and saw [my son] going after him."

In a flash, Alec had leaped over the guardrail and rushed toward the deafening torrent.

"Nobody has ever survived going over the fall," said Yosemite park ranger Kari Cobb.

Alec sprinted out into the river over the rocks on a beeline trajectory to intercept the child at the lip of the falls.

"I got half my body in the water and kept the other half out," Alec said. "I grabbed the kid twenty feet from the edge of the waterfall and pulled him back onto the bank."

"If he'd hesitated, the kid would have been gone," said Alec's uncle.

Alec said he didn't think of the danger. It had been an instant reflex, but he now realizes how lucky he was. Many rescuers have gone to their deaths in precisely the same situation at that treacherous spot.

"Don't think about what could have happened," he said. "Be grateful for what did happen."

The LIFEMORTS triggers define the best in human beings: struggling to the death to preserve life and limb, intolerance to insult from others or to one's expectations of themselves, selfless protection of family, preserving our home environment, willingness to sacrifice anything for our mate, placing our social organization above our self-interest, keeping what is rightfully ours from the hands of thieves, defending our own people against any threat, never giving up—the struggle against an overwhelming impediment. These things, which well up inside us in an explosive emotion we call rage, are the noblest characteristics of humanity.

"I have forty dead contacts in my cell phone, all good friends of mine," a veteran of SEAL Team Six told me.

I think about that often when I look at the contacts on my own cell phone.

# 6

# The Flavors of Threats

Fear, indeed, is the mother of foresight.

Sir Henry Taylor, *Notes from Life in Six Essays*

Without even seeing the robber in my peripheral vision during that attempted pickpocketing in Barcelona described in chapter 1, I had snared the thug by the neck and thrown him to the ground. How had my brain taken in all this information entirely unconsciously as my attention was fully occupied by the enormous challenges of negotiating my way through a strange new environment? This situational information had penetrated deep into my brain to reach neural centers guarding against the threat to which I was consciously oblivious. The response was automatic and apparently unchecked by rational thought. It had been a blind snatch-and-grab, but as will be revealed in this chapter, new research shows that in times of danger we have powers long considered the stuff of magic.

## Living Dangerously

Cool, alert, and hyperfocused, he sipped his black coffee while his eyes locked on mine with an intensity that could penetrate any subterfuge, periodically turning his head around to automatically scan behind him and monitor every corner of the room as we chatted. Even on a coffee break, this was an entrained habit and a lifesaving one. The human brain is wired for threat detection and defense, but some people have honed these skills to such a high art that threat detection is their profession.

When I first spotted him as he rounded the corner of the high school coffee shop on a drizzly December afternoon in Washington, DC, I thought he could have been an athletic coach—too short for basketball, maybe track or wrestling. But as he approached, the clues to his true identity began to emerge. Even without the white wire coiling discreetly up his neck from his shirt collar to the ear bud that gives him and his colleagues away while at work, the pin on his black soft-shell jacket branded him as a Secret Service agent. He walked sprightly toward me and greeted me by name even though we had never met. Special Agent Scott Moyer had agreed to meet me for coffee to talk about his chosen line of work.

Some of the most elite bodyguards in the world are the Secret Service special agents who protect the president of the United States. The threats they face range from foreign political enemies to local psychopaths. By coincidence Special Agent Moyer and I were meeting on the first school day after the horrific Sandy Hook Elementary School shooting in Newtown, Connecticut. Special Agent Moyer's demanding and dangerous line of work is protecting President Obama's daughters at school and elsewhere. The gruesome tragedy in Connecticut hit the agent close to home.

"You know that you would have to possibly lay your life down to protect the president's daughter, but that's not what I go to work thinking—that this could be the day. That's the furthest thing from my mind. You realize it's part of the job," he said matter-of-factly.

Secret Service agents must react instantly. This requires rigorous training and innate ability, both mental and physical. "In a lot of the situations that we would become involved in you react totally subconsciously—you wouldn't think about it," he explained. "When we Monday-morning quarterback, sometimes we would think, 'Wow! That was lucky. I should have done this, I should have done that.'"

In the background of the café the muted flat-screen TV rolled captions over a scene of a press conference with the by-now familiar Newtown police spokesman Lt. J. Paul Vance feeding the media's endless hunger for details of the sickening mass murder of innocents.

"I've heard enough. . . . We don't need to know where all the bodies were laying and what the kids were wearing. This only invites copycat crimes," Special Agent Moyer said, dismissing the TV screen in disgust.

Few people could do what Moyer does. These skilled agents bear tremendous responsibility, a responsibility in which a split-second personal action carries history-making consequences. Secret Service agents must confront threats with selfless and unhesitating action, including violent or deadly force.

Special Agent Moyer has experienced this instinctive reflex many times in law enforcement: halting his car and bounding out to protect a woman who was being physically assaulted on the street by a domestic partner; reading the malicious intentions of a suspicious person's body language who had been stopped by fellow officers, and in a flash sprinting ten paces to tackle the dangerous suspect when he refused to show his hands. . . . These were instantaneous actions to neutralize a threat, not consciously deliberated behaviors, and they were triggered at the appropriate point when someone was in immediate danger or being injured by someone. Other people would flee, ignore, or avoid the risk in precisely the same situation.

The answer for why different people respond differently to the same threatening situation involves a combination of personality and training, genetics and environment, specific triggers and predisposing events, but in meeting Special Agent Moyer it was obvious that some individuals excel in this respect. Their selfless instinct to protect is a matter of personal character, a deeply ingrained aspect of personality as fundamental to the person's psyche as the bones are to his body. The Tribe and Organization triggers are especially strong in such individuals. These triggers of violence in defense of society and its laws become the core identity of people drawn to a career in law enforcement and of members of the armed forces.

"I had my school-year scuffles like a lot of kids growing up. It wasn't that I ever sought out trouble, but I always felt bad for the kid who got bullied. Sometimes I'd stick up for that kid," Moyer told me when I asked about his early years. Usually the kid getting picked on would run away, leaving Moyer to handle the bullies himself.

During the Sandy Hook massacre, first-grade teacher Kaitlin Roig had implored her students to be quiet as they huddled in a bathroom. She'd barricaded the door with a bookshelf. "I told them we have to be absolutely quiet. There are bad guys out there now, and we need to wait for the good guys to come get us out." Special Agent Moyer is one of the good guys.

Secret Service agents must constantly scan the environment to extract subtle cues of a potential threat to the president of the United States in a complex and fluid situation. "A lot of it's training; to always be aware of your surroundings. I don't even notice that when I'm at the mall I'm constantly scanning, looking," he says, until someone like his wife points out his behavior to him.

"I look back over my shoulder when we are walking to the car. It drives my wife crazy a lot. I'll lock eyes with someone, and my wife will pick it up and ask, 'What's wrong? Have you dealt with that person?'

"You know, a lot of bad guys carry themselves in a certain way, just as we in law enforcement carry ourselves in a certain way. We have that confidence, and they have that street smarts or swagger. It's like telepathic communicating. He's recognized me and I've recognized him. We don't have to say anything."

## Reading Minds

With all the emphasis we place on the conscious mind—intelligence, deliberation, learning, and thinking—we easily overlook how much of what the brain does goes on without any conscious awareness or conscious control at all. Our brain unconsciously controls every aspect of physiology necessary to maintain life, simultaneously controlling multiple bodily functions and regulating them precisely over an enormous range of timescales and through widely changing conditions. Our body temperature is held to within a fraction of a degree regardless of ambient temperature through a complex system of sensors and actions to heat and cool the body—all of them controlled automatically and unconsciously by the brain with a level of precision that exceeds the performance of many electronic control systems, including the thermostat in your house. From the precision second-to-second regulation of heartbeat to the rhythm of breathing to daily cycles of sleep and the cycles of life, from the monthly menstrual cycles of women to the gradual progression from childhood to adolescence to old age—all of these myriad functions are controlled by the brain regulating hormones and other neurophysiological processes unconsciously.

Perhaps it is reasonable that maintaining bodily functions, such as temperature, thirst, hunger, growth, and aging, would be isolated from conscious control. But is it possible that many of our seemingly voluntary behaviors are also controlled by unconscious mental processing? Consider: As you speak with another person, how much of what you say is fully constructed consciously in your mind before you articulate it verbally and how much emerges automatically as the dialogue unfolds? You may hear your innermost thoughts and arguments for the first time when you hear yourself speak them. As you write, how much text bubbles up to consciousness already formulated before your conscious mind scrutinizes what the unconscious brain has synthesized and offered up? Consider the volumes of complex data handling going on unconsciously to provide real-time situational awareness in a dangerous and rapidly changing environment to enable us to drive a car through traffic seemingly automatically while engaged in conversation. Because so many functions of the brain operate without conscious thought, we tend to overlook how much the brain is controlling our actions and behaviors. Indeed, a large fraction of the brain of animals and humans is devoted to detecting and responding to danger.

## Trusting Your Gut

"Absolutely! If I get a crazy feeling, there's a reason for it. Something's off. . . . I don't think anybody gets crazy feelings if something is not wrong. My gut instinct has been a hundred percent accurate so far," a member of one of America's most elite special-operations forces tells me. I asked this veteran of SEAL Team Six, who conducted several heroic tours of duty in the Middle East, if he could cite an example of when his gut instinct proved lifesaving.

"Oh, man! I've probably got a hundred of them, just approaching buildings. Based off what you are seeing from reports from drones above and all the other information you are getting. You know: This doesn't seem right— like this house is rigged to blow."

Despite their sense of imminent danger, the SEALs, unstoppable and unwavering, advance into the threat of death. They skillfully and silently work their way into the enemy hideout. Armed to the teeth and executing treacherous entry and room-clearing procedures with precision and skill, they enter a building. A piece of cardboard is on the ground. The SEAL crouches down and lifts it up. "It was covering a five-hundred-pound unexploded American bomb that they had rigged to blow in the floor."

There was no reason that anyone can pinpoint, but the situation just didn't seem right and the SEALs' sharpened attention to possible hidden dangers was piqued by their gut feelings. Otherwise, they might have trodden right over the innocent-looking cardboard trash on the floor as they searched for "real" threats—guys with guns, grenades, RPGs, and suicide-bomb vests. But, no, he stopped to pick up and peek under a piece of cardboard. It just didn't seem right.

"The guy in the other room found a cell phone that was rigged to some stuff and the guy at the front door found some wires rigged to the front gate. Any one of those things could have set it off . . . and we ended up living through it," he said, because of their hyper state of vigilance boosted by alarming feelings rising up from their unconscious brain. This part of the brain cannot use language. It uses feelings—multicolored emotions to influence our conscious mind; if it will listen.

"I've been in houses before where you are clearing it, and it is too quiet: There's nothing in here. What's going on? We were told some people would be here.

"And when you get in, you see some drums of fuel in the living room. You know, like, that doesn't make any sense.

"[So we say] Let's just get out of here. We don't need to continue clearing this target.

"Turn around. As you are walking out the front door, the house blows up.

"That one happened several times," he says nonchalantly.

We place great emphasis on conscious deliberation and reasoning to guide us, but in fact the conscious mind has a very limited capacity and speed of operation. Long division is hard to do in your head not because the process is complicated—it's not—but simply because we can't hold the intermediate answers in our conscious brain long enough to execute the next simple arithmetic step. The capacity of our conscious brain is horribly feeble. But when you consider that all other animals rely entirely on nonverbal and unconscious brain function to act appropriately in complex situations, including avoiding threats, it is clear we should give greater respect to the massive data crunching going on beneath our awareness in our cerebral cortex. To understand how the brain integrates complex information unconsciously, we need to explore the frontal lobes in more depth.

The frontal lobes are comprised of three parts. The top and sides, like the hood and front fender of a car, are called the *dorsolateral prefrontal cortex* (Latin for "back and side"). Considering that the brain is a bilaterally symmetrical paired organ that looks something like a shelled walnut, imagine breaking the left and right parts of the walnut meat in half. The parts hidden in the cleavage—the vertical walls at the front of the walnut—would represent the *medial prefrontal cortex*. The frontal lobes of humans have expanded so much that they curl under a bit beneath the forehead like a boxer's glove. To visualize the third region of the frontal cortex, imagine the area at the tips of the fingers of a boxer's glove. This is the *orbital prefrontal cortex*, easily remembered because it is close to the eyes, which are called *orbits* in anatomical terminology. Indeed, this part of the brain is all-seeing, but subconscious. Each of these prefrontal cortical regions carries out very different functions. All of them are associated with very complex behaviors, as opposed to the simple reflexes associated with lower brain regions closer to the spinal cord, which are more robotic, such as the discrete sensory-motor processing functions that are required to perceive the world and move about in it. The prefrontal cortex deals with complexity and uncertainty.

The orbital prefrontal cortex regulates the emotions of anxiety and arousal. This is the brain region that helps sharpen our focus of attention even to the point of tunnel vision in a threatening situation, but the orbital cortex is at work in any intense activity where distractions must be blocked out. It accomplishes this in part by inhibiting the medial prefrontal cortex, which influences our behavior based on emotion and motivation. (More on this later.) The orbital cortex is what gives us that ticklish feeling of familiarity.

"You are getting a bad feeling about something," the Navy SEAL explains, describing how his unconscious brain signals to him that something is not right as they approach a target.

"There is a *reason* for it. Something in your subconscious mind says, I've been here before, this doesn't look good. Or, I've been taught that this situation is bad maybe. . . ."

The orbital prefrontal cortex recognizes that familiarity.

Olfactory signals (smells) mix with taste signals to create a unique sensation—an abstract creation of the mind that we call "flavor." Coffee does not taste the way it smells. Neither does a cigar, but the combination of taste and smell formed in the orbital prefrontal cortex creates these flavors. Touch, sight, sound, and pain also feed into the orbital prefrontal cortex. More abstractly, no matter what the individual sensory signals may be reporting to his conscious mind as he approaches the Taliban hideout, in the SEAL's unconscious mind they are combining inside his prefrontal cortex to give him the experiential flavor of familiarity and danger. The bit of brain tissue mines all that incoming data and instantaneously computes an analysis that it conveys to our conscious mind the only way it can—as a hunch or feeling.

The function of this part of the brain is very difficult to study. Not surprisingly, this brain region is very highly developed in humans but only rudimentary in lab rats. Studies on monkeys indicate that the orbital prefrontal cortex is very important for learning and for directing behavioral change when things in their environment change.

"Let's just get out of here. We don't need to continue clearing this target."

Unconscious decision-making utilizes the orbital prefrontal cortex, and this information processing can cause us to snap, either appropriately or inappropriately.

Although this "gut-feeling" function of the orbital prefrontal cortex is involved in snapping, it is vital for many different situations. Not only for SEALs in combat but for elite athletes and adventurers who must synthesize enormous and often ambiguous or conflicting streams of data coming at them, and react instantly and appropriately. Imagine a racecar driver or a downhill skier hurtling through a race—things are happening faster than the mind can think consciously.

## Extreme Free

Fresh powder overnight had blanketed green pines and high-pitched rooftops. A brilliant patch of sunshine pierced a hole in the northern sky, where billowing clouds floating on a dark menacing ooze were slowly pursing closed the blue circle. There would be a snowstorm. But for the meantime, the slopes were good.

Her face bundled up against the cold wind on the slope of Crested Butte in Colorado, she was difficult to identify among the throng of skiers crowding around the starting gates at the top of Paradise Run. Approaching closer, her athletic physique and strawberry-blond pigtails sprouting out beneath her wool hat gave her away. Her identity was confirmed by dappled tabby-cat freckled cheeks and green eyes penetrating the golden-mirrored lens of her ski goggles. Wendy Fisher greets everyone with open, friendly enthusiasm. She speaks rapid-fire, like she's slamming moguls. A former Olympian on the US ski team, her career diverted abruptly into an even more dangerous pursuit that blasted women's skiing beyond the bounds of what anyone had thought possible.

Back in the 1992 Winter Olympics held in Albertville, France, Austrian skier Sabine Ginther had been favored to win the combined downhill, but she crashed. Her hopes were dashed at a spot on the Roc de Fer (rock of iron) downhill course that Wendy found herself screaming toward at terrifying speeds. Speeds that only a few people on the planet will ever or could ever attempt with skis strapped to their feet. Ginther injured her back in the crash and she was out of competition for the rest of the season. That treacherous bump claimed four more women hurling themselves down the icy mountain at speeds of seventy miles per hour, each one competing in the race of her lifetime. All of them were badly injured.

US women's coach Paul Major explained that to negotiate the dangerous bump, nicknamed "Noodles," it is critical to hold a very precise skiing line, or disaster awaits.

"Noodles is sharp, but it does have a slight curve to it so you can pressure your skis forward and move your center of gravity forward as you go over," he said. "The left side of the bump falls very sharply and the right side stays kind of even."

If the skier does not hit the bump precisely along the right edge, she will be catapulted up to 140 feet down the slope, impacting where the run flattens out. "If you're jumping that far, you'll have a very rough landing."

"She must have flown about a hundred and twenty feet through the air," US team spokesman Tom Kelly said. Wendy's head slammed into the rock-hard surface and she slid down the slope like a rag doll until coming to rest on a flat area. Her helmet may have saved her life, but she suffered a concussion. Her thumbs were broken and both knees injured. It had been a practice run just before the Olympic race, the race that had been the focus of her entire life. Suddenly, so close to achieving her lifelong goal, the chance to live her dream was snatched away.

After recovering from the accident, Wendy gave up competitive downhill skiing. Not to sulk but rather to undertake an even more challenging goal. To become one of the first women big mountain free skiers. These are the daring extreme skiers, nearly all male before Wendy, who climb or helicopter to the top of the steepest and most isolated summits in the world. Then they launch themselves on skis down the terrifying mountain, flying over terrain—often through the air soaring off cliffs—that no human being may have ever traversed before. Training and competing for the Olympics was a drudge by comparison.

"I stopped liking skiing because of racing. I was super burnt-out," she says, explaining her decision to quit competitive downhill racing.

"The whole free-skiing scene—catching air! Skiing with all these guys doing cool lines, being in the middle of nowhere, was so fun and exciting, and an adventure. I was really drawn to really wanting to be a part of it all. Now you were with friends [rather than out to defeat other competitors]," she explains. "It is exciting and more individual."

Extreme big mountain free skiing is dangerous, far more dangerous even than downhill racing.

"Every time I was on top of a run I was always scared. Every time I stood on top of a mountain, I thought, I definitely could die today."

She elaborates: "There are consequences of avalanches, and your slough. [Slough is when the top layer of snow is disturbed by the skier slicing through pristine terrain and it releases, billowing down the mountain like liquid fog engulfing everything.] I would get lost. It is a big, *big* mountain. I would think, I know where I am, and then ski over and not really be in the right zone where I thought I was.

"My big fear was to have that sudden avalanche take me out.

"I trust my gut," she says as we ride the chairlift back to the top for another run with her seven-year-old son Aksel who is competing in a ski race today. She describes a time, caught on film, where her gut instinct suddenly made her alter her course and embark on what any rational decision-making process would conclude was a fatal mistake. "I trusted my gut that this was going to go."

In the film Wendy is seen screaming down the dreadfully steep and rugged slope alone in a landscape that has never before been touched by skis. Suddenly she disappears into an explosive plume of snowy smoke as the avalanche engulfs her.

"It was gigantic."

Miraculously she maintains control and, drawing on years of experience as a downhill racer, out-skis the avalanche and traverses to safety at the precipice of a cliff.

"My game plan was to stop at the edge of the cliff because I thought all the snow was going to go away."

But despite the deliberate planning in the midst of this deadly situation that she always feared might take her life, she abandoned the rational course and skied back *into* the avalanche, a seemingly suicidal act.

"I traversed the hill and got back up on a spine. All the snow was running next to me and I could hear it. It was right at my level but not getting me. So I end up skiing right back into everything that was running next to me. A lot of it had already gone, but it still took me out.

"I had to make a split-second decision: Am I going to stick to what I had been planning or should I abort into what I already know is happening?

"I tumbled. Still had to fight not to be pushed upside down, but then I was fine."

When it was all over, the consequences of her decision became clear. The original route would have taken her below cliffs where tons of snow from the avalanche would have buried her deeply as it cascaded over the precipice.

In an instant, threat-detection and situational-awareness circuits in her brain had taken in all the information available, and drawing on a lifetime of knowledge and experience (via connections to the hippocampus), these circuits considered the various contingencies, calculated the odds, and set her off instantly on a definite course of action.

"So I end up skiing right back *into* everything," she says in disbelief as if only an observer. In fact she was. The conscious mind works too slowly. At 60 mph you travel 88 feet every second. You don't have anything like "a second." Hesitate, deliberate, and you are dead. No one without a high-functioning orbital prefrontal cortex could have done what she did. Without this supercomputing gut-instinct neurocircuitry, she would have been killed.

Too much conscious processing can undermine performance in many situations. This is especially evident in elite sports, such as Major League Baseball or the NBA. Head team physician for the New York Yankees Dr. Christopher Ahmad explains how this is a critical issue for batters:

"So-called choking—they are bringing too many conscious elements into their swing, for example. Then what I think happens is they are trying to consciously work on some fundamental aspect of their swing and then they start to slump rapidly. The more they hit poorly, the more they are trying to work consciously on their plate appearance, trying to do something."

"There's nothing harder [than batting a ball in the major leagues]," says Jayson Werth, outfielder for the Washington Nationals with a .318

career-high batting average in 2013. "If you can hit, you can do anything, because it's the hardest thing to do. There's nothing harder. There's nothing harder in the galaxy."

Science tends to back up that perception. A baseball thrown by a major-league pitcher traveling at nearly 100 mph from 60 feet away comes zipping straight at the batter too fast to see in detail. Hitting a fastball requires instinctive reflexes that must be honed into a skill over years of practice. "When you go to swing, it's violent. It's everything you got," Werth says.

Elite athletes must become skilled at avoiding "paralysis by analysis," as Dr. Charlie Maher, team psychologist for the Cleveland Indians, put it. It may be a surprise to learn that MLB teams frequently have a team psychologist like Dr. Maher on staff. He is not there as a fringe benefit to players; he is there because he wins games. Psychology is a major part of elite performance in baseball and in other top-level athletic competitions.

"Players are under enormous pressure, particularly at the major-league levels, but in all sports. The ability to cope with the demands of the sport, to travel, dealing with the media, dealing with the ups and downs of their performances, dealing with expectations of other people like the owners and general manager who give them contracts. All of that can get to them, and if you want to add in the personal side, things that might be going on in your personal life, it is a very, very stressful environment. That's why it is a very risky environment for them. Because they are in such a high-risk environment, for some of them it is very easy for them to make mistakes," whether on the field, on the road, or bringing the stress home, Maher says.

The psychology of choking in sports happens when a player departs from the precisely automated mental operations that must be executed rapidly and accurately, and begins to consciously overanalyze the situation, Dr. Maher explains. "They are doing that for the most part unconsciously," he says, performing with lightning reflexes despite enormous stress and the demanding complexity of the situation. "The term they will use is 'automatic pilot.' Now, when something happens that takes them out of that kind of mode, that's when they run into problems."

Maher elaborates: "That shows up behaviorally as being tentative. A pitcher who's supposed to pitch inside—brush a guy back—doesn't do it. So he becomes very uncomfortable and he'll throw another kind of pitch and the batter will take advantage of it.

"The same in football," says Dr. Maher. He should know; he has worked as team psychologist for twenty-five years in the NFL and NBA as well as in the MLB.

Dr. Ahmad of the New York Yankees says he thinks pitchers and batters in the major leagues definitely have more of this unconscious rapid analytical ability than other people.

"And not only do they have more of it, they make it more of a *positive* performance feature for them," he says. "Whereas some people might make it a negative performance. A bad driver who is on the road might have that rapid reaction, but it is more of a negative thing [such as road rage or aggressive driving]. Whereas athletes who experience it on a regular basis somehow develop a way to modulate it, to know how to make it psychologically—not on a conscious level—but they make it effective for them."

Interestingly, Dr. Ahmad observes the same effect in a realm far more important than any game. "It is the same thing that surgeons do. There are a lot of parallels between [being] surgeons and athletes and being chess players and musicians."

Performance on the field is one thing, but consider the operating room, where Dr. Ahmad must perform flawlessly under intense pressure. He explains:

> You are losing your patient type of thing . . . it happens quite a bit. It may not be as severe as you are losing your patient, but like, *Hey, if you don't get things going, then you are going to have to amputate his leg, or you are going to have to reoperate tomorrow and explain to the family that you, uh, messed up the surgery.* That kind of thing.
>
> I know plenty of surgeons who panic, and when they panic their performance completely deteriorates. It crumbles to zero. They become almost infantile. They would not know how to put their mask on to get into the operating room.
>
> Then there are other surgeons who no matter how dire the situation is, they accept it, roll with it, deal with it—somehow. They are, of course, the people you'd be wanting to operate on you.
>
> I've been thinking a lot about that and how people are able to get that positive mind-set in the most critical situations. How to do it? In my opinion some of it is probably natural, but you have to have rehearsed terrible situations like [how] a pilot is able to get in a simulator and rehearse disasters. Then when a real disaster happens, he

is able to follow protocol, is able to go through his processes that he has been trained to do. The military may be the same thing.

But as surgeons we don't routinely practice terrible situations. Instead we do everything we can do to avoid them! We don't go out and do them on purpose to better ourselves!

[Instead], I mentally rehearse a framework so that in the most dire situation I have a conversation with myself where I say, *This is the time where you have to step it up and take advantage of what a great opportunity this presents to overcome something that could turn out terrible. Rise to the challenge, to be positive about it and say, "I've got a great opportunity to switch this all around and make it perfect." Instead of saying, "Oh, this is terrible. I'll never get this right."*

Performance under extreme stress is paramount in combat, and the military seems far ahead of the medical establishment in this regard.

"I have had multiple friends die in parachute accidents," a Navy SEAL says, to illustrate this point. "The reason they died is they got overwhelmed, panicked, and then it went to irrational decision-making."

I asked the SEAL Team Six member if he had read Mark Owen's book *No Easy Day*, about the operation that killed Osama bin Laden, and he had. The book describes the extensive planning and endless drills in which the assault teams practiced infinite contingencies to respond to every imaginable disastrous scenario possible, and they did it over and over inside a detailed mock-up of bin Laden's compound. When their dangerous mission inside Pakistan did in fact get off to a horrendous start, when their helicopter crashed into the compound and their well-rehearsed plan to fast-rope onto the roof went up in smoke, the team was unfazed, and pushed ahead with an alternative approach to the assault.

The book describes a heroic moment when the assault team finally battled its way to the entrance of bin Laden's third-floor bedroom and the point man encountered two women shielding bin Laden. The point man shoots his target and then instantly the SEAL tackles both women, presuming that they are wearing suicide vests. He acts selflessly, committed to die shielding his comrades from the blast. Was that an unconscious reflex? I asked.

"I guarantee he thought through every single step," the SEAL ventured.

SEALs, he explained, work hard to develop all the lifesaving threat-detection responses our brain has evolved—all of them, including the conscious and the unconscious abilities—and to refine them to the highest level of performance possible. Most important, they work to rapidly integrate the two brain threat-response mechanisms—utilize the rapid subcortical networks but integrate them with the cortical networks of impulse control and reason. The SEAL continued with his expert opinion about the split-second reaction of assaulters at the threshold of Bin Laden's bedroom:

> I guarantee you that point man read every single thing in the room from the way they are dressed to what are the odds they have a suicide vest bomb underneath based on the way they are dressed. Is it baggy? Enough clothing to conceal it? Do they have a gun? What are the odds that they have a gun behind them? Typically women do not have guns.
>
> It's like a car crash, right? In a car crash [time dilates and] you remember everything coming up to the last minutes before it because you focus on things and survival instinct just kicks in.
>
> In that car crash, the more you are able to sit back and think and slow things down the better.
>
> Like special-operation forces in all branches of the military, SEALs are gifted with natural talent and ability, but they also train very hard to improve their threat-assessment and -response capabilities.
>
> One of the drills that we do is called the hooded-box drill. (It is non-lethal. You shoot paint cartridges.) They bring you into a room—hooded. Then the instructors set up different scenarios in the room. Boom! They pull the hood off and there are any number of different scenarios, threats, non-threats, you name it, in the room. Boom! Your hood comes off and you've gotta analyze all that information instantly—life-or-death—and make the decision. There may be a hostage situation. There may be unarmed women and kids. There may be men coming after you. There may be people with guns. There may be someone right behind you getting ready to grab your pistol. Whatever it is, hood comes off, you analyze everything, look at and prioritize everything immediately, and then act.
>
> We don't want the guys when the hood comes up to freeze. We don't want the guys when the hood comes off to start blasting people. That's not what we do. What we *do* want is when the hood comes up: analyze—prioritize. Put them in steps one through ten. OK, ten

may be the least important. [You tell yourself] *I'm not even going to worry about that. I'm going to act on the top three.* And then, reprioritize after that, because it is always a changing environment. The hooded-box drill is an excellent drill to train your mind in decision-making at that stressful level.

The balance between reflex and reason under intense stress is the key to peak performance. Too much of one and you choke; too much of the other and you may snap.

# Tunnel Vision

The dorsolateral prefrontal cortical region is involved in working memory and bringing information to your consciousness for immediate action; that is, concentrating on a problem. Thus, working memory and attention are linked by the dorsolateral prefrontal cortex. You must pay attention to solve a math problem or to respond to a threat where analysis and heightened mental focus are essential. Consider Secret Service Agent Scott Moyer holding up in his memory the faces of potential suspects as he scans the crowd pressing against the ropes to shake the hand of the president of the United States. A brain stroke involving the anterior cerebral artery feeding this part of the frontal lobe would make it impossible to perform this difficult task at the high level of performance required in his job.

This region of the cortex also sends commands to the thalamus, the brain's relay point for sending information from the incoming senses to the cerebral cortex and thus to our consciousness. When someone says, "Wow! Did you see that?" at work here is the selective shunting of sensory information to our conscious mind. Perhaps it was a move on the basketball court in the middle of a game or the flight of a woodpecker through the trees that grabbed your attention. Conversely, when you see that same person yacking away obliviously on his cell phone in public, you are seeing the opposite action of this brain region as it throttles back sensory information from reaching the cortex. This connection to the thalamus shuts out the rest of the world as the person's awareness narrows to encompass only the person on the other end of the phone. The minute he hangs up, the caller may look around and be startled to realize that everyone is staring at him as his dorsolateral prefrontal cortex begins to readmit the sensory input from his surroundings to reach his conscious awareness. "Oh! Sorry . . . I was talking to my daughter . . ."

When it comes to moments of snapping, we call it "blind rage," and the neurocircuitry in this brain region is responsible for tunnel vision blindness.

"A batter gets hit by a pitch, and then he charges the mound," Dr. Ahmad of the New York Yankees describes, giving an example of how the Life-or-limb trigger of rage gets tripped. The batter throws down his bat and charges the pitcher with clenched fists. "You know he's not thinking rationally, how he normally thinks; [instead] he's reacting.

"I recall a player who got hit by a pitch and when the trainer went out and repeatedly was trying to speak with him—to find out if he was truly injured or not—the person was totally in a daze. Like not responding to him. He was just lost."

Sudden rage can overwhelm all reason and blot out everything around you except the target of your fury. On August 9, 2014, three-time NASCAR champion Tony Stewart was racing neck-and-neck with Kevin Ward Jr. in a nighttime dirt-track competition at Canandaigua Motorsports Park in upstate New York. Pulling out of a turn at full throttle, Stewart crowds Ward's car up against the wall, cutting him off and spurting ahead. The two cars brush lightly as Stewart passes, causing Ward's vehicle to career into the outside wall, spin out of control, and come to a halt facing backward on the track.

Immediately, Kevin Ward storms out of his vehicle and marches on foot against the ferocious race traffic, oblivious to the oncoming hail of auto-mobiles and deafening roar of engines hurling the racecars at high speed toward him. The Stopped trigger of rage had been tripped when Stewart crushed Ward's chance of finishing the race.

An oncoming vehicle swerves abruptly, narrowly avoiding hitting Ward as he's marching against the traffic. As Tony Stewart's car laps once around the track, coming back upon the scene of the accident, Ward thrusts out his arm, points at Stewart, and starts yelling. Ward storms directly into the path of Stewart's onrushing racecar. Apparently unable to process the startling image of a person appearing on the racetrack, Stewart strikes Ward with his racecar at high speed, catapulting the twenty-year-old driver to his death. Controversy immediately surrounds the accident because of Stewart's infa-mous reputation for his aggressive temper and sudden snaps of anger on the track, but clearly Ward's hotheaded reaction was foolhardy. Anyone who would step onto a racetrack in the middle of an automobile race risks causing mayhem and death. But that thought never reached Ward's cerebral cortex.

The orbital prefrontal cortex regulates arousal; the dorsolateral prefrontal cortex regulates awareness. Together the two act in partnership to respond appropriately to the world around us. This is why people who have had a prefrontal lobotomy are so impaired even though their cognitive capabilities are not diminished. However, while the dorsolateral prefrontal cortex helps us in a crucial way by focusing all of our mental capacity onto one difficult problem, it also impairs us by causing tunnel vision, and it can decrease human performance by overthinking.

A recent scientific study performed at the University of California at Santa Barbara and published in 2013 demonstrates this convincingly with a clever experiment. Participants in the study were shown a blitz of kaleidoscopic geometric images for about a minute. After a one-minute break, the participants were presented with a series of images and asked to state whether they had seen the images before. But here's the interesting twist that separates this study from so many other studies of memory: The test subjects were required to say how certain they were about having seen each image previously. That is, to state whether they remembered it in rich detail, or whether they only had a vague impression of having seen it, or whether they were guessing blindly. Paradoxically, the subjects did better when they said they were just guessing.

Why? From what we have just discussed, it could be that when people in the study focused on the object consciously and committed the details of an image to memory, they were actually impairing their performance in the test overall. "Paralysis by analysis," as Dr. Charlie Maher would say. In contrast to the sluggish conscious mind, the high-speed unconscious brain can easily keep up with the blizzard of kaleidoscopic images hurled at it. But this brain region can only communicate its output to the conscious brain by emotion— by provoking a sense of recognition or a hunch. "Trust your gut." This rapid unconscious processing must be blazing away inside the brain of a racecar driver speeding around the track in a blur and relying on rapid unconscious reflex to keep his speeding car on the edge of control.

The researchers went on to test their hypothesis. Remember that the dorsolateral prefrontal cortex is what focuses our attention on one thing amid a hail of complex sensory input. If this explanation of why guessing produced better recognition scores is correct, then inactivating the dorsolateral prefrontal cortex should improve performance by letting the high-speed unconscious circuitry do its work without interference from the conscious brain as it attempts to focus the mind on analyzing the details of each image. What the researchers needed to do was take out the dorsolateral prefrontal cortex and then see if performance improved. There is a harmless way to do this. If very powerful electromagnetic pulses are delivered through the skull, activity in brain circuits at the focus of the pulses can be suppressed. This technique is called *transcranial magnetic stimulation* (TMS).

Using TMS the researchers suppressed activity in the dorsolateral prefrontal cortex while showing subjects the kaleidoscopic blitz of images. Once participants in the study were unable to focus mentally on any one image, they would have to go with their hunches, just as skier Wendy Fisher did in her split-second decision to avoid the avalanche danger. Discussing the results, Dr. Taraz Lee, the lead author on the research paper, said, "If we ramped down activity in the dorsolateral prefrontal cortex, people remembered the images better."

Dr. Lee relates this situation to the common breakdown that occurs on the golf course: A professional golfer who holds the lead at the eighteenth hole with only one easy shot on the putting green left to win falls apart under the pressure and misses the easy putt. "You just can't think about that sort of thing," he said. "It just doesn't help you. Activity in this part of the brain hurts you," he says, when performing under pressure and when needing to rapidly analyze a stream of complex sensory input.

Connections to the dorsolateral prefrontal cortex are refined during childhood and adolescence. Synaptic connections are pruned away to develop brain circuits that give each of us our unique personality and individual character. This synaptic pruning through experience enables our brain to become ideally suited to coping with the environment we are reared in. Integration of emotion and sensory input makes this brain region especially critical for processing complex social settings and for conducting normal social interactions. Hostile experience in early life can wire it to put aggressive behaviors on a hair trigger and lead to impulsive violence. Fundamentally, though, this is an adaptive behavior; it is what you would expect evolution to provide. Wiring these hair triggers for aggression is likely to help a person survive in a hostile environment, but in a different environment impulsive violence will be detrimental. A similar effect can be experienced by a serviceman returning home after combat or in the debilitating fear that can remain unsubdued long after a traumatic experience. The effect in that case is called post-traumatic stress disorder (PTSD).

This adaptation and variation occurs at a genetic level too. Many recent studies show that impulsivity, risk-taking, and aggression have a strong genetic component, and genetic risks are associated with attention deficit hyperactivity disorder (ADHD), antisocial personality disorder, borderline personality disorder, intermittent explosive disorder, suicide, drug abuse, and violent crime. Mental illness and crime are outside the scope of this book, but in exploring the biology of why we snap, we brush up against this fascinating body of genetic research in considering impulsivity, individual variation, and the work of the prefrontal cortex in focusing attention and controlling impulsive action.

Moreover, the same neurotransmitters (dopamine and serotonin), and therefore some of the same genes, are at play in both rapid responses to danger ("good" and "bad" snapping) and violent crime. Given the high-level executive functions of the prefrontal cortex it is obvious how many psychiatric disorders, from depression to post-traumatic stress disorder to schizophrenia, would involve this brain region and the neurotransmitter systems that regulate it. Drugs for treating psychosis, depression, and schizophrenia work by regulating the amount of dopamine and serotonin in the prefrontal cortex. It is important to remember that human behaviors and traits are the result of multiple genes working in combination, and that there are many influences on brain function from environmental experience. It is remarkable, though, how influential variation in even a single gene influencing communication between the prefrontal cortex and the amygdala can be.

Dopamine is an important neurotransmitter in the prefrontal cortex. High levels of dopamine facilitate concentration and working memory; lower levels are associated with increased impulsivity and inability to stay on task. Dopamine signaling is what gives us the sense of reward when we achieve a goal or have a pleasurable experience. Motivation, cognition, drug addiction, and psychiatric disorders are all strongly dependent on dopamine signaling.

An important source of dopamine comes from neurons located in the midbrain (*mesencephalon*). If the brain can be likened to a mushroom, the midbrain region corresponds to the area where the mushroom stalk joins with the bell. Within the midbrain there are two clusters of neurons, called the *substantia nigra* and the *ventral tegmental area*, which contain neurons that use dopamine for communication. They extend their axons broadly throughout the brain to reach the cerebral cortex, the limbic system, and the striatum. Stop for a moment and consider what this anatomy tells us. This tiny spot in the middle of our brain (operating well below consciousness) has an enormous, wide-ranging influence on the brain at large, including the cerebral cortex and thus behavior.

The striatum is a brain region important in controlling movement, and the loss of dopamine supplied by substantia nigra neurons is the root cause of Parkinson's. Substantia nigra, incidentally, is Latin for "black substance," because again, anatomists had no idea what these neurons did, but this spot in the brain is darkened by melanin pigment inside the dopamine neurons. Much of the cerebral cortex receives dopamine from neurons in the ventral tegmental area, including the cortical regions that are of interest with respect to fear, threat detection, and impulsive action (namely the *nucleus accumbens* and *ventromedial prefrontal cortex*), but also the amygdala and hippocampus.

Individuals with the so-called warrior gene display higher levels of impulsivity and aggression in response to provocation. The gene encodes an enzyme, monoamine oxidase A (MAOA), which breaks down dopamine, norepinephrine, and serotonin. We will want to consider these three neurotransmitters together, not to form an encyclopedic "laundry list" of neurotransmitters but because these three neurotransmitters work in concert to control the behaviors we are interested in here.

Norepinephrine reaches the cerebral cortex from neurons located in a region of the brain stem called the *locus ceruleus*—even lower down the "mushroom stalk," in a region called the pons—and this promotes arousal. Thus, this neurotransmitter is critical for arousal, response to stress, anxiety, panic, and the fight-or-flight response. Norepinephrine is also released into the blood from the adrenal glands. These neurons look dark blue, thus the name *ceruleus*, meaning "blue," and *locus*, meaning "spot" in Latin. The axons of these norepinephrine neurons reach far and wide. They influence almost the entire brain, including the hypothalamus, cerebral cortex, and amygdala. It makes sense that a system designed to elevate arousal would need to have a wide-ranging network to rev up activity in the entire brain.

Serotonin also comes from neurons in the lower brain regions (the pons and mesencephalon). Serotonin regulates emotional states, and thus is involved in many behaviors, including aggression, impulsivity, attention, decision-making, reward, and (when disrupted) in schizophrenia, drug addiction, and autism.

Humans have different variations of the gene that metabolizes these neurotransmitters at different rates. Boys with a particular variation of the MAOA gene are more likely to join gangs and also to be among the most violent members. The same effects are not seen in girls. This is because the MAOA gene is located on the X chromosome. Females have two X chromosomes, whereas males have one X and one Y chromosome. If a male inherits the "violent" MAOA variant, there is no other copy of the gene to moderate its effect as there is in women, who have two copies of this gene. A study of prisoners in Finland found two gene variations were associated with extremely violent prisoners: an MAOA genotype associated with low dopamine turnover and CDH13, a molecule on neurons that controls brain development. The statistics show that five to ten percent of all severe violent crime in Finland is attributable to these two genotypes.

Other genes also regulate levels of these neurotransmitters. Two variations of the COMT (catechol-o-methyltransferase) gene have been associated with becoming either a worrier or a warrior. Differences in the MAOA and COMT gene have also been associated with depression and other mental disorders or traits including panic disorder, OCD, and neuroticism. (As an aside, I checked to see if I might have the "warrior gene." I sent a saliva sample to a company that for a fee returned a detailed analysis of thousands of genes and gene variations in my personal genome. I was happy to learn that I did not have the warrior gene. Instead, I have inherited a gene variation producing elevated dopamine in the prefrontal cortex, which promotes concentration and working memory—very helpful things for book writing. For this I am grateful to my ancestors and struck by how much of our fate is channeled by a genetic dice roll.)

Acetylcholine, which comes from neurons in the basal region of the brain (the diencephalon and basal nucleus), also stimulates arousal. Acetylcholine increases responsiveness to sensory stimuli and focuses attention. The loss of these inputs is seen in Alzheimer's disease, and this contributes to lethargy and cognitive decline.

The question is, could someone like Tony Stewart or Kevin Ward Jr. have become a champion racecar driver if they did not have an extraordinary ability to focus, to rely on reflex, and a willingness to impulsively risk everything to win? To blot out everything except achieving the immediate goal of crossing the finish line first? To fearlessly accept danger, to risk a horrible fiery death in a dangerous situation where your fate depends on split-second reactions? Clearly not. Any spectator feels the sudden surge of fear and adrenaline rush through their body as the engines rev up and whine with terrifying power. Most of us would be paralyzed with fear on a professional racetrack. But without the prefrontal cortex to moderate the powerful survival instinct driven by subcortical circuits of rage, fearlessness can suddenly slip into recklessness or snapping in a blind fury.

## Making a Move

The *dorsomedial prefrontal cortex* is important for motivation and initiating activity, and thus it is critical in regulating impulsive behavior, such as snapping or responding to threats. It becomes activated by emotional processing and eye gaze. People with damage to this region become apathetic, listless, lack spontaneous movement, and often fail to respond to commands. This region communicates with areas of the cortex that control bodily movement (motor cortex) to enable it to evaluate and regulate information from the body's senses about the external environment as well as the internal state of the body, and to launch an effective response.

You may not be a Secret Service agent, a member of SEAL Team Six, or a professional athlete, but you can see the unconscious mind at work in threat detection in your own brain. Just as in exposing a magician's sleight of hand, all is revealed by careful observation of everyday behaviors that ordinarily go unnoticed.

On July 31, 2012, there was a near disaster in Washington, DC. A collision between three jets carrying 192 passengers was narrowly averted by an air traffic controller with only 1,650 yards of separation remaining between the planes screaming toward each other at a combined speed of 436 miles per hour. This near disaster highlights something we take for granted: how our brain constantly monitors our surroundings and automatically acts to prevent collision as we weave effortlessly through throngs of pedestrians and circumvent obstacles in our path without giving it a thought.

We are all familiar with the experience of approaching another person and, instead of passing each other gracefully, you both move abruptly in the same direction, putting you on a collision course. Instantly you veer in the other direction, but the approaching person makes the same dodge and you come to within a split second of colliding head-on. You both halt, releasing a little laugh at your spontaneous body-language breakdance.

We laugh for two reasons: Such navigational dilemmas and near misses are rare, and secondly, all of this behavior and maneuvering is carried out unconsciously. It's amusing because both of you are just spectators.

But how? Collision avoidance is an extremely difficult problem. Enormously complex systems, sophisticated communication mechanisms, and strict adherence to standard protocols guard against collision at an airport or in any transportation system. Yet in walking around freely there are no codified rules for collision avoidance. If we followed a formalized rule to safely negotiate passing another pedestrian on foot—always pass on the right, for example—there would be no problem, but there are no such rules. You and the person approaching you are free to pass each other either on the right or the left. There is no requirement to communicate or negotiate your preference. If left to chance, this would result in near collisions 50 percent of the time, but in fact collisions between people are infrequent. Watch the throngs of pedestrians weaving in all directions through swarms in Grand Central Station, where collisions are rare. How can this be explained? Does each person somehow read the other's mind and know which way to pass? Without the slightest conscious thought our hidden cerebral guidance system solves this problem instantly and constantly so that you are free to engage your conscious mind in other matters.

Unconscious mind reading is very much at work here, because people use a different unconscious strategy for collision avoidance depending on whether the obstacle ahead is an inanimate object or a human being. If the obstacle to be avoided is an inanimate object, regardless of whether it is stationary or moving, we simply divert our course (and speed) onto a new trajectory to safely avoid colliding with it. If the object ahead is a person, though, this strategy is not used, simply because the other person might suddenly change and move in any direction—speed up, slow down, stop, or step right or left. In addition, it is possible to communicate unconsciously with the other person by using body language. The "steering a safe trajectory" approach to maneuvering around an object is the strategy we use while driving a car to avoid a squirrel in the road, and we all know how badly that can end.

After analyzing video footage of subjects avoiding collisions between pedestrians, researchers A. H. Olivier and colleagues at the Université de Rennes in France have found one mathematical variable that will accurately predict the collision-avoidance maneuvers we perform to prevent "pedestrian roadkill." The factor is not a variable unique to either pedestrian's behavior, such as a change in speed or direction of either individual, it is a factor that *combines* the actions of both parties—a reciprocal interaction of both people—known as mean predicted distance (MPD). This is the anticipated distance of separation between two people at the point ahead where their current trajectories would bring them to intersect with each other. If that predicted distance of separation is less than one meter, *both* subjects alter course in a way to increase the distance of predicted separation at the intersection to one meter or more. If the MPD is greater than one meter, neither person alters their stride or direction.

The unconscious mutual guidance maneuvers are executed in seconds or fractions of a second. An observational phase begins when the two walkers first see each other, and realize (unconsciously) that the current trajectories place them on a collision course (MPD less than one meter). This observational phase lasts only one-third of a second—remarkably quick, considering all of the situational awareness and subconscious calculation, analysis, and decision-making that must be executed.

Next the walkers begin maneuvers (again without conscious deliberation) to *coordinate* an increase in the MPD if necessary. The course correction is based on feedback from what maneuvers the other person executes. The analysis and response typically lasts only three seconds.

Clearly people weaving through crowded places like the streets of New York do not simply maintain a one-meter distance of separation. Researchers find that in a final regulation phase of collision avoidance, two walkers fine-tune their trajectories and speed to allow them to pass each other even closer than the minimum distance of one-meter clearance the instant they pass. This frugal last-minute adjustment occurs in the final 0.8 seconds before the crossing point. This is about the time it takes to make one stride. Both walkers can now permit the clearance distance to tighten because the collision-avoidance problem has been solved and the other person's trajectory can no longer change in the final instant before passing. Unfortunately, this melding of minds does not happen very well between a squirrel's brain and the brain of an automobile driver. The human brain is built to meld with other human brains.

We are pretty darn good at this, and we do it all subconsciously while our conscious mind is fully engaged in other activities such as contemplating where we are going and what we will do when we get there; turning over a pressing problem; conversing with a companion; noticing an attractive person or a suspicious stranger in the crowd and divining their intentions; or puzzling out where we are in an unfamiliar environment.

Collision avoidance while simply walking around illustrates how our unconscious mind takes in and constantly analyzes an enormous amount of information outside of our awareness, but it also suggests that an apparent unconscious telepathic communication is going on between the brains of two people. Indeed, that is the case.

Researchers from Princeton University used functional MRI (fMRI) imaging to see what part of the brain is activated when two people communicate through speech. They found that when two people communicate, neural activity over wide regions of their brains becomes synchronized, with the listener's brain activity patterns mirroring the same patterns of activity sweeping through the speaker's brain, albeit with a short lag of about one second. If the listener, however, fails to comprehend what the speaker is trying to communicate, their brain patterns decouple.

Interestingly, in part of the prefrontal cortex in the listener's brain, researchers have found that neural activity *precedes* the activity that is about to occur in the speaker's brain. This only happens when the listener is fully comprehending the story and anticipating what the speaker will say next. "Communication is a joint action, by which two brains become coupled," Uri Hasson, the lead scientist on the Princeton study, explains. "It tells us that such coupling is extensive [across many brain areas]." Language binds brains together, and this melding of minds forms societies.

The Princeton research team is interested in determining if nonverbal communication (rather than language) similarly causes mirrored brain activity in the recipient's brain. One can only imagine if an fMRI machine would show Special Agent Moyer's and the bad guy's brain activity locked in synchrony as their eyes meet at the mall.

Recall my encounter with the robber in Barcelona. In the silent stalking that preceded the thief's grabbing my wallet, nonverbal information was all that my subconscious mind would have had available to deduce the potential danger. For one thing, these unconscious navigation-control circuits would have detected that the individual was violating the MPD to get close enough to snatch my wallet. That may have been the final threat alarm tripped in my unconscious mind as my conscious mind tried to take in the complexity of a new environment that was opening up ahead as my daughter and I emerged from the subway. Even though that potential danger had not yet been sent as an alarm to my cerebral cortex through the brain's relay point (the thalamus), my unconscious brain "knew."

Research shows that vision and language can meld minds unconsciously, but so can sound. You walk into a bar and music is thumping. All heads are bobbing and feet are tapping out the same precisely coordinated movements. Somehow the rhythmic sound grabs control of the brains of everyone in the room, forcing them to operate simultaneously and perform the same behaviors in synchrony. How is this possible? Is this unconscious mind control by rhythmic sound only driving our bodily movements, or could it be affecting deeper mental processes?

The mystery runs deeper than previously thought, according to psychologist Annett Schirmer. Rhythmic sound "not only coordinates the behavior of people in a group, it also coordinates their thinking—the mental processes of individuals in the group become synchronized," she says. This helps explain how drums unite tribes in ceremony, why armies once marched with bugle and drum into battle, why worship ceremonies are infused by song, why speech is rhythmic and punctuated by emphasis on particular syllables and words, and perhaps why we dance.

EEG recordings of electrical activity in the brain showed that waves of brain activity (alpha and beta waves) become synchronized around the auditory rhythm. That is, brain waves become phase shifted so that the peaks of the brain waves always occur at a precise point relative to the next beat in a drum rhythm. Rhythmic sound synchronizes brain waves.

What are the consequences of shifting brain waves to follow the beat? To find out, Schirmer and her graduate student Nicolas Escoffier from the University of Singapore first tested subjects by flashing a series of images on a video monitor and asking them to quickly identify when an image was flipped upside down. While participants focused on this task, a synthetic drumbeat gently tapped out a simple four-beat rhythm in the background, syncopated by skipping the fourth beat of each measure.

The studies showed that when an inverted image was flashed on that missed beat, the subjects identified the image as being inverted much faster than when the upside-down image was flashed at times out of synch with the beat or when all the images were presented in silence. Somehow, the brain's decision-making was accelerated by the external auditory rhythm, and cognitive ability was heightened at precise points in time with the beat. An analysis of electrical activity in the brain revealed how this boost in performance on the anticipated missing beat was happening.

The EEG measurements showed that not only does the auditory rhythm *couple* the brain waves of everyone in the crowd, but the phase-locked brain waves alter visual perception. Like ocean waves crashing on the shoreline, brain waves deliver sensory information to the mind in synchronized pulses. The brain wave measured at the back of the skull over the region where vision is processed peaks each time an image is flashed on the screen. This is called a *visual evoked potential*, meaning that vision evoked a rising electrical potential in a visual cortex as the sensory input from the eyes reached the brain. But the experiments revealed that when an image was presented simultaneously with the missing drumbeat, the electrical brain response evoked by seeing the picture is bigger than when the image is presented out of rhythm or flashed on the screen in silence. The bigger electrical response means that these visual circuits are more responsive when the image appears precisely in synch with the auditory rhythm.

This region of the brain (the visual cortex) processes the earliest steps in the process that gives us vision, the arrival of visual input to the cerebral cortex, rather than brain waves that reflect analysis and processing of that visual input. This means that our *perception* of the external world entering our mind through our eyes is "gated" by the rhythm of what we hear. Something seen at a point precisely on the beat with an auditory rhythm is more likely to be perceived than if it appears out of synch with the rhythm.

This gating of visual input by auditory rhythm does not require prolonged meditation on the rhythm to cause the person to enter into some sort of a trancelike state; the effects are nearly instantaneous. "Within a few measures of music your brain waves start to get in synch with the rhythm," Schirmer says. "Rhythm facilitates our interpersonal interactions in terms of not only how we move but how we talk and think," she says. "Rhythm facilitates people interacting by synchronizing brain waves and boosting performance of perception of what the other person is saying and doing at a particular point in time."

Rhythm, whether the lyrics to a song or the meter of a poem, facilitates language processing, she concludes. "When people move in synchrony they are more likely to perceive the world in synchrony, so that would facilitate their ability to interact."

Tap out the first line of Lincoln's Gettysburg Address using the fingers of both hands. Do it four times: "Four score and seven years ago." Take a minute to try this and it will become obvious why Lincoln did not begin his speech with "Eighty-seven years ago." The rest of Lincoln's short speech is equally rhythmic. His brief oratory synchronized the brain waves of those who heard Lincoln's words that day, and the words resonated throughout history in the minds of all who read them.

". . . of the people, by the people, for the people, shall not perish from the earth." Analyze the rhythm: The first three phrases, if set down as a musical score, would be three measures of four quarter notes in 4/4 time with metronomic regularity:

Of the people (one, two, three, four . . .)

Then in the final measure, the timing abruptly changes. As the new research shows, whatever was said on that missing beat would have had greater impact on the brain: "Shall not perish . . ." (one, two, three, four) . . . and then: "from the earth" (one, two, three—ending on a missing beat). That message was the nugget of Lincoln's entire speech: that the country now torn by civil war will survive. The same information could have been conveyed by omitting the vacuous "from the earth" bit. How else can one perish except from the earth? But those three words were there for the syncopated rhythm that synchronizes brain waves and amplifies the sensory input into the minds of all hearing or reading his speech.

Lincoln's two-minute address was preceded by a two-hour oration by politician Edward Everett, not a word of which is remembered. Lincoln's address proved unforgettable.

Contrast Lincoln's rhythmic oratory with Everett's tin ear. Both speeches convey the same message, but the two would have had completely different effects on the listener's brain waves and thus perception. Consider the last lines of Everett's speech:

But they, I am sure, will join us in saying, as we bid farewell to the dust of these martyr-heroes, that wheresoever throughout the civilized world the accounts of this great warfare are read, and down to the latest period of recorded time, in the glorious annals of our common country, there will be no brighter page than that which relates the Battles of Gettysburg.

This newly discovered phenomenon would have been working against my pickpocket in Barcelona. To snatch a wallet from the cargo pocket above my knee, the thief had to synchronize his movements and pace precisely to match mine. That synchrony would have put him more prominently on my unconscious radar, because the thief and anything else operating in synchrony with my own periodic peaks in sensory perception and decision-making would have elevated that sensory input to my brain more powerfully than, say, the rest of the crowd or his accomplices moving out of synch with my own footsteps. When you consider that it is impossible to focus simultaneously and consciously on everything in your environment, gating sensory information to your cerebral cortex in synchrony with your own movements and brain activity makes sense. Indeed, it came as a shock for me to learn that the pickpocket was not acting alone but rather operating with a team of criminals. Had he lifted my wallet as we sat still on the subway car, the outcome might have been very different. In fact, I once had my pocket picked while sitting on the subway and I never noticed a thing.

Much of human communication takes place nonverbally, through our eyes and ears, but touch also provides a rich and universal means of communication. Touch, after all, was the trigger for my snap reaction to the pickpocket. This communication was instantaneous, and it conveyed a clear message in a complex situation.

Every political campaign appearance starts and ends with vigorous handshake sessions with as many potential voters as possible despite the danger candidates face when wading into a crowd of strangers peppered with possible adversaries. Some campaign appearances are completely focused on pressing the flesh—a fleeting grip of palms accompanied by little or no verbal exchange of substance, but for politicians that brief handshake is priceless—worth risking even death. That is because handshakes and body language in general communicate powerfully and deeply about the internal state and intentions of other people. Through the handshake we establish the level of trust between ourselves and the other individual. We divine earnestness, truthfulness, and we measure the depth of friendship or deceit in the other party. In all formal social interactions and in business, the handshake in Western society precedes discussion and seals the deal at the end.

Body language can communicate and persuade even more powerfully than what is said. After two presidential debates in 2012, in which the candidates crammed endless facts into their speeches and rehearsed to hone their rational arguments, what everyone remembered and was talking about afterward was the body language of the candidates—Obama's grimace, Biden's smirk, Ryan's composure under fire. The same is true of the classic Kennedy/Nixon presidential debate. Does anyone remember what point the two candidates were debating while Nixon lost the argument (and election) appearing so uncomfortable, sweating and nervous in the spotlight? These political debates always start and end with handshakes between contestants and often with the moderator. The handshake touches neural circuits inside the brain that predispose a person toward positive feelings of competence and trustworthiness, and it opens a relationship of positive cooperation while suppressing negative feelings.

Much higher ratings of competence, interest in doing business, and trustworthiness are found in psychological studies if an encounter is preceded by a handshake. It activates a part of the brain called the *nucleus accumbens*. The amygdala (part of the limbic system) activates when a handshake ends a successful business meeting. Such handshakes leave us markedly less stressed. Our interest in further interactions is boosted.

The nucleus accumbens is a central component of the brain's reward pathway that is linked to positive experiences and emotions such as excitement. Casual social interactions as well as getting to know people over the long term require ascertaining the other person's internal state of mind and character. Nonverbal communication via body language and touch, even as formalized a touch as a handshake in Western society, allows our mind to tap into the other person's mind. It is a rich, unconscious, and universal language that transcends words.

# 7

# Extrasensory Perception?

If you don't like something, change it. If you can't change it, change your attitude.

Maya Angelou, on her Facebook page, February 11, 2013

Do the subconscious modes of communication discussed in chapter 6 sound too much like pseudoscientific mental telepathy and ESP? Well, prepare yourself. The science gets considerably stranger—and it seemed to chase me down.

A reader called me to compliment another of my books. Then he confided the true reason for his call: He wanted to share with me an extraordinary change in his brain and ask for my neurobiological insight.

"After having a stroke I found that I could read other people's minds," he said.

*OK...* I thought to myself, poised to find footing on an uncertain precipice.

I explained to the caller that I am a basic research scientist, not a medical doctor. (I was thinking psychiatrist, to be specific.) "All of my 'patients' have fur and tails," I joked. "I'm not able to offer medical advice."

He was not deterred.

"After the stroke I could not speak for a long time, but strangely I could still sing. I'm a songwriter and musician."

He went on to describe how after suffering a stroke that left him mute, he could read the private thoughts and intentions on the faces of the hospital staff and people around him. Quickly it became obvious to me that this person was not suffering a psychiatric illness, nor was he describing a hallucination rooted in street-drug "pharmacology." The man was completely lucid, articulate, and very intelligent. What he was experiencing, I concluded, had developed from a neurological disorder.

As he laid out his life history, I was reminded of how many people who lose a sensory ability sometimes develop heightened abilities in the other senses. The blind can develop extraordinary abilities to hear, for example. Some blind people can perceive speech sped up to three times the rate at which a sighted person can comprehend, understanding with ease what sounds to everyone else like babble. Others can echolocate to navigate the world without vision. Ray Charles and Stevie Wonder are famous examples of people for whom extraordinary musical ability appears to have been enhanced by the loss of vision and increased reliance on hearing.

As far as mind reading goes, I appreciate how much of what the brain does takes place deeply below the level of consciousness. I know that humans utilize a rich vocabulary of body language that works almost entirely subliminally. The world bombards us with a blizzard of signals. Our brain takes them all in, evaluates them, synthesizes them, makes decisions, and influences our conscious mind through subtle feelings and urges. Only a sliver of this perception and analysis reaches the level of consciousness.

So I speculated that if the loss of verbal communication associated with a stroke had forced the caller's brain to rely more heavily on other channels of perception that were previously suppressed by the predominance of the conscious mind, perhaps he had developed them to a greater extent than before. This seemed to make sense to both of us.

As we closed the conversation I asked, "You mentioned that you were a musician and that I might know some of your songs." He revealed the name of the hit he wrote, which had made him and his rock band a legend. A delusion of grandeur? I wondered. He could claim to be Jesus Christ and thoroughly believe it, I reminded myself. But as we discussed the musical intricacies of the piece, it was obvious that he indeed knew the score intimately. His revelation was no fabrication, as I later confirmed. However, he related this story to me in confidence, so I don't have the right to reveal his identity. The world is a small place.

I was reminded of this phone call when I read a new study from the

University of Bologna, Italy, by psychologists Bertini, Cerere, and Lavadas, writing in the journal *Cortex*. The researchers studied people who were blind in one eye, and their experiments revealed that the sightless eye was able to communicate to the brain unconsciously.

It sounds like pseudoscience, like ESP. But as I told the rock musician, the brain is able to take in much more information subconsciously than we can possibly hold in our conscious mind. Neuroscientists are beginning to map out the hidden brain circuits that are responsible for this.

The researchers in Italy flashed pictures of faces on a computer screen and asked the subjects to identify whether the faces were male or female and whether their facial expressions were happy or fearful. What the subjects did not know, because they could not see through their blind eye, was that the researchers were also flashing pictures simultaneously to the sightless eye. The surprising result was that when they flashed a picture of a person with a certain expression to the sightless eye simultaneously with an image shown to the good eye, the subjects were able to identify the gender and facial expression of the picture shown to their good eye much faster. How could showing the blind eye anything at all affect how quickly a face is seen through the good eye?

Here's the most important clue: The accelerated reaction time only occurred when the researchers showed the blind eye a picture of a fearful face while showing a happy face to the sighted eye. Fear or alarm on the faces of others, even though it was impossible to see visually through a blind eye, somehow alerted the subject's mind to danger and elevated their mental performance to a heightened state of vigilance.

You see, these people were blind in one eye as a result of a brain stroke or traumatic brain injury to the part of the brain that gives us the sense of vision—the visual cortex located at the back of the skull. The eye itself was not damaged. Recent research shows that there is a pathway from the eyes to part of the brain responsible for vigilance, detecting threat, and fear—the amygdala. Although the subjects could not see through their blind eye (that is, they had no vision through it), the unconscious pathway to the brain center responsible for detecting threats remained intact.

Fear on the face of someone else means that there is good reason for you to fear or become alert to danger as well. In many threatening situations, sending signals to the visual cortex and consciously evaluating the danger would take much too long. So nature has given us a high-speed shortcut for alarming visual information to reach the part of the brain that is always on guard for deadly threats. In many life-or-death situations, the regal conscious brain is just too slow. This is becoming a familiar story, no?

We react instantly to evade an onrushing car and then, after jumping to safety, gasp in amazement—"What was that?!" Then the fight-or-flight response kicks in as the adrenaline surge hits our bloodstream and our heart races, skin moistens, muscles tense, and we shake with wound-up energy to fight to the death or run like hell, but in leaping out of the path of the onrushing car we have already reacted instantly and decisively to the danger. The threat has passed before the fight-or-flight response even started. That's what leaves us shaking afterward—we are primed to fight or run, but we can do neither one.

All the new strange sights and difficulties of negotiating an unfamiliar environment in a country where my language was foreign fully taxed my conscious mind when I responded to that thief in Barcelona. I was simultaneously engaged in delightful and excited conversation with my daughter on our sightseeing adventure, but my unconscious brain was working just as feverishly taking in situational information through multiple sensory circuits (sight, sound, touch, smell) all below the level of awareness and evaluating them for danger. I was not aware of the potential threat, and I had not taken any particular notice of the crowd pressing around us as we ascended the steps out of the Metro station looking for directions to the Gaudi cathedral, but the parts of my brain that are dedicated to danger and threat detection must have been monitoring everyone around me. Even though I was consciously blinded to the thief as he plucked my wallet from my pants pocket, just like the fearful face shown to the blind eye of people in the experiment described above, visual information was reaching the subcortical centers of my brain in the limbic system, and it must have been alerted by what they perceived. When I felt my pocket empty, my limbic system, coiled to spring into action in its heightened state of alarm, released my defensive response, activating my hypothalamus and preprogrammed motor skills to take down the robber, probably within one-third of a second, just as fast as we do every day in unconsciously avoiding collisions with others while walking.

Those motor skills were not trained through martial-arts classes or military experience; they were innate behaviors for self-defense and the defense of my offspring, coupled with whatever training I had as a school student on the wrestling mat decades earlier. When I found myself on the ground struggling with the thief and my conscious brain began to apply itself to the situation, it was these lessons in wrestling from forty years prior and from watching my son and my other daughter wrestle in high school that came to my mind and guided my actions in pinning and immobilizing the robber as the fight-or-flight response ignited in my body. The course had been set by the time the adrenaline hit—it was a fight. My mind and body were fully prepared and engaged.

As we were chased like prey through the foreign city, my daughter and I were not overcome with feelings of fear. Rather, we were gripped by an intense heightened state of awareness and purpose. Our minds raced with elevated mental acuity as we outwitted our pursuers. We were hyperalert and focused. The sensation of being pursued was very familiar, however. After we were safely in our hotel reliving the experience, Kelly and I both commented that the sensation of being chased by the gang felt exactly like rock climbing when negotiating a very difficult and dangerous move. This reaction, like me drawing upon my high school wrestling lessons, suggests that prior experience must have a role in how individuals will respond to different threats and situations that trigger aggression or fear.

We have taken a brief glimpse into the enormous flood of new information on nonverbal and subliminal communication, body language, threat detection, fear, and aggression. These new insights are emerging from fMRI research that enables cognitive neuroscientists to see the neurocircuits in action inside the human brain in a way that psychologists of the past, who probed these same questions by devising insightful behavioral tests and sticking electrodes into animal brains, could only have imagined.

This neuroimaging research shows that there are two distinct routes through the brain for detecting and processing information related to threats, and that each one triggers a different kind of behavioral response. Conscious awareness is not only unnecessary for you to detect and react reflexively to a threat, body language and threatening facial expressions that enter the brain unconsciously evoke *faster* reactions.

It is no wonder Special Agent Moyer says that he and other Secret Service agents "react totally subconsciously" in many threatening situations. To add to the evidence already noted, cognitive neuroscientist Julie Grèzes and colleagues reported fMRI studies in 2012 showing that "threat signals trigger defensive responses independently of what observers pay attention to," and that very different parts of the brain are activated by threats involving conscious and unconscious processes.

The first route of threat detection proceeds through circuits involved in early emotional responses. These include part of the amygdala, hippocampus, and hypothalamus, but they proceed independently of attention to the threat. The second route flows through the cerebral cortex, and this requires our conscious attention. The first route of unconscious threat alarm triggers a reflexive behavioral response; the second evokes a voluntary behavior. In my case the first neural circuitry was what caused me to instantly grab the robber by his neck, and the second circuitry, which was activated after my attention was raised to the threat, was responsible for me working to gain hip control on the ground and deliberately cinch the choke hold to vanquish the perilous attack.

It is difficult for us to comprehend all of the complex threat-detection and situational-awareness activity in our brain, because so much of our sensory ability and the constant surveillance of our environment is unconscious. However, we can obtain deep insights through the study of truly exceptional individuals who have extraordinary minds and extraordinary sensory abilities. Meet the most extraordinary research subject I have ever encountered: Tricia.

## Willing to See

The first part of her I saw was her brain. It was stunning.

Looking back, that fMRI brain scan that so excited and intrigued me scientifically has diminished into a paper-thin representation, a two-dimensional slice of a richly multidimensional person I came to know.

She speaks with open enthusiasm. Energetic, ebullient, almost childlike in her honesty and unguarded sharing of emotion conveyed by dynamic inflections of her voice. She speaks with glee. Effortlessly articulate, her linguistic prowess shines with the intelligence of an inquisitive, sharp mind that is well read, filled with wonder and fascination, but empowered with fierce determination; constantly seeking, her sentences are prone to ending on a rising pitch. She is a lovely young woman, charming, with chestnut-brown eyes and brunette hair attractively styled with two sprigs of burnt-orange tinted ends framing her face like colorful feathers at her ears, but it is most fitting to know Tricia by her voice. That is what we can share. Tricia is congenitally blind—that is, blind from birth. I hasten to add, sensing Tricia's ire rising at my last statement, that while genetics has robbed her of functional photoreceptors in her retina, Tricia lives in a visual world of her mind's own creation. The absence of a sense that she has never known is a source of ongoing discontent. So she has willed herself to see. In the process she has rewired her brain to create vision. Strange science, indeed.

If you have ever explored a cave, you invariably reach a point where you decide to shut off your flashlight beam to experience the momentary thrill of absolute darkness. That sudden plunge into blackness is frightening and disorienting. Suddenly you feel lost, vulnerable, and helpless. If you have not had this experience, you have no doubt experienced awaking in darkness in a strange room, momentarily alarmed and lost until your eyes open and faint shadows you can barely discern anchor you. Ah, there you are.

It is common for people to be afraid of the dark. How do people who are blind cope with the many dangers and threats lurking invisibly in the darkness? Research in this area might reveal how I was able to identify and grab the robber behind me without actually seeing him. Do the blind have a heightened sense of awareness and rely on a means of threat detection that most of us who depend on vision lack? What better way to probe the brain's many sensitive and unconscious mechanisms of threat detection than to speak with someone who cannot see? An encounter with Tricia reveals the vast untapped potential of the human mind, but more profoundly, meeting her illuminates our limited and artificial perception of reality and exposes the hidden clockwork of the brain.

"I tend to be hyperaroused or hyperaware," Tricia told me when I asked her about threat detection and fear in the absence of sight. "I can hear in people's voices when something sounds threatening. Sometimes I'm not accurate; I have a thin trigger. I might think someone's mad at me or upset when they often aren't.

"I have intuitions about people usually," she said. "Sometimes when I touch people's bodies I can make comments about the kinds of pain that they are feeling or what that pain might be linked to, but this hasn't been empirically tested. I'm careful with it because it's kind of scary."

When one sense is diminished, the other senses become more acute. People who are deaf can discern speech by watching a person's lips move. This shift to strengthen the remaining senses can be observed by neuroimaging methods as changes in both the structure and function of the cerebral cortex, and importantly, by changes in connections linking different cortical regions. This adaptation to sensory deprivation illustrates not only the capacity of the brain to self-modify its structure and function adaptively, at a more profound level it reveals the vast potential of the human brain to extract information about the environment through subtle stimuli that we are not consciously aware of, but that nevertheless contribute to our perception. These subtle, unconscious "perceptions" operate together with our conscious perceptions to form an intuitive analysis of other people's intentions.

For most of us, many subtle sensory inputs are ignored. This is not surprising, considering how even strong sensory input is often completely ignored. The conscious mind can hold only a tiny fraction of all the sensory input it receives in a momentary state of awareness. If we attend visually to one feature of our environment, our awareness of other sensory input (sound, touch, pain) becomes diminished. This filtering and focusing of sensory input can be seen by monitoring the flow of electrical activity in the brain and by functional brain imaging, but it is readily apparent from everyday experience. A nurse may distract a child momentarily to quickly remove a bandage on a painful wound. While the child's brain is focused on the distraction, the pain stimulus never reaches the conscious mind. Or you may be rudely awakened by an angry horn honking when you failed to see a red traffic light turn green while you were intently engaged in conversation. Even though you are staring right at the traffic light and all the neural systems are delivering that visual information in exquisite detail to your brain, the brain shuts off this stream of input to your conscious mind (higher cerebral cortex) to devote more of its limited computational resources to comprehending sound.

Now consider the situation with a person who is congenitally blind. One-third of the human cerebral cortex is normally devoted to vision. This is an enormous amount of computational power. Only a sliver of the cerebral cortex controls all of our bodily sensations and movements. It's called the sensory-motor cortex and it runs in a narrow strip down our cerebral cortex much like the Rocky Mountains on a map of the United States. A tiny patch of cerebral cortex known as Broca's area on one side of the brain (left side usually), about the size of a fifty-cent piece or a silver dollar, is devoted to the enormously complex and uniquely human capability of speech. Imagine if all of the territory of the United States from the eastern seaboard to the

Mississippi River were completely uninhabited, devoid of towns, cities, roads, factories, and farms. This would be akin to the situation in the cerebral cortex of a blind person.

For people who are congenitally blind, in contrast to people who lose their sight at some later point in life, vision is an abstraction. They know nothing of color. They cannot easily imagine or comprehend shapes, visual textures, or the motion of objects from sensing the way light reflects off surfaces. The concept of a horizon, stereo-optic space and perspective, and the ability to discern foreground from background or to recognize an object— for example, a cup no matter how it is oriented (projected on the retina as a circle, a cylinder, or any rotation in between)—is difficult for a blind person even if vision is restored later in life.

Without the experience of vision while the brain is wiring itself in early life, the necessary connections between visual neurons in the cerebral cortex that are required to analyze visual information cannot develop. In experiments with animals that are sightless from birth, if vision is restored, say, by removing an eye patch on a cat's eye that was applied when it was a kitten, the cat remains blind in that eye. Light sensations are received after the eye patch is removed, but the brain can make no sense of the information.

In our analogy of the cerebral cortex to a map of the United States, restoring visual input to the brain later in life would be like building a highway from San Francisco up to the west bank of the Mississippi River but having all the territory beyond it be undeveloped. All of that development we see now east of the Mississippi River, the centers of commerce and the complex systems of bridges, highways, railways, airlines, and communication networks linking them, were formed over time according to the needs and activity of our country during its formative stages of development. A new road being opened into the vast territory east of the Mississippi River that lacked Chicago, New York, and all of the developments between these and other cities would be useless. This predicament is a serious problem in the attempt to restore sight to congenitally blind people through new medical interventions, for example using gene therapy to restore retinal function.

A congenitally blind person, though, has always inhabited a different sensory world that lacks visual input, and therefore does not typically feel a sense of loss for something they have never known. People who are congenitally blind (or deaf) face obvious challenges, frustrations, and difficulties because of societal structure and people's attitudes toward them, but the visually impaired and deaf utilize their other senses to cope with these impositions. Congenitally deaf people may refuse to learn American Sign Language or have a cochlear implant, because they contentedly self-identify as being deaf. They may have no urge to transform into something that they are not; their set of sensory abilities are in the minority, but they feel ownership of them. Tricia is different.

"Do you have a favorite color?" I asked Tricia when I first met her in her home in Berkeley.

"Burnt orange. And turquoise, like the *Caribbean* color of the sea. I also like to think about *whiteness*. Like *cheap, horrible* whiteness versus, you know, whiteness that has a *warmth* and more of a *cream—and brown* to it." Her tone of voice is dynamic and dreamy, conveying so much emotion and appreciation of color. She sounds like an artist savoring color or a stage actor who can instill feeling in others through precise emphasis and inflection in speaking her lines.

"Certain colors that are shown to me will actually *hurt* my eyes. Like, *Oh*! I don't want to look at *that* anymore. Like *really insane* hot-pink. Or insane *purply* colors, like very, very bright purples. It's like, *Oh* my God! *Goodbye!*"

Medical tests confirm that Tricia's retina is nonfunctional. She has a genetic mutation that destroyed the photoreceptors in her retina. Tricia, unlike the typical person who is born blind, has never self-identified as blind. Vision is so important to Tricia, she has overcome the absence of a functional retina, rewired her brain, and taught herself to see—with her fingers. Yes. A challenging notion.

Dermo-optical perception (seeing by touch) has been in the scientific literature sporadically for centuries. There are periodic accounts, primarily of blind individuals who can perceive images printed on a piece of paper by sweeping their fingers lightly over the surface. Some can reportedly read text this way. The ability to "see" with fingers has been dismissed because of charlatans and "mentalists" who use trickery to perform the feat as entertainment. An influential paper published in the journal *Science* in 1966, and written by the renowned science writer for the magazine *Scientific American* Martin Gardner, concluded that there was no scientific evidence to support dermo-optical perception. Gardner debunked the entire field and ascribed accounts of dermo-optical perception to cheap trickery accomplished by peeking through blindfolds. Essentially no scientific credence or scientific research has been given to the subject in more than fifty years.

I count myself among these scientific skeptics. Only two years ago while dining with a biophysicist in Tel Aviv, where I had been invited to give a lecture on my research on brain plasticity, my host shared with me new theoretical research that his colleague, also a physicist, had just performed. His calculations proved that—in theory, at least—it would be possible to perceive images by photosensitive cells in the skin. Even without a lens to focus light into an image, the brain, in theory, is capable of extracting visual input from the skin and reconstructing images in the brain by using complex analysis and mental calculation. It is a fact that many plants and animals do have photosensitive cells in the superficial layers of their skin. Plants, for example, track the sun's path across the sky. Flower petals open and close according to ambient light. Many sea creatures—scallops and the giant clam, for example—react quickly to a shadow cast upon them by clenching their shell closed. Our skin detects the sun's ultraviolet light and produces melanin in response.

But "seeing" with your fingers? Despite the theoretical possibility based on mathematics and physics, I was skeptical; dismissive, I must admit. I never thought about dermo-optical perception again until my colleague Dr. Manzar Ashtari, a neuroscientist at the University of Pennsylvania, shared her recent experiments with me and told me about Tricia. To my chagrin, she also sent me a copy of the physicist's paper on dermo-optical perception that I had rejected in my mind over that dinner in Tel Aviv. It all sounded too much like ESP. Maybe Tricia was a fake or fooling herself.

I had just given my invited lecture at the University of Pennsylvania in November 2014, and Dr. Ashtari was walking me briskly back to Penn Station to catch my train back to Washington, DC. As we walked she eagerly shared the results of an experiment she had done the day before on a young woman who was a graduate student at UC Berkeley. I listened intently as we rushed into the station.

Dr. Ashtari is an expert in neuroimaging, and her current research, in collaboration with her colleagues, is using gene therapy to restore sight to people who have a certain genetic mutation that causes blindness. The results are extremely exciting. The blind subjects in the study are regaining limited but very useful vision, and by using sophisticated human brain imaging, Dr.

Ashtari can see the visual circuits in their brain rewiring after retinal function is restored. This rewiring and restoration of vision is happening despite the large body of research on animals in the lab showing that vision cannot be restored after a certain age—the point in early life when the visual system must wire itself correctly in cortical circuits guided by visual experience.

Unfortunately, Dr. Ashtari's study subject, Tricia, was congenitally blind due to a genetic error in a different gene from the others in the study. Tricia was not eligible for the gene therapy, but Dr. Ashtari was so intrigued by what Tricia claimed to be able to do, she decided to test Tricia in a functional MRI machine.

As we waited in the noisy station for my train to arrive in a few minutes, Dr. Ashtari popped open her laptop to show me the brain images. Oblivious to the crowd of onlookers surrounding us, we pored over the images excitedly and in amazement. First Dr. Ashtari had tested Tricia by flashing images of different colors, intensity, and contrast to her eyes through goggles while she was in the brain scanner. Dr. Ashtari pulled up the results on her computer for me to see.

"Absolutely nothing!" Dr. Ashtari exclaimed, pointing at the cold X-ray-like image of the woman's brain. "Right eye, left eye—nothing. Absolutely nothing."

Normally when an image is presented to a sighted person, the visual cortex becomes actively engaged in analyzing the visual input. This appears on the fMRI in a rainbow of colors, where warmer colors indicate higher levels of brain activity. The back of the cerebral cortex would normally appear orange or red on an fMRI when visual stimulation is applied to the eyes through the goggles, but Tricia's brain scan was completely black-and-white, indicating that there was no visual information reaching her cortex that could be detected by brain imaging. Tricia is indeed blind. Electrophysiological measurements of her retina and genetic analysis had already told researchers that much.

Next Dr. Ashtari handed Tricia a postcard and said to her, "I want you to do your thing—whatever you do."

The blind woman did not know that half of the postcards she was being given to probe with her fingertips had no image on them. Dr. Ashtari would compare Tricia's brain's activity while her fingers scanned postcards with images on them to when she unknowingly scanned blank postcards as a control.

Demonstrating her ability to see without vision before this brain scan, Tricia had swept her fingertips lightly over the surface of a postcard like a treasure hunter waving a metal detector over beach sand.

"I can pick out basic shapes, some colors, and contrasts between colors, but I never know precisely what I am looking at."

But what is "color" to a person who has never seen one?

Tricia marvels at how so many of the pictures she "sees" with her fingers have a common theme: a horizontal gradient slicing across the bottom of the pictures. If you had never seen one before, could you possibly imagine a horizon?

During the experiment Tricia's head was resting inside the enormous doughnut-shaped magnet of the fMRI machine and she wore the same goggles that the researchers use to display images to subjects with sight. Dr. Ashtari instructed Tricia to extend her arm out to the side and touch the postcards with her fingers. Tricia lightly scanned the series of postcards using her fingertips as the fMRI machine peered inside her brain.

My inbound train was about to arrive as Dr. Ashtari and I stared at the results of the blind woman's brain scan on the computer screen. Tricia's brain is ablaze with activity. In contrast to the cold black-and-white image of Tricia's brain as pictures were flashed on her retinas, her brain was now splashed with the orange and red color signifying intense activity throughout her brain while she probed postcards with her fingers.

"Amazing! It's all over," Dr. Ashtari exclaims. "Her whole brain is involved."

Dr. Ashtari jabs at the computer screen, pointing out regions of intense brain activity: "Prefrontal cortex. So much cerebellum! She's drawing so much out of cerebellum. That's very interesting! Motor [cortex], sensory [cortex], sensory association area, parietal lobe, occipital! That's a very specific area for object naming," she says, pointing to an orange splotch of cortex. "Amazing!"

We discuss the technical aspects of the measurements to assure ourselves that we are not being fooled by weak signals or noisy measurements. There is no question about it. The signals are strong. Tricia's brain is "seeing" through her fingers.

"It is as if she is forcing her eyes to see. I can't explain this. I cannot explain this at all," Dr. Ashtari says, shaking her head.

This woman's brain had apparently rewired itself to utilize her visual cortex to extract sensory information from her other senses. Analysis of Tricia's brain connections, even in a resting state, showed that her visual cortex was wired to the somatosensory cortex (touch). Touch would thus have the enormous computational power of one-third of the cerebral cortex that is normally devoted to vision, but instead applied to analyze tactile input. Her auditory cortex was also strongly connected to the visual cortex. This is often the case in blind people, and it enables them to apply increased computational power to analyzing sounds, but the amount of interconnectivity between touch, vision, and auditory input was profound in Tricia's brain.

How would the brain perceive the input from the fingers when they are connected to the visual cortex? Would a touch produce an image? What's more, Tricia told Dr. Ashtari that when she uses cannabis her power of dermo-optical perception is greatly amplified, and that under the influence of marijuana she can actually begin to see something with her eyes. She excitedly urged Dr. Ashtari to repeat the experiments under these conditions, but it would be impossible because of the restrictive laws on cannabis, even for scientific research. Medical marijuana is legal in some states, but in Pennsylvania, possession of marijuana is a crime. Dr. Ashtari did not doubt that marijuana could possibly have the effect Tricia was reporting, knowing how the active component in marijuana, THC, can amplify synaptic transmission in brain circuits to enhance sensory perception, but unfortunately, this is something that cannot be studied because of legal restrictions. Possibly, if what Tricia says is true about being able to see light under the influence of marijuana, new drugs might someday be developed to help the blind see better.

The brain scan that shows how Tricia's cerebral cortex is not divided up into the five senses the way the cortex of sighted people's would be strongly suggests that Tricia must experience profound synesthesia, a mixing of senses. Tricia's perception of the world could be quite different from what most of us experience. The language we have developed to describe our sense of the world would likely be inadequate to convey the cognitive experience of a person with this degree of synesthesia. Touch might provoke an image or some combination of tactile and visual sensation.

That brain scan proved to me that dermo-optical perception is real. Tricia's brain was picking up very feeble tactile differences from the printed image on the postcards and working hard to make sense of it. Slight differences in temperature of different pigments, minute differences in the conductivity of heat sucked from fingertips touching different pigments, and possibly even effects of light on skin, although this is not yet proven, were being collected, analyzed, and reconstructed in the visual cortex to reconstruct a sense of vision. The intriguing thing is that all of this sensory input is potentially available to all of us—literally at our fingertips! But we ignore it. I had to meet Tricia.

I flew to Berkeley on a Saturday in December to meet her at her mother's house, an attractive middle-class home in a nice neighborhood. I rang the doorbell and a small dog started barking behind the double doors. It took quite a long time before the door opened, and a small curly-haired white dog burst out, springing past the young woman's legs in a flurry of confusion as Tricia, wearing a turquoise cashmere sweater, greeted me. The little dog dashed back inside and returned immediately with a red stuffed toy in its mouth to play. Once inside I noticed that the floors were all hard surfaces of stone, noticeably affecting acoustics, and Tricia led me to a round two-person granite table at the end of the kitchen counter. It was a bright and airy home. On the day of my visit her mother, an artist, was away showing her work. It turned out Tricia has discovered her dermo-optical ability quite recently by helping her mother with her painting and other artwork.

Tricia is in graduate school, and she was stressed about an important term paper due that coming Monday. I asked how she reads and writes, and she explained that she uses text-to-speech software to read and write on her laptop. She demonstrated by lifting the lid on her Apple laptop, releasing a rapid stream of babble from the speakers. The rate of speech was sped up three to five times faster than a sighted person can comprehend. It sounded something like Stephen Hawking's electronic voice on a reel-to-reel tape jabbering incoherently backward on rapid rewind.

Tricia snapped closed the lid of her laptop. When I asked, she said yes, she can also read braille.

"I only understand speech that is sped up like this with this particular voice," she said in a self-deprecating way.

To me, it was utterly incomprehensible. There was no white cane in sight. She does not wear dark glasses or shuffle unsteadily about using her hands to guide her. She knows her way around her home and she confidently but cautiously navigates in part by echolocation. "Although I'm not very good with echolocation," she said, again dismissively.

"When I use echolocation or when I hear sounds, *everything* gets localized to my eyes. I don't consciously use my ears to navigate very much—I mean, I do to some extent, but the kind of mental imagery of space, like, the working memory, tends to be centered around my eyes."

Tricia's eyes constantly wander and roam, unable to fixate on me or on any object. Sighted people take for granted that our eyes rest quietly and move in a coordinated manner to settle effortlessly on objects, but this requires visual input from the retina to enable the brain to fix our gaze and pursue objects. Blind people cannot do this, and her constantly wandering eyes (nystagmus) frustrate Tricia.

"Right now without marijuana I am fighting against the nystagmus," she said, eager to show me the beneficial effect of cannabis on her blindness. "Like, it's *really* hard to control my eye muscles, and my proprioception of my eye muscles is diminished, because I don't have the marijuana. So the feeling that I'm getting now is—imagine an arthritic hand. The fingers are clenched; you don't have a lot of mobility. The first thing that I noticed when I used the marijuana was an increased elasticity. I think it's the nerves [controlling eye muscles better under the influence of cannabis], because I am aware of them, particularly toward the back and sides of my eyes. The nerves suddenly become something that I can manipulate. They become like these strings that I can *push* and *pull* and *stretch* and do whatever I want with."

Tricia launched eagerly into a monologue about her condition and about her thorough search of the scientific literature on the CB1 and CB2 receptors in the retina that would be activated by cannabinoids. This could explain how her vision is improved under their influence. Her detailed, accurate, and up-to-date knowledge of the scientific literature was impressive. When I realized that she has learned all of this, and everything else, without being able to see, I was even more astonished. This is complicated stuff, and every scientific paper I have ever read is filled with figures and illustrations. Scientists love and depend on graphs. I doubt that I could have comprehended the papers without seeing the figures. Clearly Tricia has a gifted mind to be able to excel at one of the best universities in the country.

Without body language to orchestrate and punctuate our conversation, Tricia was speeding away on a monologue and I had to break in. "Before we go further," I interrupted, "I would like to know more about what it is like to perceive the world without sight. For you the sense of vision is touch," I propose, recalling her fMRI and the connectivity between her somatosensory and visual cortex.

"Yeah. It's touch with my eyes."

Elaborating on a concept that is strange to a sighted person she explained, "Let's say I'm looking at a white wall [while on marijuana, when she says she can see faintly]. I'll have a sensation that my eyes are slipping and sliding up and down that white wall. This is very hard for people to imagine, but imagine that your eye is a marble, and that you are holding that marble, and that you are sliding that marble up and down a sheet of ice. That will be what it feels like when I'm looking at a white surface.

"I discovered this with my mom and I said, 'Why is it, like my eyes are slipping up and down the wall?' And she said, 'Because they are white walls.' I said, 'Oh! OK. I get it now.'

"You know, the main difference between touch and vision is that you can only touch what is in your parapersonal space. *However,* vision is *different!*" she exclaimed with rising excitement in her voice. "What happened to me is that I had to learn these eye movements that would basically *push* my eyes away from my face and get them into more *distant* space. So then I started developing *metaphors* to account for that experience, because I can't reach ten feet away from me with my hands, right? So I started to develop metaphors to describe what my eyes were doing. They include things like *walking*," she said in a singsong delight. "*Dancing*; that sort of thing. Marbles rolling—*leaping* into the air. Another common one is a limb outstretching, because obviously you can touch more if you stretch your arm than if you can't. So this is why I think eye movements are part of my synesthesia. I push my eyes consciously away from my face, it creates an illusion of distance or of space.

"Because I have synesthesia, it's impossible to tell what's coming from sound and what's coming from vision. It seems like it's not just a simple matter of vision to touch, there's a whole kinesthetic component that is at play, which is that different colors seem like they impact my eye movements in different ways."

"Can you see color?"

"No. On *pot, then* it's exciting! But without the marijuana, no I don't. I don't have access to colors. I have access to—let's say I am walking and there's a bush on one side. Different objects as I am passing by kind of come to me as different textures, like the bush might feel softer, like around my face, and the wall will be harder, but again it's hard to know how much of that is vision and how much is echolocation. Sometimes when I'm walking it feels like I'm trailing something with the back of my hand and sometimes it feels like I am touching something with the front of my hand. I'm not touching anything! I'm just moving my eyes around. It all seems to have some kind of mapping with hand and finger positions. My friend who uses echolocation a lot says that she does *not* experience those kinds of textures."

"Let me ask you some questions to help people better understand what it's like for a person without vision, because vision is so important for other people—" I began.

"But wait!" she interrupted forcefully. "Back up!"

"Vision is *super, super, super, super, super, super* important to *me*! Which is *why* I am doing all of this. Because *my* perception of the world, unlike other blind people who've been blind since birth and they somehow manage to adapt—*I* have—I mean, *I* can get around. I'm going to Berkeley, but I *still* feel a gap. I am still *palpably* aware of my missing sense. *That's* why I'm spending all my time and energy on all of this, because I *want* to find a way to recover as *much* of that as I possibly can, because vision is *so important* to me."

I felt bad that I had made a misstep and upset her. She is fiercely driven by her goal. In my research for this book I have had the immense privilege of meeting many extraordinary, strong-willed people who are driven with a passion and who have achieved extraordinary heights through relentless devotion to their cause, and through hard work to shape their mind and body to succeed beyond what others could hope to achieve—members of the Special Forces, elite athletes, and others—but to my mind, Tricia stands shoulder-to-shoulder with any of them.

"People think visually," I explained.

"Right, I do [too]."

What about dreams and imagination? I asked her, because both are such visual processes for most of us.

"It's very hard to tell you if there are visual components of my dreams, because in the waking state I have to spend so much cognitive energy focusing and attending to vision to make sense of it, so in a dream state I don't have that conscious control. So I think my dreams tend to be more auditory or they tend to use other senses.

"*Imagination* on the other hand," she said with the delight of a creative artist, "and *thinking* are a whole different story. If I'm thinking of ideas about a paper and, like, organizing it, I'll often do these abstract paintings in my mind of some ideas on *one side*, and some ideas on the *other*, and I can't even *put* that kind of thinking into words for you, because it goes *beyond* language. It is a very *non*verbal process, and that's partially why I say when I think *visually*, it *feels* very visual to me. It *feels* like it's coming from an area that isn't a language-based area, and that thinking that I have has developed a *whole* lot more since the use of cannabis. I didn't use to be able to do that."

She gave me an example. "When I'm reading a book and there are colors mentioned in my book, often I will stop and I will *imagine* the color that I'm reading about in the book, so that I can practice simulating what it would be like to *see* that color.

"Or if I am thinking about a specific character, I could think about *looking* at that character. My repertoire of visual images is *still* so limited, which is why I think I am so frustrated, so *that obviously* has an impact on imagination. So something like, 'She was pouring the milk into the tea and the tea became like a taupe color.' You know, I *imagine* that. I *imagine* the tea *lightening* in color."

"How do you think dermo-optical perception works?" I asked.

"There are a couple of different theories. One of them is that *all* that's happening is that I'm picking up on things like the texture of the paint or the temperatures of the paint. So I'm not *actually* doing anything that's all *that* special. I'm just particularly *aware* of subtle, subtle, subtle changes in texture. That's one explanation.

"Another explanation accounts for that sometimes I am able to do this while the picture is on a computer screen. So then there goes your *ink* and your *paint*. But I can *definitely* feel different textures on the computer screen when there are different colors. It's just that I'm less consistent and it's harder to map them."

"But as you are doing this with your fingers, your sensation is touch. It's not an image?" I asked.

"No. It's a touch."

Clearly words invented to describe the world in five senses are going to be inadequate to describe the kind of synesthesia that allows Tricia to see. Different colors under her fingertips evoke different tactile sensations, textures, and visual images: bumps, smoothness, sharp pointed objects, resistance, and silkiness—all of it a different tactile sensation in her mind derived from differences that colors on the picture somehow convey to her fingertips.

"I'm aware of a sort of touch coming from my eyes that is more in line with what my pointer fingers would do. As I'm touching something, suddenly I'll just have this awareness—Oh! The line of sight goes from here to there! I'll just be like, Oh! That's the way your eye is meant to move."

"Do you think anybody could do this?"

"Yeah. I do think that. I think that the difference is that I'm trying to use whatever this is to help me understand vision. It's helping me understand how someone who's seeing with the eyes could understand color and how colors interact with each other to delineate edges and forms and dimensionality."

"What's so interesting to me about this is the untapped potential of the mind," I said.

"Yeah!"

"Do you think that we have all of these unconscious abilities?"

"Yeah!"

"That they get overlooked?"

"Oh, yeah! So, for example, the way that I understand a piece of art . . . I was doing this two nights ago, actually. I was under the influence and I was using both my eyes and my hands and actually that's when I'm most effective. Right now I can't use my eyes. They are not working." She laughed. "In general, *seeing is, like, really* rewarding, like my dopamine is like *boom, boom, boom!* Like *yes, yes, yes!*" She laughed again. "Yes! I'm seeing, I'm seeing!"

I took an eight-by-ten color photograph out of my backpack and placed it on the table. "Will you try one for me?"

"Sure!"

She began sweeping the index finger of her left hand over the image—not cautiously or analytically probing the surface and straining to extract some meager signal out of the glossy photograph but oscillating fluidly and rapidly over the surface of the photo, the way you would feel the difference in texture of velvet when your finger moves with and against the grain. But there is a striking difference. As I watch, her finger is sweeping along the edges of objects in my photo.

"Yeah, so this right here—" she said. "I notice, like, a contrasting sensation. Like there's like a change, uhmm . . ." She probed contemplatively in silence for a second.

"Wait! Watch!" she exclaimed "Maybe just watch how my finger moves," she said, as if we were both just observers.

"Oh! *Definite* boundary!"

I sat in stunned silence. I don't want to make a sound to give her any unconscious clues, but her finger had found a prominent feature in the glossy photograph and she was swirling around the edges of the object in the picture as if it were an object sitting on the table.

"That! I found—uhmm—a boundary! Like some *smooth* color. *Really* dense and resistant." (Her finger was probing a large rectangular black object in the photo, made of granite.)

Sweeping her finger around the photo she sang with glee from high to low pitch, "Smooth, smooth, smooth," as her finger traced the unseen object accurately before my eyes.

"I don't know what it is, but I *like* it. *Really* soft—and smooth—uhmm," she said, savoring the sensation of a white-colored image.

"I think, like, a line, but I'm not entirely sure, like another boundary. Ooh! Like there, this whole area, like there it's so smooth and soft. I *really* like that."

There was no question that her fingers were tracing objects in the photo.

"Oh and there! Right there there's a disturbance," she exclaimed. I was stunned as I saw her finger discover another object in the photo. "There's, like, a bump, like a change or a disturbance in the colors.

"Something else that's *really* interesting is that when I touch an area like *this* that is *very* smooth, it gives a feeling of *openness* or of *space*." I was biting my tongue in disbelief.

"Uhmm, this *whole* area— So here and there's a boundary. There's this whole area that's smooth on this, like, top edge." In fact, her finger was tracing along the top edge of a cliff.

"There's another area here. There's something on *this* side and then there's something on *this* side." Her fingers were leaping across a valley between two mountain ridges in the image.

"And then there's this whole area here that's smooth. Like this might be rectangular. [It was.] But I'm not exactly entirely sure . . . like, it's very hard to tell.

"What is it?" she asked as she probed the rugged edge of the rectangular cliff.

"Jeez, I wish I had videoed this," I said, exhaling and speaking for the first time.

"Oh, boundary *there*," she delighted in a singsong voice as she continued to scan the photo with her finger.

"That's amazing," I said.

But I was tongue-tied. The photograph is of my son and me after we had climbed to the top of Half Dome in Yosemite Valley when he was thirteen, following a technical climbing route called the Snake Dike. Half Dome is the iconic massive granite mountain in Yosemite Valley that was sliced in half by a glacier, creating something that looks like a scoop of ice cream sliced in half. Its sheer granite face plunges from the 9,000-foot summit to the valley floor 4,000 feet below. But where to begin? How do I describe Half Dome or Yosemite Valley to someone who has never seen a mountain? How do I describe the dizzying sensation and majesty of looking up at the sheer granite walls of Yosemite Valley towering above you, which strikes you with humility and awe and makes your palms sweat just to look at them? Also I was conflicted. I had grabbed this photo at random and suddenly I regretted it. I might be describing a sensation of beauty that I fear could accentuate her pain.

I tried to explain the photo, but it was difficult.

"Do you know Half Dome?" I asked.

There was no reply.

I reached out and guided her fingers across the photo. "These sharp boundaries; it's a cliff that drops—"

"Oh! Cool!" she exclaimed as the puzzling image that she has never seen in real life snaps into her mind.

Excitedly she asked, "What were the *very, very, very* big, smooth—like, was it *green* stuff? What was it?" Her fingers were tracing the rectangular patch of green: the meadow of Yosemite Valley.

I tried to explain. "Yeah, so here." I reached out again and guided her fingers along the edge of the cliff and other features that she had been tracing. I was so moved by what had just happened, I broke the tension by changing subject.

"By the way, you only used your finger on your left hand." That meant that she was using her more intuitive right brain rather than the more analytical left brain. This is typical of most people who are reported in the scientific literature to do dermo-optical perception, regardless of whether they are right-handed or left-handed. Tricia is right-handed.

"I know. But I *could* have used both."

I went back to describing the photo. "OK. So this here is just the top of the mountain."

"Yeah."

"And right in there—"

"Yeah! What is that?!" Tricia's wandering eyes were scanning the ceiling as her head tipped back while we probed the images in the photo with her fingers.

"That is like four thousand feet to the bottom of the valley," I tried to explain.

"So it is like air?"

Indeed it is. When her finger traced this area of the photograph earlier she had described it as giving her a feeling of space.

"But what's this particular area?" she asked, jumping her finger from the cliff on one side of the valley to the mountain on the other side.

"Well, that is the mountain looking across—"

"But then why is it smoother? Like, why is it like less dense than the surrounding area? Is it just because of the contrast?"

"Well, um," I stumbled. She was right; it *is* a sharp contrast in color. "So it's hard for me to describe," I said. Clearly she saw the images in the photo using her fingers, an image she had never experienced.

I tried again to do a better job of explaining the peculiar image. "So this is the top of a mountain," I explained, moving her finger along the ridge.

"Yeah!"

"And this is, you know, space. That's what you said. You said it was space."

"Oh!"

"Because—"

"Yeah!" she exclaimed.

"It's four thousand feet to the valley floor."

"Oh! So it's the drop! It's the drop!"

I moved her fingers to the part of the image that they had circled repeatedly on their own. "And then down here, down here—feel this part—is the valley floor four thousand feet below. And the other side of the valley has got this other mountain, and this is the ridge of the other mountain," I explained, returning her fingers to the places I had watched them trace when she explored the photo herself.

"So what colors would you say are there?" she asks me.

"OK, so it's not . . . It—it's mostly dark brown—"

"Uh-huh."

"And black because, you know, it's a mountain—they are rocks."

"Yeah."

"So there are heavy shadows."

"Right."

"And the space is deep blue."

"*Ohhh!* OK!" she exclaimed.

"And the valley floor here is green," I said, moving her fingers back to the meadow.

"OK."

"You guessed that it was green."

"I *did!*"

"And then up here—"

"Well, that's because of training!" she said triumphantly. "That's because I've touched—like, my mom and other people have shown me that, and I've touched that enough times and I've been able to make that association. That's one of the few [colors] that I've been able to do that with."

"Then up here at the top, this transition right here that you are feeling up here, that is the horizon that—"

"Oh, right!"

"And it's just pure white above that—"

"Right! Oh, right, right there," she said, sweeping her finger across the brilliant sky.

"Because it's just bright sky."

"Oh, yeah! I can feel . . . Oh, yeah! OK."

"That's amazing," I said, taking the photo and replacing it in my backpack.

"So I'm *not* crazy," she said, having dispelled my possible skepticism.

"You might have to convince other people that you are not crazy, but you forget that I've seen your brain scan, so you don't have to convince me."

"The *problem* with all of this, which is *of course* not *your* problem, it's *my* problem. . . . Emotionally it is very difficult to deal with—this sort of thing. . . . Because I feel it's *cool* but it's also *extremely* isolating. . . . Because I just, I want . . . I'm *doing* all of this to get *out* of this situation that I'm in, and I feel like I have to make that *really clear* to people. . . .

"Because I think there's a temptation where someone is like, you know, people tend to do this with blind people—it's like an exoticization of it like, '*Oh*, that's so *great* that you can do x, y, and z.' And *yeah*, it *is*, but it's *not* as useful to me as it would be if I could actually *see* that picture.

"So what I was getting from that picture—the *smoothness* and *roughness*, and *that* sort of thing—is *great*, but it didn't allow me to *connect* with you.

"Like, let's say you were my friend and you were showing me that picture; it only allowed me, in a limited way, to have an emotional response to that picture. And as someone who *wants* to be an *artist* . . ."

I heard the pain in her voice.

"I of *course* want to *maximize* my ability to emotionally respond to things, and to have them be *meaningful*."

"I think I understand," I said, "but why visual arts?"

"*I. Don't. Know!*" She punctuated each word with her fist on the table.

"*You* saw me. When you made that comment, '*Visual* images are important for *sighted* people.' *You* saw me. I cut in right there. The *reason* that I cut in right there was because when other people are talking about me as *not* a sighted person, I feel *weird* and *strange* and *bizarre*. *Why* would I not be able to just *accept* when someone talked about me as a blind person? I am *not*. I have a *visceral* reaction *every* time that happens. Why do I want to make this art that I can't even see? It's *just* the way it is. *That* is why I want to see—so that I . . . That is one of the many reasons," she reflects.

"Yes, but you are obviously a very creative person," I offered.

"Yeah," she agreed.

I asked her why she does not exercise her creativity in another way, by writing, for example: to paint with words.

I'm *not* going to stop writing! I'm *going* to write. But what you just *said* is a *line* that I've been fed for my whole life. *Any* creative person *can't* explain or *rationally describe* the type of creative expression that they prefer.

I discovered that yes, writing is *wonderful*. I am *never* going to give up writing. I *love* what I can do with words. But there is this *other* form of creativity where you almost enter this sort of meditative state where there *are* no words. And that's where I want to go.

The reason I feel so compelled toward the visual is that I love the *uncertainty* and *confusion* and the *newness* of this place where I am experiencing things that I can't fully articulate or put them into words nor *should* I, because once I put them into words, in some ways they are diminished. And I just have this *incredibly* strong desire to *push* myself and to be able to express myself in this *other* way, where it's just an *image* that's up for interpretation.

I am living in two realities simultaneously. One of them is a reality in which I am sighted, and one of them is a reality in which I am not. In some ways both of those realities are equally valid for me, so it's not a matter of *longing* for something I don't have. It's actually a matter of making something that I *do* have manifest.

No matter how many things I can say to you about my ability and I can show you—"Oh, let me do this with a color . . ."—I'm still always going to be dealing with that disconnect between what my mind is doing and what my eyes are doing. I am trying through the development of either a pharmaceutical intervention or through a prosthetic . . .

I'm trying to minimize that disconnect, because *that's* my disability, actually. It has caused me a *severe* amount of psychological distress.

"Anybody could understand that. I totally understand that," I said.

"There is something fundamentally different about me that is different from every other person I have ever talked to who has been blind since birth, which is that I am extremely motivated to learn about visual things and assign meaning to them, and basically construct this ability to simulate vision. And other people that I have talked to, even if they have a little bit more vision than I do, have no *idea* what I am *talking* about, and they are *OK* with being blind."

"Absolutely," I said. Tricia is right about her uniqueness in this respect.

"They might get annoyed or they might get frustrated, but usually those annoyances and frustrations have to do with societal attitudes, whereas mine is a dysphoria."

"Your brain is sitting there saying, 'I don't want to be blind,'" I suggested.

"'I didn't ask for this. I take all the input I can—every input that I can get—and I'm going to make vision out of it.'"

Tricia giggled. "Yup! You see, I *like* it when you say it like *that*."

"Well, we can see it in your brain," I told her. "Your whole brain gets involved when you are doing dermo-optical perception."

"My concern is that the dermo-optical thing is a *tiny, tiny*, tiny, tiny corner of things that have happened to me—that I have. I think that the visual experiences that I've had [while under the influence of marijuana] that are *not* dermo-optical are *tons* more interesting. The tough part is that I can only use it [marijuana] when I am not functioning in the world. Vision isn't something that I just want to go visit once in a while; it is something that I want to have access to and actually just have it not be climactic.

"So that's actually the problem, that marijuana is only somewhat of a solution. I am desperate for another solution, either through a neuroprosthetic— I don't know what it would look like, but to have all this work that my brain is doing, combining the senses—to have it happen more naturally, and have it happen *without* the marijuana, so that I can access [vision] all the time."

Tricia had been eager to show me the effect of cannabis on her vision since I arrived. She has a prescription for medical marijuana, so I felt that for me to observe the effect of this prescribed medication on her condition was not unethical. I was, however, concerned about interfering with her day and especially disrupting her work on her term paper as the deadline was approaching, but she assured me that her paper was in good shape and the adverse effects would only last a couple of hours.

"If you have complex things to ask me," she said, "you should ask them before I smoke marijuana. I'm not going to have a lot; I am just going to have some." And she walked into another room to get her prescription. The clicks of her heels on the stone floor resonated and trailed off as she walked away. In a minute the footsteps rose in crescendo and she returned to the kitchen with her prescription inside a baggie, and a vaporizer to administer it.

"Certain strains [of marijuana] just don't work," she said. "Different strains actually affect different aspects of vision. So I can have *one* strain that my *color* perception increases, and I can have *another* strain in which my *motion* perception and *contrast* perception increases, and rarely do I get both at the same time."

She set the vaporizer on the island kitchen counter across from me, turned it on, and inhaled the vapor through a clear tube. She held the smoke in her lungs and then exhaled.

"Will this take effect quickly?" I asked.

"Yes. The visual effects only really work for the first, short amount of time of the high. And then they go away."

She drew the smoky vapor from the tube again.

"It will just take a second—here . . . OK, I'm already noticing some effects.

"What's happening is, like, my eyes felt drawn to move forward, away from my face. I'm aware of their movement."

She inhaled the pungent vapor deeply into her lungs again and held it for a second. Clouds of blue-gray explode from her mouth and nose in a series of short, violent coughs.

"Sorry, this strain is, like, harsh," she said. "*Oh!* I'm aware of . . . See? This is my *favorite* thing that happens. I'm *aware* of *something* up here."

She turned and faced the window at the end of the kitchen.

"Like the light or something like that."

She stepped toward it.

"Then the line of sight, like what I am drawn to do is look over there for *whatever* reason, I'm, like, *drawn* to this area of the kitchen."

She took three quick steps directly toward the bright sunlight streaming into the kitchen through the window.

"I'm drawn to start looking at it! Like my eye just *roooooolls* in that direction," she said with delight.

Chills and goose bumps flashed over my skin as I saw a veil of blindness begin to lift.

"And I learned that . . . So here; so here's the table, right?" she said.

For the first time since we met, Tricia's eyes stopped their endless chaotic roaming. Suddenly they are seeking. She faced the table and stretched both her hands out toward it, seeing it for the first time since I arrived.

"So if I take my eyes—what are they doing now? And I *make* them go along this side of the table . . ."

Her eyes and gaze were tracking the rounded edge of the table.

"Like that makes more sense! It makes more sense spatially."

As I watched, a vision of her fMRI comes to my mind. I could see how her entire brain is now actively engaged in making sense of the first dawn of light in her mind.

She commented on the remarkably delightful sensation of her eyes seeming to be moving on their own, not wandering but purposely and automatically probing the kitchen.

"*Oh!* And when I am walking here"—she took a series of quick steps excitedly along the island kitchen counter that is separating us, back to the small table that we sat at all afternoon—"again I am *so* aware of, like, a *light,* or I don't *know* if it is a light or, like, the table. I am aware of this object in space and I'm thinking about its relation to my head and then I'm turning *this waayyy.*" She sang with delight as if she had just lifted her arms and taken flight.

"I have been in here before, but I am *really, really, really* aware of the light that's coming from over there."

She began to pace around the kitchen with a sense of rapture toward different bright features in the room.

"Color? Can you see color?" I asked.

"This doesn't seem like it is doing that," she said. "This is not an example of when it is working very well, unfortunately. Oh! I just caught a glimpse of something. . . . Oh! I guess that was the computer!"

She pivoted and stared directly at her gray Apple laptop resting on the granite table.

"That was funny! So this was cool! Classic blindsight! So I thought I caught a glimpse of something metallic, and it was the computer! That was really fun! Because what happened was that my eye was up there"—she pointed to the ceiling—"and then it went *boom, boom, boom,* and it went down and then it hit this, and I was like, What *is* that?" Her finger jabbed from the ceiling to the computer on the table.

Highly aroused, Tricia took several more excited paces.

"OK! I'm walking—this is an interesting phenomenon that happens to me that I don't entirely understand. So I'm walking toward the fridge and all of this stuff, and again, maybe it's the tracking sensation, because I got, again, this strong desire to *look* that way. I'm like, *Why?* It's like the fridge is tugging at me."

She is seeing the fridge and the image is sucking her in.

"One of the weird analogies that I use to make sense of my visual perceptions is, you know how when you have a street there are two lanes of traffic? And there's a near lane and there's a further lane? You know, there're multiple lanes not just two."

"Yeah," I said as I began to realize that what she was perceiving is a sense of depth—three-dimensional visual space—as her eyes focus on nearby objects and then distant objects. This is something that is beyond the reach of two-dimensional tactile sensation.

"Sometimes when my eyes are traveling up and down objects it feels like it's in a *far* lane. But then sometimes it feels like it's in a *closer* lane. And that has to do with . . . um, I think where the visual image is being projected on my body? I think? Or how . . . I don't understand why this is happening." Tricia started to retrace her steps back and forth in front the fridge, exploring the effect.

"Now I am walking. . . . It looks like all of this cabinet, or whatever is over here, when I was walking toward it?"

I observed her working to form the image of the kitchen cabinet in her mind.

"Like, talk about *threat* detection!" she said. "It felt like someone was coming closer and closer and closer to my face. I was like, What is *happening* there?"

She kept pacing in front of the cabinets, looking at them. "I was like, *Why* is that happening?

"So! Do you notice how I was walking and I *thought* that I detected *threat*? Like I detected a moving object. The object wasn't moving. *I* was moving! But I can't tell the difference."

She turned to face me. For the first time all afternoon, her two brown eyes looked me straight in the eye. It was as if I were seeing her face for the first time.

"Essentially what I want, Doug, if anyone can make this happen, is for someone to teach me how to see. Because it is *really* frustrating trying to figure it out by myself."

I suddenly realized that part of the reason it can feel uncomfortable to be in the presence of a blind person is that it can make you feel like a voyeur. You can see them, look anywhere you wish, but they can't see you. It is a guilty sensation. It is alienating. But she broke through the curtain and looked me straight in the eye. I was moved by a sense of intimacy and recognition.

"People are literally blind to the beauty of the visual world," she murmured to herself dreamily.

She stood drinking in the light and savoring the sensations of vision. I had the feeling that something precious was being shared with me.

"I'm sitting here talking to you and I'm drawn over there? What's so interesting that I'm drawn over there? That's the eye-movement cue! So I get cues from my brain to tell me which way to move my eyes, right? And then I move it so that something happens—something interesting, like it is a reward! Like I get the cue. Like it's a little nudge. And then I go with it and I'm like, *What* is it? It's the sense of surprise! Of having my eyes move on their *own*, which is such a *lovely* feeling."

"I think you may have it backwards," I said. I speculated that she is making sense of this new visual input from her point of reference that relies on touch to derive sensation. "Probably you are picking up a sensation that is not conscious, and like any sound, it is orienting your eyes to that object. So in other words, if you hear a sound to your right, you turn towards it."

"Yeah."

"But none of that's conscious."

"Yeah."

"That's what I think is happening to you," I said. "You are getting unconscious spatial cues and then your eyes are moving to it. And they are moving far away if it is in the distance and you have this sense of depth."

"So it's spatial," she said.

I believe the marijuana gives Tricia a heightened sense of synesthesia, which makes her visual/tactile/auditory perceptions of her environment flow and combine in her mind even more freely in her altered state of consciousness. It was also clear to me, as she believes, that the cannabis is also acting on any remaining photosensitive cells in her retina that are normally inert; with cannabis they respond, however feebly, to light. The light sensation reaches her brain, and her facile mind eagerly strives to make sense of it.

All of the aspects of vision that we take for granted took us years of experience from infancy to sculpt our visual cortex until we could see objects, track them, distinguish foreground from background, identify colors, and all the rest that we are now able to do completely automatically, but Tricia is working this all out from first principles to make sense of it. She is building bridges over the Mississippi.

Another important thing that Tricia's experience reveals is that many of the photoreceptors in our retina have little to do with what we normally think of as vision. They do not form images. Some retinal cells sense light, but they are connected to the hypothalamus to regulate our twenty-four-hour sleep cycle, for example. Other cells in our retina sense sudden movements in our peripheral vision. These retinal photoreceptors have very broad fields of view, unlike the densely packed photoreceptors at the center of the retina, where light is tightly focused to form a detailed image. The narrow field of vision of these cells in the center of our retina gives us high-resolution images, just as with a camera that has a large number of megapixels. These photoreceptors in the peripheral field of vision cannot form an image, and they do not connect to the region of visual cortex where image analysis is performed (called VI). They connect to an adjacent area of cortex. These cells and this cortical region are specialized for threat detection. They detect sudden changes in light intensity and blurs of motion in our visual space that are sent very rapidly to the brain to orient our eyes, body, and attention toward the source of potential threat. All of this detection and rapid eye movement toward the "unseen" but clearly perceived visual stimulus is automated and unconscious. Some of these photoreceptors have survived in Tricia's retina and, under the influence of cannabis, they have become more sensitive to light.

She says, "You have to remember, you are seeing degraded, degraded, degraded functioning. You are seeing the spatial dimension; you are not seeing the color dimension [with this particular strain of marijuana]. Those are the two big ones. Now you are just not seeing the color because the color one isn't happening." Tricia went on to explain how wonderful and how useful the ability to perceive color is in helping her form a visual map of the world.

I am frustrated that for political reasons the laws prevent me and other scientists from exploring the compounds in the marijuana plant that could provide new cures and new drugs to relieve human suffering.

"*Oh* my God!" she said. "I'll tell you one last thing I was getting also"— she pivoted her body back and forth in front of the island counter in the center of the kitchen with obvious excitement—"I love *occlusion* because I couldn't figure it out for the *longest* time. But *now* I understand *exactly* how it works. Because I was looking at this line of things and then behind it is, like, I don't even know what lines I'm looking at, but it's like *something* and then the *wall*, and then it's so *interesting* how things are at different *heights*, and like you can see *some* things at *one* height, but then at the height *behind* it you *can't*—you can only see the *top* of it. It's really interesting! I *love* that I understand what occlusion is."

"Yeah," I say. "I mean, it's amazing."

"What?"

"You just willed yourself to see, is what you've done."

"Yup!" She laughed again. "Yup! That's what I've done."

"Yup," I said rather meagerly, feeling somehow inadequate.

Her eyes became calm and focused. A broad smile broke out on her face and she reached out slowly with her hand.

"I keep wanting to look at the flowers!" she said. "Sorry, I am distracted by the flowers."

The small vase was setting on the table next to my right elbow. I hadn't even seen it until she mentioned it.

Tricia touched the delicate cream-colored petals gently with her fingertips, looking directly at the flowers with an impassioned smile.

They are lovely.

# 8

# Heroes and Cowards

Courage is resistance to fear, mastery of fear—not absence of fear.

**Mark Twain,** *Pudd'nhead Wilson*

Anxiety, attention, stress, fear, anger, gender, personality traits, and social context all have strong influences on suddenly awakening the beast within. However, no one could snap to action and attack an adversary if they were gripped by fear. Fear seems to vaporize when one of the LIFEMORTS triggers is tripped. Does fear really play no role in the moment we snap? Does this tremendously powerful emotion have no input to our rage circuit? Where in the brain does fear come from anyway? New research is defining precisely how and through which brain circuitry these factors play out.

I asked a courageous member of SEAL Team Six who survived a career in the most nightmarish situations one could imagine about fear.

"Anybody who's been in combat who says they weren't scared is lying to you. Anybody who has truly been in a situation where it's like, Damn! I'm done. . . . If you are truly in a situation like that, you absolutely will experience fear."

SEALs act fearlessly and ferociously when necessary, but they are not immune to fear. SEALs train to use fear, because fearlessness is dangerous.

It was ten o'clock at night on a dark street. Sally Martin was walking alone. She approached a small vacant park where a shabby man, looking "drugged-out," as she would later describe him, was sitting on a bench leering at her as she approached. No one else was in sight. She could have crossed to the other side of the street to avoid him, but instead she walked right past the seedy man. As she did, the man called to her. He motioned for her to come closer.

Sally walked right up to the guy. When she was within arm's reach, he suddenly bolted up, pulled her down to the bench by her shirt, and stuck the sharp blade of a knife to her throat. "I'm going to *cut* you, bitch!"

Sally remained completely calm. She felt absolutely no fear. Not the slightest sense of panic.

In the distance the sound of a choir singing inside a church drifted through the cold night air. "If you're going to kill me, you're gonna have to go through my God's angels first," she told him.

The would-be robber and murderer suddenly released her, startled by her fearless reaction to the blade at her throat. Instead of running away, Sally simply stood up and walked casually back to her home.

The next day, after surviving a near-deadly attack that would have terrified anyone else and induced PTSD in many, Sally walked past the very same park with no feeling of fear.

Sally is not her real name. This forty-four-year-old woman, normal in every other way, including IQ, memory, language, perception, is known to the medical literature as SM. Her medical abnormality is that she completely lacks fear. In experiments, when researchers brought her up to a cage of snakes or tarantulas, she eagerly started to handle them. Even after being warned sternly that the snakes were not safe and could bite her, Sally experienced not the slightest sense of fear. She has never known fear since about the age of ten.

The same scientists tested her responses inside a haunted house. She showed absolutely no fear reaction at all to any of the hidden monsters that provoked other members of the group to jump out of their skin, shrieking in fright.

When she was shown film clips of scary movies, scientists found that she lacked any fear responses to scenes that set hearts racing and palms sweating in everyone else. Her responses to other emotional events in the film were normal, inciting the appropriate emotions of disgust, anger, sadness, happiness, surprise, amusement, but no fear. She did, however, find the horror film and the other dangerous tests very exciting.

SM's medical condition is caused by a brain defect—the amygdala in her brain was destroyed by an infection when she was ten years old. This rare brain disorder shows that the experience of fear depends in a critical way on the amygdala.

People commonly refer to those who engage in dangerous or heroic actions as "fearless," but this is rarely the case. If you ask them, they will tell you that they *do* feel fear, but they persist by determination in spite of it.

"Every time I was on top of a run I was always scared," women's extreme free skier Wendy Fisher says. "Every time I stood on top of a mountain, I thought, I definitely could die today."

"Fear is a force that sharpens your senses," writes Navy SEAL Marcus Luttrell. "Being afraid is a state of paralysis in which you can't do anything. It's critical to understand the difference."

Fear and the turbocharged brain and bodily responses it fuels are life-saving. Fear is vital for successfully operating in any dangerous situation. Fearless SM, on the other hand, lacks the heightened vigilance, boosted mental acuity, and supercharged bodily function that the fear response triggers, and she has suffered greatly because of it.

SM has very poor threat-assessment ability, so she overlooks even obvious threats in her environment. She does not learn from experience to avoid dangerous situations. As a consequence, she has experienced a harrowing number of life-threatening encounters. She has been held up at knifepoint and at gunpoint. She has been physically accosted. She was nearly killed as a victim of domestic violence. She has been threatened with death many times. In all of these deadly encounters she never felt fear and she did not behave with any urgency or sense of desperation, according to police reports.

Justin Feinstein of the University of Iowa, who studied SM says, "It is quite remarkable that she is still alive."

Imaging in fMRI shows that activity in the amygdala increases when someone experiences fear and anger. Patients with anxiety disorders, social phobia, and PTSD have heightened amygdala reactivity compared with normal subjects when shown emotional faces. Therefore, whether one is the anxious type by nature or a cool cucumber relates in part to how responsive their amygdala is. This fits with the studies on fearless SM. The amygdala in SM's brain has been destroyed, which confirms that the experience of fear depends in a critical way on the amygdala, but this does not mean that fear resides *in* the amygdala.

We have all seen pictures of the brain divided up into tiny sectors, each one having a name describing the specific function supposedly performed by that brain region. A key concept that has become clear from newer methods of studying brain function, however, is the importance of brain *circuits*, in contrast to the older idea of brain regions. Insight into the circuit model comes from studying people like SM who have had damage to certain parts of their fear and threat-detection circuitry. By tracing these circuits connected to the amygdala, researchers predicted that they could find a way to scare the devil out of SM to give her a sensation she has never felt—intense fear and panic.

Suffocation causes panic in almost everyone. It is the basis of the torture technique known as waterboarding. Scientists have used the body's heightened anxiety produced by mild suffocation to probe fear, by having a person inhale $CO_2$. This is harmless, but anxiety quickly rises as the subject's body becomes oxygen-deprived and $CO_2$ levels in the blood increase sharply, signaling suffocation to the brain. This sensation is genuinely alarming, because if suffocation is not resolved in a matter of minutes, you will die.

The surprising finding is that SM and other people who have extensive damage to the amygdala, and who are otherwise fearless, nevertheless panic in response to the suffocation test. This test sounds frightening, and it is, but the subjects were all fully informed and they willingly volunteered for the experiment to further scientific understanding (and, perhaps, to better understand themselves). "They were scared for their lives," says Justin Feinstein, lead author on the study. SM and the others without an amygdala panicked immediately when they started to inhale the gas. They gasped for air; their heart rate shot up; they tried to rip off their inhalation masks. This study helped prove definitively that the amygdala is a critical switch point in the circuitry of fear, threat-detection, and anger, but the sensation of fear can be induced without it. In the case of $CO_2$-induced panic, this is caused by the carbon dioxide in the blood activating part of the autonomic response circuitry downstream from the amygdala inside the hypothalamus. The amygdala is the gateway, but the mental state of fear comes from a larger circuitry in the brain, including the cerebral cortex.

# Fear in Your Body

Remember the last time you were afraid? I mean really afraid—*terrified*. It might have been a close encounter with a robber on a dark street, or quaking just before speaking in front of a crowd, or a personal phobia, such as fear of heights or flying. Recall the sensations of that fear—the racing heart about to burst, panting, intestines twisted into knots, knees shaking and hands trembling, cold sweat oozing out of your palms and beading up on your forehead.

Now imagine exactly the same scene, but without any of the bodily sensations that fear brings. No sweaty palms, heart rate and breathing calm, your muscles relaxed and your stomach content. Are you still afraid?

What would fear be without the body? Can fear exist only in the mind?

If the bodily sensations produced by fear are instead evoked artificially by direct stimulation of nerves to set the heart racing, for example, you will feel afraid. You would experience something of a panic attack initiated by the body rather than unleashed by the brain. This is what happens to the people in the $CO_2$ experiment as carbon dioxide rises in their bloodstream, signaling suffocation—even if they do not have an amygdala.

It is only the cerebral chauvinism of our conscious brain that makes this seem so surprising; but as we have discussed, much of what goes on inside our head is unconscious; much communication among people (and all communication among animals) is nonverbal; and much of what our nervous system does has its origins outside the brain. The practice of tai chi and yoga are based on the body controlling the brain rather than following the presumption of our conscious mind that it must always be the other way around. Manipulating the body is how our mind becomes relaxed, and soothing hormones are released by massage. It is how art and music, taste, touch, sights, and smells move our mind and behavior. We close our eyes or cover our ears to quell fear. It helps, at least at the movies.

The fact is that most animals get by just fine with no brain at all. Their "brain" *is* their body. Creatures like worms, insects, and snails all eat, reproduce, sense the environment, and defend themselves without any difficulty in the absence of a brain. Some, like ants and honeybees, have complex social structures and sophisticated means of communication and they can even learn, all without having a brain or even a spinal cord. The nervous system of animals without brains is a knotted network of neurons connected together and cast throughout their body. Often neurons are wadded together and stashed next to the body structures they operate, such as the bunches of neurons clustered like grapes next to each segment in a lobster's articulated tail. A similar neuroanatomy is true of people, but human beings, with their emphasis on conscious, intellectual function, easily overlook this and assume that the brain does it all.

Many neurons in your body—in fact, most of the neurons associated with producing the sensations of fear—are not located in your brain at all, or in your spinal cord for that matter. They are stashed in clusters throughout the body close to where they are needed. All of the sensations of touch and pain, for example, come from neurons tucked next to each vertebra in our backbone. The neurons that set our heart racing, our skin sweating, and our stomach churning are draped in a network throughout the body cavity next to the organs they control—completely outside the brain and spinal cord. So it should not be surprising that the body itself in stimulating these neurons would affect the brain and the conscious and unconscious mind, which would in turn affect hormones, mood, mental function, and behavior. When we manipulate our body, be it a dance, a smile, or a posture, we also manipulate our brain. This basic fact of neuroanatomy may hit some like a blow to the solar plexus (that knot of neurons in the belly that controls breathing and other organ functions).

Interestingly, many studies have shown that a low resting heart rate correlates strongly with decreased fear and a greater propensity for seeking novelty and adventure. Even after correcting for other factors that influence heart rate, the association remains strong, especially for males. People with heart rates below 66 beats per minute tend to be tolerant of fear and much more likely to be thrill-seeking. Such people become bored easily and they are averse to routine. They enjoy intricate and challenging activities, in contrast to those with a heart rate above 95 beats per minute, who seek comfort in routine activities and are averse to danger. Higher aggression and criminality are also strongly correlated with low resting heart rate, but the effect is twice as strong for men than women. One explanation is that men tend to show more direct aggression, whereas women engage in more indirect aggression.

One theory to explain this relationship between heart rate, fearlessness, and novelty seeking is that such people seek more exciting experiences to elevate their heart rate because low heart rate may be physiologically unpleasant to the body. Alternatively, a low heart rate may simply convey less sensation of fear in a dangerous situation compared to someone whose heart is already beating fast and starts racing at the slightest provocation. Regardless, the correlation between heart rate and fearlessness shows that the nervous system, in part by acting on the heart and other responses controlled by the sympathetic and parasympathetic nervous system, determines an individual's tolerance for and attraction to danger, adventure, thrill, and novelty. Genetic factors, nervous-system differences, and fitness activity that lowers heart rate will dictate a person's desire for novelty and adventure, and tolerance of danger.

Astronaut Edward White was almost removed from the *Gemini 4* spaceflight because of doctors' concerns that his low heart rate (52 beats per minute at rest and 70 beats per minute after exercise) could predispose him to fainting—in fact, he may have been the fittest astronaut ever. Today it is recognized by many fitness trainers and medical experts that low heart rates are typical of extremely fit athletes. Resting heart rate in the low 50s is not uncommon in military personnel. (White was a West Point graduate and fighter pilot in the US Air Force.) The commonality may be not only a consequence of physical training, but might possibly result from self-selection of people joining the military who are drawn to adventurous and dangerous professions and activities, and thus naturally have a lower heart rate at rest.

White was killed in the horrific *Apollo 1* cabin fire on the launch pad during testing that also killed astronauts Virgil "Gus" Grissom and Roger Chaffee as they were strapped in their seats on January 27, 1967. Heart-rate monitors indicated that White's heart rate increased abruptly from the mid-30s to the 60s right after the fire was reported. Such an increase could have been produced simply by rising from his seat. In the midst of the blaze White did unstrap himself and get out of his seat. He turned around and wrestled with the stuck escape hatch for at least sixteen long seconds before he was overcome by the inferno fueled by the pressurized 100 percent oxygen atmosphere in the cabin.

We speak about emotions as coming "from the heart," and indeed there is some truth to that. It is the body's response to stress that gives us the sensation of fear. If a person has a lower cardiovascular (and other sympathetic nervous system) response to stress, or if those nervous-system responses in the body can be suppressed, the sensation of fear in the person's mind will be quashed. Fear and threat detection and response are all about interconnected brain regions and the connections between the brain and body.

# Needles

The wicked black blade shot out of the stiletto and locked into place with a solid snap. Reaching above his head, he stabbed and firmly embedded the deadly knife into a wooden pillar. Next the Navy SEAL grabbed a bag of lactated Ringer's solution and hooked it onto the handle of his stiletto. Stretching out the clear tubing draining the bag, he handed the woman an alcohol wipe and laid his right arm out on the bar top. Clearing aside beer bottles, he told her to wipe the crook of his inner arm and then asked her to tie the blue elastic strap tightly around his biceps muscle to engorge his veins with blood. Then he patiently instructed her how to stab the needle into the vein to give him an IV, something the woman had never done before.

It did not go well. The Navy SEAL never flinched as he encouraged her to try again. She poked the needle in again, moving it this way and that, raising and lowering the angle of the needle, probing and digging to find and then pop into the vein. It seemed to take forever. Finally after much trial and error she punctured his vein and a small stream of blood seeped out, dripped down his arm, and simultaneously tainted the clear IV fluid with crimson backwash.

This was not a medical emergency. The SEAL and his team members were out for some R & R in the backwoods of West Virginia between combat missions in the Middle East. When a young woman mentioned that she had recently received a Wilderness Emergency Medical Technician certificate and hoped to become a trauma nurse after college, the medic on the SEAL team immediately offered to share his knowledge. Beyond sharing his medical skills with an interested student, an important part of his role on the SEAL team, he explained, is to teach anyone to give a person an IV in an emergency. His ability to teach this skill could be lifesaving in the event of multiple casualties. Just as a marksman practices his skill constantly, so does this SEAL in perfecting his lifesaving medical techniques. He is responsible for delivering emergency medical treatment in battle, and he prides himself on being able to get an IV into any vein in any part of an injured person's body in a hurry, and to train others to do it on the spot. I have heard that in nursing classes they teach this skill by first using an orange as a surrogate arm, but not Navy SEALs. On the battlefield there is no time for oranges. And the battlefield is not the place to practice.

Next he showed the woman a three-and-a-half-inch-long, 10-gauge needle—slightly thicker than one-eighth of an inch in diameter. Such a needle had saved his life, he said. SEALs don't discuss their operations in detail, however. His description was focused on the needle and its lifesaving potential, not on how he had stabbed one into his own chest after being shot by a 7.62 mm bullet fired from an AK-47 automatic rifle. That's all the information he gave, other than very detailed instruction on how to locate the precise spot in the second intercostal space in the chest in which to jab the needle up to the hilt, how you feel a "pop" when the needle enters the chest cavity and then hear a hiss of escaping air. The needle works because when a person is shot in the chest or suffers a penetrating injury to his rib cage that allows air to seep into the chest cavity, the victim will quickly suffocate. The trapped air prevents the lung from inflating and the lung collapses uselessly (a tension pneumothorax). This needle carried by soldiers in combat has saved many lives.

This experience underscores the intriguing neuroscience of fear and, in some people, phobias. Some people faint at the *sight* of a needle. People can be fearless in other endeavors but terrified of needles. The fear of needles keeps them from going to doctors, and they avoid immunizations. This is very strange; the same person may stoically brush off a bloody injury caused by almost any other accident, but the sight of a needle sets off profound fear and even fainting by activating an unconscious part of their brain beyond their rational control. The actual injury caused by an immunization is trivial, and people with a fear of needles know this. They know that an immunization will in fact save them from disease and death, but their conscious awareness of the benefit and the rational need for an immunization cannot override their body's automatic response to the terror evoked by the needle. Even watching another person get an IV or an immunization can set off the same panic response, even though there is no threat whatsoever to their own body. Others, like this Navy SEAL, either don't have this fear or they completely suppress it, and they also completely suppress the pain and autonomic physiological response to having an IV needle miss its mark, something that frequently puts people into mild physiological shock.

The next morning, this SEAL had a tennis-ball-sized hematoma where the woman had bungled the IV—a purple bulge caused by blood leaking into and inflating the surrounding tissue. When his buddy asked what had happened, he replied casually, "Might have bruised my arm." Then they all went mountain climbing.

Some fears seem innate, such as fear of snakes or heights, but others are learned. All of our fears have a biological basis in being lifesaving. People who faint at the sight of a needle are suddenly overwhelmed with anxiety, their breathing becomes rapid and shallow; they become light-headed; their pulse accelerates and becomes shallow; blood pressure drops; their skin becomes cool and clammy; and they become confused, dizzy, and they faint. You will recognize all of these physiological responses as the symptoms of shock. Shock is not an emotional state. It is an automatic physiological response triggered by the body's sympathetic nervous system as a lifesaving response to severe injury. The cardiovascular reaction and fainting in shock reduce blood loss after trauma. You only have eight to ten pints of blood in your body if you are a man, five to eight if you are a woman. It does not take long to pump that small volume out of the body through a wound if the heart is pumping at its maximum rate. No wonder women tend to be more prone to fainting. If not for this automated shutdown, your heart would be pounding furiously from the intense experience of having just sustained a serious wound.

Fear of needles is irrational. It is a phobia, but the biological basis of this phobia is thought to be that a puncture injury to the body will result in blood loss. Through the course of evolution, those who fainted immediately after a stabbing wound were more likely to survive the blood loss. It becomes a phobia when the lifesaving response gets set on a hair trigger in some individuals.

Fear is at the heart of any threatening situation, but much as different people have different pain thresholds, fear varies enormously among different people. Also, very different kinds of dangers provoke different levels of fear in any individual. Even the bravest people may have fears and minor phobias for specific kinds of dangers. I have spoken with "fearless" law enforcement personnel who willingly confront the most frightening individuals in deadly situations in dark back alleys, but they are afraid of heights. A Secret Service agent insists that she doesn't have the slightest hesitation about "taking a bullet for the president," but she is terrified of caves. The thought of screeching bats snarling in her hair in a dark cave sets her skin crawling. Where do our individual fears come from? Our DNA?

Most people are spooked at the sight of spiders and snakes. Lynne Isbell, professor of anthropology at the University of California–Davis concludes that primates are hardwired to fear snakes. Modern mammals and snakes evolved at about the same time, about 60 to 100 million years ago, which

Isbell believes could have resulted in evolutionary pressure for nonhuman primates (monkeys and their relatives) and humans to fear snakes. You rarely get a second chance if you are bitten by a deadly snake, so it might be impossible to learn to fear them through experience. Individuals who naturally feared the sight of a deadly viper lived to pass on their genes. We inherited the fear of snakes from our ancestors.

Collaborating with neuroscientists who are experts in recording electrical activity in neurons in the visual system, Dr. Isbell's research team showed rhesus monkeys images of snakes and other objects while probing the monkeys' brains with their microelectrodes. They found neurons in the visual cortex that responded strongly to pictures of snakes, but fired only weakly when the scientists flashed a different object on the screen. Remarkably the snake-sensitive neurons were more common than neurons that recognized other objects, including images of other monkey faces or hands or even simple geometric shapes.

"The results show that the brain has special neural circuits to detect snakes, and this suggests that neural circuits to detect snakes have been genetically encoded," says Hisao Nishijo, a neuroscientist who participated in the study.

The monkeys being tested were bred and raised for experimental research in an enclosed colony. They had never encountered a real snake. "I don't see any other way to explain the sensitivity of these neurons to snakes except through an evolutionary path," Isbell says.

It is remarkable to contemplate that our individual fears and behaviors could be programmed inside the strands of DNA we inherit from our parents. What greater fatalism than accepting that the biology that will dictate our fears and behaviors is already set at the zygote stage when our life began as a single cell? This seems to shatter the illusion of free will, but Susan Mineka's research at Northwestern University counters this argument. Her research shows that monkeys that are raised in the lab are not afraid of snakes, even though they may have special neurons to detect the sight of them. She also found that laboratory monkeys can learn to fear snakes very well, much better than they learn other information.

A subsequent study on humans provides a new twist on where our fear of snakes originates. In one experiment they showed infants as young as seven months old two side-by-side videos. One was of a snake and the other was a different nonthreatening animal, like an elephant. The babies showed no signs of fear in response to either image, which runs counter to the idea that we have inherited a fear of snakes through natural selection. But interestingly, the babies did spend more time looking at the videos of the snakes.

The research team then tested three-year-olds by showing them nine images on a screen and asking the children to pick out one of the pictures upon command. The three-year-olds identified snakes more quickly than all the other images, which included flowers and other animals that look similar to snakes, such as caterpillars and frogs. Some of the young children at this age were afraid of snakes and others were not, but that made no difference. Both groups of children identified snakes faster than any other object.

This body of research shows that snakes are different in two ways from other objects, says Vanessa LoBue, who collaborated on the research. "One is that we detect them quickly. The other is that we learn to be afraid of them really quickly."

So we do have a propensity to develop fears for certain dangers more readily because of our genes, but it is our interaction with the environment that makes our genetic predispositions turn into specific fears. There is no need to fear snakes if you live in an environment where you will never encounter one. This is the reason that individuals have different levels of fear for different kinds of threats. We all have a different genetic makeup, but we all also have a different life experience. This combination of genes and environment is a theme that carries through all of human behavior, including fear, bravery, thirst for adventure—and snapping into a rage.

## Building Courage

A person can become incapacitated by fear and freeze, unable to take any action against a threat or powerless even to run away. On the other hand, fear can sharpen one's wits, sensitize the senses, speed thinking, and fortify the body's strength. Fear can alert you to a potential impending danger or motivate a vigorous response to a sudden dangerous situation. Different people respond to fear differently.

Navy SEALs, for example, train to use fear and train to overcome it. "I don't like heights. I never have," a member of SEAL Team Six says. "I didn't like skydiving when I first started. The idea of it was cool. All these SEALs that I looked up to skydived and did all this stuff, but when you open the door of an aircraft and you're gonna jump out—it's scary! It's not normal.

"I forced myself. I jumped right out the first time they told me to do it," he said.

The personal challenge drove him, just as it does for so many others who operate in dangerous situations.

"Skydiving scared me, so I volunteered for every skydiving trip I could go on. I volunteered to jump the tandems. I volunteered to jump these seven-hundred-pound containers, because I said, *OK, that's part of being a SEAL. I want to be the best SEAL I can be, and if that's what it means to do it—I'm gonna do it!*"

Wendy Fisher, the extreme big mountain free skier, said that every time she looked down from the steep summit to begin a run she would say to herself, *Let this be the time I don't feel fear,* but every single time she did feel fear. Regardless, she was committed. Even after becoming a mother and many people thought her days of risk-taking and adventure were over, she continued to pursue extreme big mountain skiing. She is a devoted mother, but adventure and skiing dangerous mountains is her identity.

"It kind of became mind over matter," she says. "I was always fearful, but I did not want my fear to control my life."

What Wendy Fisher and the member of SEAL Team Six are describing is something beyond fear: persistence despite fear—that is, courage. Science is beginning to trace the neurocircuitry for this honorable trait.

Antonio Rangel, an associate professor at Caltech, has used neuroimaging to study people with strong self-control in comparison to people who have less self-control, as determined in psychological tests. "In the case of good self-controllers . . . the dorsolateral prefrontal cortex becomes active," he says. His research using food temptation in a self-control test is backed up by another study on self-control that reinforces this finding. Researchers at the University of Minnesota placed subjects in an MRI scanner and then had them perform two different self-control tasks one after the other. The first task required test subjects to ignore words that flashed on a computer screen, and the second test involved choosing a set of preferred options. The subjects had a more difficult time exerting self-control on the second task, a psychological phenomenon called "regulatory depletion." This means that self-restraint slowly drains away (something any parent who has been subjected to unrelenting repetitions of a request of some sort from a child knows). The dorsolateral prefrontal cortex became less active during the second test of self-control as the subjects' resistance to temptation waned. (OK, fine! You can have another chocolate! Your dorsolateral prefrontal cortex is pooped out.)

The constant pestering of a child's begging is annoying, but persistence is one of the rare traits that make winners. Persistence is what it takes to survive the physically and emotionally intense demands of training to become a SEAL. Only recruits who are exceptionally persistent will make it through training. Persistence is what makes Wendy Fisher pick herself up off the slope after a horrendous fall that knocks the wind out of her and wrenches her joints painfully, and then head straight back to the top of the mountain to tackle the challenge again. Persistence is what kept Thomas Edison searching through endless failures to finally find a filament that would make an electric light.

Researchers have found a tiny circuit in the human brain that fuels persistence, and they discovered that they can turn it on with a few electrical impulses. Dr. Josef Parvizi of the Department of Neurology at Stanford University studied two people with epilepsy who had electrodes implanted in their brains to help doctors learn about the source of their seizures. When doctors stimulated neurons in one spot in the anterior cingulate cortex, both patients described feeling the powerful expectation of an imminent challenge coupled with a determined commitment to surmount it. Their heart rate increased and they reported a physical sensation in the chest and neck, bracing themselves with determination to succeed. This is a complex but powerful and immediately recognizable human emotion—determination to fight for a goal. Stimulating the brain only five millimeters away from this spot did not provoke the same response at all.

"That few electrical pulses delivered to a population of brain cells in conscious human individuals give rise to such a high level set of emotions and thoughts we associate with a human virtue such as perseverance tells us that our unique human qualities are anchored dearly in the operation of our brain cells," says Parvizi. But like everything else about our brain and body, different parts are developed somewhat differently in everyone. "These innate differences might potentially be identified in childhood and be modified by behavioral therapy, medication, or, as suggested here, electrical stimulation," he says.

New research illuminates how the brain decides when to stop and rest. From how athletes pace their running in competition to when people take breaks during work, two brain regions constantly monitor the cost/benefit of sustaining effort, adding into the calculation the incentive anticipated upon reaching the goal, to determine when to stop and take a break. Neuroscientist Florent Meyniel and colleagues at the Hôpital de la Pitié-Salpêtrière in Paris, asked participants to squeeze a handgrip to win a given amount of money. The cash payoff was proportional to the time spent in exertion above a given force level, versus time spent resting. The size of the incentive was always displayed to the subject just before each trial, but the difficulty of closing the handgrip was unknown until the subject began to squeeze it. In this way, researchers could measure precisely how the difficulty of the task affected how often breaks were taken, and for how long. Then they could gauge the influence of motivation on how we allocate break time against effort by changing the monetary payout.

Naturally, the subjects spent less time squeezing the grip as the difficulty of the task increased. Likewise, they spent more time squeezing and less time resting if a higher monetary reward was offered. But how does the brain decide when the cost/benefit balance reaches a point where taking a break is the best decision?

What the researchers saw using brain imaging was that as the effort required to squeeze the grip increased, activity in certain brain regions also increased, and the activity accumulated with prolonged effort. This included both the posterior insula, which is a region of cerebral cortex known to be involved in somatosensory function, as well as the ventromedial thalamus, deep inside the brain.

These two brain regions are part of a network known to be activated in response to physical pain. The new data showed that these regions continuously signal the costs of the effort and rejuvenation provided during the rest periods. It is not clear exactly what physiological or sensory input these brain regions monitor to determine the "cost"—that is, the pain associated with prolonged effort. What is known is that direct electrical stimulation of this brain region induces a painful sensation.

Interestingly, monetary incentives slowed the accumulation of the cost signal in these brain regions during sustained effort, and they speeded the dissipation of the "cost signal" during the break. The authors speculate that this might reflect a motivational signal being subtracted from the cost. They propose that these positive signals could arise from other brain circuits that are involved in reward processing (for example, the ventral striatum).

The range over which the cost signal fluctuates was also adjusted by the incentive. This could be related to the psychological phenomenon that, when motivated, we literally push back our limits, allowing our body to work closer to exhaustion. The ability to do this may contribute to the superhuman strength of a mother lifting a car off of an injured child. Even a mountain climber who must be able to push themselves to the limit to carry heavy loads up high mountains in an oxygen-starved atmosphere will have something more in reserve if the incentive is great enough. The climber can become thoroughly exhausted and collapse in his tracks but would instantly spring to his feet with newfound energy upon hearing the thunder of an avalanche. Placebos (sugar pills that a patient believes to be medicine) have been shown to reduce responses to painful stimulation in these brain regions. Therefore the brain can adjust the sensitivity of its pain circuits depending on expectations.

One wonders if these new findings could lead to new kinds of performance-enhancing drugs that strengthen the brain instead of the body. (Imagine if there were a drug that could push endurance rather than doping blood with erythropoietin, or EPO, to absorb more oxygen as Lance Armstrong admits he did.) "We are currently testing drugs on this paradigm in our lab," Mathias Pessiglione, a coauthor on the study, says. They are testing analgesic drugs (painkillers) such as morphine on the theory that this might slow the cost accumulation in the pain regions, and testing drugs that boost motivation, such as dopamine enhancers. "Amphetamine, for instance, may push back the bounds of cost accumulation," he speculates.

Pessiglione suggests that this new information might be exploited without resorting to performance-enhancing drugs. "Perhaps people should listen to their brain signals! That is, instead of planning breaks in advance, monitor their fatigue online, and have a break when it reaches a given threshold. The other possibility is to increase the incentive."

Courage, by definition, is perseverance despite great risk and fear. The neural mechanisms of courage are being revealed by an experiment that might accurately be called "snakes on a brain-imaging machine." Participants had to choose whether to move an object closer to or farther away from them while their brain was being scanned inside an fMRI machine to let the scientists see what part of their brain became active during the challenge. The scientists used either of two objects: a teddy bear or a live corn snake. The researchers had already divined each participant's fear of snakes by using a questionnaire before the imaging session.

What they found was that bringing the snake closer stimulated the body's fear response as measured by increased skin conductance (cold sweats) in their test subjects. However, some subjects admitted feeling afraid and others did not. The subject's denial of fear did not always match what their body's heightened arousal response was showing the researchers. That is, some of the brave subjects reported little fear even as the snake came closer while their body started to sweat more from the stress. In the brains of the people who chose to react courageously, the anterior cingulate cortex increased its level of activity.

"Our findings delineate the importance of maintaining high anterior cingulate cortex activity in successful efforts to overcome ongoing fear and point to the possibility of manipulating anterior cingulate cortex activity in therapeutic intervention in disorders involving a failure to overcome fear," one of the researchers, Dr. Yadin Dudai of the Weizmann Institute of Science, Rehovot, Israel, concludes.

Different people react differently to a sudden threat. Some snap to action instantly; others become helpless. Certain types of people accept dangerous professions and enjoy exciting, even risky, activities; others cling to routine and security. One strategy is not better than the other; both have their advantages. This is why there is such genetic diversity in our species to predispose individuals in the population toward either fearlessness and adventure or timidity and security. Where would our species be without those among us who crave novelty, exploration, and adventure? There would be no Lewis and Clarks, no astronauts. At the same time, where would our species be had there been no monks in the Middle Ages, patiently transcribing the world's knowledge and literature with endless care, patience, and prissy penmanship? How would we have clothed ourselves had there not been people who draw pleasure from routine, knitting and weaving threads into fabric?

Interestingly, some professions are so diverse they attract and need people from both ends of the spectrum. In science, for instance, there is no award for second place. Either you make a new discovery before anyone else, or you don't. To succeed, research scientists share the same drives as an Olympic athlete or a racecar driver (except that in the latter activities there is at least a second- and third-place prize). Other scientists—anatomists and museum curators, for example—succeed through patience, mastering routine, and assimilating and cataloging information without haste. Who would want a tax accountant to get bored with routine and take shortcuts? Who would want a first responder to become flummoxed in a complex, novel situation that is outside the routine? Any automobile can be utilized for transportation, but different makes have different advantages and limitations. In a race between a Porsche and a pickup truck, which one would win?

The answer would depend on whether the race was held on a racetrack in Daytona or off-road in Baja.

## Adventure

Billowing clouds engulf the small blue-and-white Cessna 208 airplane. Through abrupt openings in the mist the pilot and passengers catch brief glimpses of breathtaking mountainous terrain below—the rugged wilderness of Alaska. As they soar through the heavenly domain of gods they feel as insignificant as a leaf blown in the wind. They buoyantly thread their way feebly through the high mountain passes, the scream of the Pratt & Whitney engine whining defiantly, spinning the propeller into an invisible blur in front of the windshield. Jagged ice-tipped crags tower on both sides as the small plane is buffeted through the sky, pressed low to the ground by the solid low ceiling of clouds. Below them a great stretch of glistening white glacier is banded by brilliant turquoise cracks, crevasses split as the ancient mass of ice flows on geological time through the mountains, grinding away stone, gouging a pathway to the sea. They are following the Davidson Glacier after a week of kayaking on Glacier Bay to reach their intended destination of Haines, Alaska. Pilot Drake Olson cranes his neck to peer through the top of the windshield in search of a break in the gray-and-black ceiling of clouds.

Suddenly my stomach carves an arc upward through my guts and a wave of nausea washes over me as the wing dips sharply. The plane pitches left and then right as the ground rises, filling the entire windshield.

"These are really interesting questions," Olson says, speaking into the foam-covered microphone that curves to his lips from his army-green headphones. He is responding to my series of questions about the appeal of dangerous activities like being an Alaskan bush pilot. "I'd like to answer you but there's a lot of stuff going on in my mind right now." All three passengers are similarly equipped with headphones to communicate with one another inside the noisy cabin.

"Sure, please . . . never mind . . ." I babble into my microphone.

Jagged mountain ridges displayed on a navigation screen at the center of his control wheel turn from green to yellow and then splash red.

"The north wind is pushing us down," he says out loud, though he's speaking to himself.

Drake's chosen profession is to transport hunters and adventurers into remote regions of Alaska that are often inaccessible by any other mechanized form of transportation. Forget about nicely lighted runways, control towers, radar, and accurate weather forecasts; an Alaskan bush pilot lands and takes off from gravel, ice, or water without the benefit of even a windsock to point the wind's direction. Days later he must return to pick up the party waiting for him in the backwoods. Regardless of weather, he will be there at the pre-arranged time unless it is impossibly dangerous.

Airline pilot is the third most deadly occupation, according to the 2011 census by the Bureau of Labor Statistics, with 56.1 fatalities per 100,000 pilots. Occupations widely regarded as risky—for example, police and firefighters—fail to even make the top ten on this list. What is the appeal of these dangerous jobs?

"It is a very basic thing," Drake says, carrying on his side of the conversation alone as he negotiates the turbulence and guides his plane through the narrow rocky passes. "It is a freedom." He is answering my previous question about why he is drawn to this dangerous profession, although I've long since stopped talking or asking questions. "Also, I get bored with the regularity of life."

Suddenly out of nowhere another light plane approaches us head-on, coming up the glacier. There are only so many ways through these mountains, and in this weather the Davidson Glacier is the best bet. Drake diverts to the left, hugging the black stone cliff streaking by our wingtip.

"Flying is a combination of intelligence and guts. So is racecar driving," he says. In the instant of danger automatic unconscious reflexes in the brain take over to rescue you from a perilous situation, but it is the conscious control of those highly honed reflexes that is the reward for Drake and others drawn to dangerous professions.

"I'm always thinking of my outs—thinking of contingencies."

On the ground or in the air Drake Olson is intense, impatient, hyperfocused, and driven by a fierce independence, quick intelligence, and curious mind. His steel-colored hair drapes over his shoulders in back as he liberally expounds his viewpoints on a wide range of subjects without the nicety of politely sugarcoating them to appease the sensibilities of people who may not share his opinions. Drake has no patience for bullshit. Classic rock music plays in the background through the headphones.

"It's a high-level chess game," he says. His voice is clipped by the metallic electronic amplification of the headphones. "Like a racecar driver it is very sensory, but I'm always thinking of stability. It is a balancing act all the time."

The balancing-act analogy was patently obvious at that moment. It felt as if we were balanced in midair on a beach ball. The parallel to racecar driving was not an abstraction either. Before moving to Alaska and starting Drake Airways, Drake Olson was a champion Formula 4 and Formula 1 racecar driver, winner of the Porsche Cup North America in 1986.

Reflecting on a near crash when his car "kicked out" suddenly on a curve, he says, "When you are going a hundred and forty miles per hour accelerating full-throttle and it slips and you catch it, there is no way in hell you thought about that."

A week earlier, his passengers had seen a demonstration of that. Drake had flown the party to Glacier Point for an ice-climbing expedition. As they approached the remote gravel strip on the glacial moraine, suddenly a complex vortex of crosswinds spilled down from the glacier and collided with winds blowing up the Chilkat Inlet, tossing the plane erratically seconds from touchdown. Drake, on the edge of his seat stretching his neck forward for a better view over the high dashboard, began rapidly but precisely rattling the control wheel back and forth like a person testing a locked doorknob, but these were quick and controlled maneuvers to keep the wings level and the plane at the appropriate angle while landing in the turbulent crosswinds. Seconds before touching down, an alarm began to sound. The wheels touched down smoothly, crunching on the gravel although stopping quite a bit farther down the gravel strip than usual.

"Well, that was exciting," Drake said casually, a cool-tempered understatement as he brought the airplane to a halt and took off his headphones.

After two days of ice climbing on the glacier, the party had been scheduled to return home, but by then the weather had soured. Gusts were whipping up chop on the Chilkat Inlet and it was raining. The party discussed the possibility of needing to camp another night or so until the weather cleared enough for Drake to retrieve them. They debated their options as they sat next to the soggy gravel strip drenched by rain and keeping cans of bear spray within arm's reach. They passed the time picking wild strawberries bursting with flavor.

Then in the distance they heard the faint hum of an airplane engine, and looking up precisely at 6:00 p.m. they saw a tiny speck appear in the angry black sky with its landing lights ablaze, cutting through the rainy mist.

It was a vision of precision work. Not risk-taking for the daredevil thrill of it but managing and dominating risk with innate skill, reflexes, and experience. And a welcome sight.

When you consider the full spectrum, from people who cannot travel as a passenger on a commercial airline because they become incapacitated with panic and fear to a bush pilot flying in the remote Alaskan wilderness utterly on his own, one wonders what accounts for the difference? Why are certain people tolerant of or even drawn to danger when others avoid risk at all costs—even to the point of irrationally fearing commercial airline travel when they know consciously that this mode of transportation is far safer than driving a car?

What happens when people on opposite extremes of the scale of tolerance of fear and danger are suddenly confronted by a dangerous encounter? Will instinct drive the fearless to battle and confront the threat while immobilizing the timid in passivity like a deer in the headlights?

Do people on opposite extremes of this scale have the same propensity to snap in uncontrolled rage, but will do so only in response to quite distinct situational triggers?

Why are some people driven by a constant thirst for novelty and seek out new experiences, while others are unsettled by change and content with routine?

Do some individuals rely more heavily on intuitive reflexive cognition, whereas others rely on conscious analytical thought, and if so how do people with these different modes of thinking respond in a threatening situation, and which types would be more likely to erupt in rage?

Let's address these questions, not in a descriptive manner or from a psychological perspective but from a neuroscience perspective. That is, how do the brains of such people differ and what accounts for the differences in their neural circuitry?

I might have expected Drake to respond disparagingly about people who are afraid of flying, but when I asked him his opinion, his immediate response was that such people might have had a bad or even tragic experience that would make them fear flying. This explanation is insightful, and familiar to us from the brave soldiers returning home from combat who are afflicted with PTSD.

"The bravest people I know," Daniela Schiller, a neuroscientist who studies fear at Mt. Sinai Medical School, told me, "are people with PTSD." This is because people with PTSD must confront debilitating terror on a daily basis. The same serviceman who may have fought with valor in combat can display even greater bravery after returning home, forcing himself to ride as a passenger in a car while his body is unleashing a full-blown panic response as a consequence of having narrowly survived a roadside bombing in Afghanistan when his jeep was blown to bits.

New brain research on fear, threat detection, and conscious versus automatic modes of cognition shows that both genes and environment will dictate how a person responds to danger and whether one is compelled to seek out excitement and novelty or will shun them for the comfortable refuge of routine. The differences among people on this metric are both innate and the result of experience, and they can now be seen in the schematic of physical and functional differences in people's brains.

As we have seen, everyone uses both conscious analytical thought and unconscious cognition (gut feelings) to make decisions and guide their behavior, but different individuals utilize and develop these two modes of "thinking" to different extents. Since snapping in rage is an unconscious reaction, are people who are more deliberative less likely to "snap," or will anyone "snap" in the right situation given the right triggers?

"Maybe it's adrenaline," Drake speculated when I asked him what the appeal was of taking up the dangerous profession of Alaskan bush pilot, but then he immediately rejected that explanation. "It's not adrenaline. It's that . . . you know something? I've realized I'm bored." Drake picked up the same theme the next day inside his hangar as he was replacing a tire on his plane. He had just separated the wheel. A glossy black and polished aluminum engine sat on the bench—spotless. Not a drop of oil in sight. I recalled him telling me during the flight the previous day that when he was a racecar driver in the 1980s, he was also his own mechanic.

"Andretti would come over—he was really good. He had everything.

People took care of the car, every detail. After a race he'd just walk away from the car, but I worked on my own car. He'd come over and be amazed. I was mechanic, driver, and I'd do everything. I slept outside next to my car sometimes."

Drake is drawn to excitement and adventure, bored by routine, but he is also drawn to mastering the most intricate and complex challenges by working independently. Research shows that novelty seeking and an interest in mastering complex tasks working independently are often coupled.

Finally placing his finger on why he is drawn to his risky business he says, "It's the fluidity of movement, the conservation of inertia that you have to have in driving a racecar. You know you are full-throttle—that thing's got seven hundred horsepower—it's a magnificent two-hundred-thirty-mile-an-hour machine, but you are still trying to coax every little last ounce of inertia out of that thing, and have a little more inertia and a little more exit time out of each corner better than the next guy. That is such a precious feeling.

"Racecar driving for me was the same thing as skiing—linking those turns [in a racecar] you are running gates; it's that preservation of energy. It's that perfect carved turn," he says, staring ahead and visualizing the perfection of motion.

"It's the same thing the surfers like. It is the same thing the big mountain skiers like. It is that you are giving it *all* you've got, but it's not arms flailing or anything, it's *perfect*. That '*whoosh*' . . . there is something about that."

He continues: "It is a spiritual thing, actually. In flying it is really great. You get that nice glassy air and it's those same things."

I am reminded of big mountain skier Wendy Fisher's comment: "*The whole free-skiing scene—catching air! Doing cool lines, being in the middle of nowhere.*"

Those who crave novelty and adventure are perplexed by those who do not have such thirsts.

"Hell," Drake says, "I went to the market this morning to get something quick to eat and I just marvel. People sitting at tables. It's not that comfortable. It is kind of hot and stuffy, and yet they will sit there for hours and it's like, fuck! Do you not see this glorious day? And there's shit to do! Hell, they will spend hours there."

So let's have a look inside Drake's head, or at least the heads of people like him who are driven to seek novelty and adventure. Novelty enhances exploration by engaging circuits in the brain's reward system, which encompasses the midbrain, striatum, amygdala, and orbitofrontal and mesial prefrontal cortex. To select the novel choice means accepting greater risk and elevated danger. Using an fMRI scanner, Dr. Bianca Wittmann and her colleagues at University College London (UCL) constructed a game that presented subjects with the choice of unfamiliar options or a safe familiar option. When subjects selected the exotic option, activity in the ventral striatum increased. This region is part of the brain's reward system that releases the feel-good neurotransmitter dopamine. Whenever we make a choice that turns out to be beneficial, a bit of dopamine is released to give us that rewarding feeling of achievement. There is a downside: "Increased novelty-seeking may play a role in gambling and drug addiction, both of which are mediated by malfunctions in dopamine release," says Professor Nathaniel Daw, who participated in the study. Other studies also show that increased novelty seeking is associated with gambling and addiction, and that both behavioral activities are also linked to dopamine function. In fact, a well-known side effect of treating Parkinson's disease using the drug L-dopa is increased risk-taking, gambling, and sexual promiscuity. Parkinson's disease is caused by the death of substantia nigra neurons that use the neurotransmitter dopamine to facilitate control of voluntary movement, and L-dopa elevates levels of dopamine in the brain.

Sensation seeking is partly in the genes. Different forms (polymorphisms) of the D1 dopamine receptor gene are associated with sensation seeking in alcoholic males. In another study 1,591 adolescent twin pairs were evaluated for differences in sensation seeking. In comparing twins it was found that about 50 to 60 percent of the total variation in sensation seeking could be attributed to genetic factors. The strongest correlations were with thrill and adventure seeking and new experience seeking. Another study showed that boredom susceptibility and disinhibition (impulsivity) correlated with aggression, but aggression did not correlate with sensation seeking in this study. These various "risky/adventurous" traits are often blurred in common usage, but psychologists have different tests for evaluating boredom susceptibility, disinhibition (impulsivity), novel experience seeking, and thrill and adventure seeking. Somewhat different (but interrelated) brain circuits are involved in each of these traits, and people may exhibit any one of these qualities without exhibiting others. For example, some adventurers who seek novelty through exploration (mountaineers, for example) may have no interest in thrill-seeking pursuits such as BASE jumping.

Another genetic study also links sensation seeking to dopamine. Jaime Derringer at the University of Minnesota analyzed eight genes related to the neurotransmitter dopamine. Of 273 genetic variations in those 8 genes, she found 12 variations that accounted for most of the difference between people who are categorized as sensation seeking compared to those who are not.

"Not everyone who's high on sensation seeking becomes a drug addict. They may become an Army Ranger or an artist. It's all in how you channel it," Derringer says. In fact, another study by Matthew Cain at Notre Dame and Stephen McKeon at the University of Oregon links sensation seeking and risk-taking in CEOs. CEOs who enjoy the adrenaline rush of flying a private airplane are more likely than other CEOs to display bold management characteristics, according to a study by the University of Oregon. "CEOs who seeks thrills in their personal lives are more likely than others to be aggressive in their corporate policies," Stephen McKeon says.

Founder of Virgin Airways, Richard Branson, is a prime example. The Virgin Group of four hundred companies, whose products span from records to vodka, has since 2004 set its sights on space tourism with Virgin Galactic. In 1998, Branson attempted an around-the-world balloon flight, which abruptly ended in the Pacific Ocean on Christmas Day. He has competed in transatlantic sailboat races, kite surfed across the English Channel, and climbed Mont Blanc, among many other adventurous activities. "These adventures are physically challenging, and mentally challenging, and technologically challenging, and that is what makes them fascinating," Branson says.

The scientists found CEOs for their study by searching the FAA database—179 corporate executives who have private pilot's licenses were compared to 2,900 non-pilot CEOs. A sensation-seeking scale, developed from psychological tests, identified the urge to fly airplanes as a very high predictor of thrill- and adventure-seeking traits, and this airplane-flying group of CEOs were found to take on greater risks in business. "These individuals take on higher leverage than their counterparts and are more active in mergers and acquisitions. The volatility of equity returns in their companies is higher," McKeon says. Risk-taking is good for some businesses but not for all. An adventurous spirit can transform entrepreneurs into billionaires, but there are some undesirable behaviors sometimes associated with thrill seekers if such individuals lack creative outlets for their burning quest for novelty. Risky behaviors can lead to substance abuse, catastrophic failures, and personal injury, as well as other negative health consequences.

Teenagers are notorious risk takers. "The reason that teenagers take risks is not a problem with foreseeing the consequences, it is more because they choose to take those risks," said Dr. Stephanie Burnett from UCL. In a study of eighty-six boys and men who were asked to play computer games, scientists measured each subject's emotional response and found that teenagers showed increased enjoyment from winning in a "lucky escape" situation. "The onset of adolescence marks an explosion in 'risky' activities—from dangerous driving, unsafe sex, and experimentation with alcohol, to poor dietary habits and physical inactivity," Burnett says. Functional MRI studies show that part of the reason for increased impulsivity in teenagers is that there is an increase in dorsolateral prefrontal cortex activity with increasing age into adulthood. The thickness of this region of prefrontal cortex correlates with differences in impulsivity seen in different individuals regardless of age. Together such studies indicate that children and adolescents are able to recognize risk and to understand right from wrong, but they have a weaker ability to implement impulse control until this brain region matures.

Men tend to score higher than women on measures of sensation seeking, but both genetic differences and socially mediated effects are responsible. Salivary testosterone levels are significantly correlated with susceptibility to boredom. A particular variation in the dopamine receptor D4 gene (DRD4) is associated with thrill and adventure seeking, again linking the brain's reward system to thrill-seeking behaviors.

The DRD4 dopamine receptor gene is also associated with people who have a propensity for one-night stands. As explained by Justin Garcia, who reported the findings of his team's research, "Some will experience sex with committed romantic partners, others in uncommitted one-night stands." After analyzing the DNA of 181 people, the investigators determined that propensity to engage in uncommitted sex correlated with a variant of the dopamine receptor D4 gene. "The motivation seems to stem from a system of pleasure and reward, which is where the release of dopamine comes in. In cases of uncommitted sex, the risks are high, the rewards substantial, and the motivation variable—all elements that ensure a dopamine 'rush.' . . . One-night stands can be risky, both physically and psychologically," says Garcia. "And betrayal can be one of the most devastating things to happen to a couple." Now we know that a big part of this intimate personal behavior is genetic.

Brain imaging shows that when high sensation seekers view arousing photographs, the brain region known as the insula shows increased activity.

This same area is activated during cigarette craving and other addictive behaviors. Low sensation seekers have increased activity in the frontal cortex when shown arousing photographs. This is interesting, given the importance of the prefrontal cortex in controlling emotions. So high sensation seekers respond very strongly to arousing stimuli, but they have less activity in the brain regions that regulate emotion. That may relate to why sensation seeking can result in substance abuse and antisocial behaviors. But the dopamine addiction is not all bad. Risk takers have lower rates of Parkinson's disease. People with Parkinson's, however, score lower on sensation seeking and risk-taking behaviors, and they show elevated tendencies for anxiety and depression.

Together these studies show that there is a genetic basis for "adrenaline junkies," and the same circuits of reward in the brain that are activated in addiction are stimulated by adventure and thrilling pursuits. Impulsivity is also linked into this network of risk and reward neurocircuitry and therefore these neural pathways are another important component in the neuroscience of snapping.

## The Fearless Leader

Certain professions require a level of cool-headedness in a dangerous situation that far exceeds the norm. Ideally it is desirable to know how a person in a dangerous profession would respond in the instant of danger before a life-threatening event is encountered. Are there biological measurements that could be taken to gauge an individual's bravery?

Take as a prime example fifty-seven-year-old US Airways captain Chesley "Sully" Sullenberger, the former US Air Force fighter pilot hailed as "The Hero of the Hudson" after safely crash-landing his Airbus A320 on the Hudson River on January 15, 2009, minutes after both engines were disabled by hitting a flock of geese on takeoff from LaGuardia Airport. All 150 passengers and the 4 crew members were safely evacuated in what was a potentially catastrophic disaster over one of the most densely populated regions in the country.

Imagine your worst nightmare as a pilot comes true. For Sully it was the most terrifying moment of his entire life. Suddenly he and everyone on board were about to crash and die.

"To have zero thrust coming out of those engines was shocking—the silence," Sullenberger recalled later.

Sully's voice remained calm as he first reported the disaster to the control tower: "Hit birds—we lost thrust in both engines."

Barely 100 seconds later, after rapidly evaluating several alternatives, he calmly and matter-of-factly alerted the emergency-response team on the ground through the air traffic controller, "We're gonna be in the Hudson."

Here's the transcript of Sully's recorded communication with the LaGuardia control tower in the midst of this disaster. Coolly in control, he is analyzing options and taking action to avert the impending crash:

> 2029:21 [LaGuardia ATC] "Cactus 1529, turn right two eight zero—you can land runway one at Teterboro."
> 2029:25 [Sullenberger] "We can't do it."
> 2029:26 [LaGuardia ATC] "OK, which runway would you like at Teterboro?"
> 2029:28 [Sullenberger] "We're gonna be in the Hudson."
> 2029:33 [LaGuardia ATC] "I'm sorry, say again, Cactus?"
> [No reply from Sullenberger; he is not heard from on the control tower transcript again.]

"He was the last one off the plane," Mayor Bloomberg announced at a press conference immediately after the safe crash landing on the river.

Sullenberger walked the full length of the cabin twice as the empty plane was sinking, to be doubly certain that every one of his passengers and crew had evacuated safely before he escaped the aircraft himself.

Many can be attracted to dangerous occupations such as being a captain responsible for the lives of others, but unfortunately not everyone will react fearlessly in the moment of disaster. Take for example the captain of the luxury cruise ship the *Costa Concordia*. The captain, fifty-two-year-old Francesco Schettino, accidentally ran his ship aground off the coast of Giglio, Italy, in the middle of the night on January 13, 2012, ripping open the hull on the rocks and flooding the ship, which rolled on its side and sank just off the shoreline with 4,229 passengers aboard. Thirty-two people were killed and the vessel was destroyed.

Schettino was later dubbed "Captain Coward" for leaving the ship while terrified people were trapped aboard the cruise liner. Witnesses described him as sobbing and confused, providing no clear information to authorities or instructions to crew and passengers. Evacuation of the vessel did not begin until sixty-eight minutes after the crash.

"At around 2:30 a.m. I spoke to the captain," chaplain Raffaele Malena told the French magazine *Famille Chrétienne*. "He embraced me and cried like a child for about a quarter of an hour."

An audiotape of the disaster that night shows a Coast Guard captain exasperated with Schettino for balking at returning to his ship after he had abandoned it before the passengers were off. It was essential that the captain help supervise the evacuation of hundreds of passengers still on board the darkened ocean liner. Schettino wasn't "lucid," according to the Coast Guard official's testimony.

Schettino admits that he left the ship in the midst of the disaster, a fact that is not in question, but he claims that he tripped and fell into a lifeboat while it was being lowered into the sea and he could not get out.

Why did it take sixty-eight minutes to start the evacuation? Why did the captain announce to passengers and report to the Italian police that the ship had only suffered an electrical problem when it was obvious to everyone on board that the ship had crashed violently and was taking on water and sinking? Why did crew instruct passengers, "Go back to your cabins; everything has been fixed"?

Later put on trial for multiple manslaughter charges, the captain denied any wrongdoing and claimed that he was an innocent scapegoat. In February 2015, Captain Schettino was found guilty on manslaughter charges for the death of thirty-two passengers killed in the disaster and sentenced to sixteen years in jail.

It would be desirable to know how fearless an individual is before placing him or her in charge of a dangerous situation where their life and the lives of others are at stake. We have already discussed one simple physiological test that requires no fancy instrumentation or exhaustive psychological evaluation to measure how fearless and adventure-seeking a person is likely to be—a person's resting heart rate.

Although a heart muscle will beat on its own rhythmically even when isolated from the body, the brain can regulate heart rate. "The frontal cortex developed [through evolution] as a way to do offline problem solving. It freed us from direct behavioral consequences of our actions so that we could ponder them. . . . Basically [the] prefrontal cortex inhibits the brain-stem response fight-or-flight behavior and typical stress responses," Julian Thayer, a psychologist at Ohio State University, summarized in a lecture on neuro-visceral integration. "The cortex [constantly] inhibits this activity so that we can go about our day, but it can be rapidly released in the presence of a threat."

This inhibitory circuit from the frontal cortex allows us to self-regulate attention, emotion, heart rate, and other stress responses. This response can be demonstrated when the brain is exposed for surgery to treat epilepsy. Infusing a drug that temporarily inactivates the prefrontal cortex causes the heart rate to jump, showing that the cortex applies the brakes to the stress-response circuit in the brain stem. Thus, by reverse engineering, it is possible to monitor the strength of the prefrontal cortex simply by measuring heart-rate variability (HRV).

When patients with general anxiety disorder are asked to just worry as they normally do, they have a lower heart-rate variability than control subjects without anxiety. The startle reflex, which springs from activity in the amygdala, is also suppressed by the prefrontal cortex, but people with high heart-rate variability are more prone to startling.

This cortical inhibition of the stress response is an important factor in performance under stress. In a study on cadets in the Norwegian Navy and with Special Forces in the United States and in Afghanistan, Thayer found that heart-rate variability decreases as the servicemen are preparing themselves, and thus anticipating performing working memory and cognitive tests. During threat conditions, cadets with the greatest decrease in HRV performed better in the test because their ability to sustain attention was greater.

In baseball, some athletes are prized for their ability to perform under extreme pressure. Charlie Maher, team psychologist for the Cleveland Indians, cites as an example the best clutch hitters. "They are able to take a non-routine play—let's just call it the World Series, two outs, all of that—and they take the adrenaline rush from that and they turn it into a very positive 'go get 'em' type of atmosphere. Somebody else, who has a negative mind-set at plate has this 'I hope I don't strike out' [mentality], is doomed as soon as they start saying that in their mind."

Being able to perform under pressure can be more important than the overall level of skill. "You see people who are typically so talented and they so-called choke," says Maher. "I think that's what happens. People are really talented and choke all the time. [There are] other people who are not so talented, but just never choke." Clutch hitters must have stronger prefrontal cortical control over their brain-stem stress response and, one would predict, greater heart-rate variability.

There are negative effects of too much cortical control suppressing the heart rhythm, however. Such people can have increased blood phobia and be prone to fainting (vasovagal syncope).

Experimental research in animals supports these human studies. For example, via selective breeding, two strains of mice have been developed that are either susceptible to fear or resistant to fear. The neuronal circuits controlling the body's response to stress and activating the fight-or-flight mechanism are different in these two strains. These circuits are also known as the *limbic-hypothalamic-pituitary adrenal axis* (LHPA). Neurons in the hypothalamus, after being activated by the limbic system, control secretion of hormones from the pituitary gland in the brain (notably ACTH), which stimulates the release of corticosteroids and other stress hormones from the adrenal glands attached to the kidneys. The fear-resistant mice have genetic differences that suppress the LHPA axis drive. Brain fMRI scans of fear-resistant mice show less limbic-system activation than in mice susceptible to fear. Also, the levels of the stress hormone corticosterone at rest are significantly lower in fear-resistant mice. Higher levels of corticosterone have been positively correlated with fear and amygdala activation in humans, and human brain imaging shows that the amount of functional connectivity between the prefrontal cortex and amygdala relates to fear-induced learning.

It is not only the specific triggers that can explain why we snap in certain situations or account for how we react to a threat; different people respond to the same triggers differently. When the "underwear bomber," Umar Farouk Abdulmutallab, attempted to detonate an explosive device on Flight 253 on December 25, 2009, surrounding passengers fled at the sound of a pop and flames, but one passenger, Dutch graduate student Jasper Schuringa, instinctively leaped over rows of seats and subdued the terrorist, putting out the flames with his bare hands.

"I didn't think," Schuringa said. "When I saw the suspect—that he was getting on fire—I freaked. Without any hesitation, I just jumped over all the seats."

In contrast to this instant reaction to a dangerous situation, in other life-threatening situations, bystanders may do nothing. In October 2011, a van struck and ran over a two-year-old girl. Security cameras show the driver pausing, then driving on, rolling over her with his back tires. The toddler was ignored by at least eighteen other passersby as she lay grievously injured for seven minutes. A second vehicle then ran over the child and drove on. The video shows pedestrians walking by the child, others riding bicycles and driving cars around the girl's motionless body lying in a pool of blood until a woman carrying a sack appears in the security-camera video ten minutes after the initial collision. Instantly dropping her sack she quickly moves the girl to safety and goes to look for help.

The woman was a fifty-eight-year-old scavenger named Chen Xianmei. "I didn't think of anything at the time," she told reporters. "I just wanted to save the girl. Blood was coming out of her nose and mouth. I didn't understand why no one else had carried her from the street."

How could people be afraid to help an injured child?

There are many examples of similar instances, which psychologists call the bystander effect or bystander apathy. Research shows that the more people who are present, the less likely someone is to respond to the danger. Why take the personal risk? Moreover, our human nature to form tribes, and thus conform to the behaviors of others in the group, compels people to do as they see others doing. In this respect, it is not surprising that the heroine in the sad incident above was a scavenger—something of a social outcast.

Despite the genetic differences and the largely automated reaction to threat and danger, our brain is equipped with cortical control circuits that can regulate these automated responses, and these circuits can be strengthened or weakened by experience. In a study by Air Force major Christopher McClernon on stress and aviation, he commented on Captain Sullenberger's ability to maintain his composure after losing power in both engines. It wasn't that Sully is immune to stress; he was handling the stress in part by deliberately controlling his body's physiological reflex to danger. Sully said that he noticed his body's stress response kick in, including tunnel vision, an increase in heart rate, and perspiration. He said that he felt an "adrenaline rush right to my core." But as his heart rate increased, he forced himself to ignore the physical symptoms in his body. "I had a job to do. I did not allow it to distract me," he says.

"I knew the situation was bad immediately—losing both engines over a highly populated area. The physiological reaction I had was strong," explained Sullenberger, "and I had to force myself to use my training and force calm on the situation . . . it took some concentration."

It must have taken "some concentration" for the Navy SEAL who unflinchingly allowed a complete stranger to poke a needle into his vein while he totally suppressed the pain, the autonomic response to injury, and fear. He was practicing.

These same general brain centers are involved in all aggressive and defensive responses to threats. This makes sense because aggressive behaviors share many common features and they rely on many of the same sensory inputs and physiological states of mind and body regardless of what initiated them. Thus, fearlessness is involved in aggressive actions in both heroic bravery and in criminality. "Low resting heart rate appears to be the best-replicated biological correlate to date of antisocial and aggressive behavior in children and adolescents," conclude Ortiz and Rain in their 2004 paper. A 2009 study confirms these findings after even after controlling for several potential confounds.

Although it seems that almost anything can set off a sudden snap of violence, thereby making news reports of people snapping violently seem so bewildering, closer analysis reveals that this is not the case. There are a few very specific triggers for a sudden commitment to violence. This also makes sense from an evolutionary standpoint. All behaviors arise from brain operations, but a behavior that risks serious injury or death in an instant must be highly regulated. It must only be initiated in response to a very discrete stimulus. Launching into life-risking violence is the equivalent of the national missile defense system detecting a nuclear threat and responding rapidly in kind. That system must maintain constant vigilance to wide-ranging threats, but it must be very highly regulated, activated only when it is absolutely vital.

Our understanding of fear, threat detection, and aggression have advanced enormously as a result of new experimental techniques to study brain and gene function, advancing insight far beyond what can be gleaned from behavioral studies alone. This is not to discount the importance of behavioral work, in animals or people, but in the next chapter we examine the rage response in animals, tracing the neurocircuitry of the LIFEMORTS triggers to their origins in the animal world.

# 9

# The Best Defense

I don't even call it violence when it's in self defense; I call it intelligence.

**Malcolm X, speech to Peace Corps Workers, December 12, 1964**

Until recently, the accepted view was that there is a single brain circuit that underlies all types of fear and aggression. We now see that this view was the consequence of analytical tools that were too blunt to resolve the intricate details. New research on lab rats and mice has identified distinct circuits for fear and aggression in response to pain, predators, attack from members of the same species, and in mothers protecting their young. Different provocations trigger different threat-detection circuits in the brain that activate different response circuits to drive various behaviors of aggression or submission.

## Momma Bear

Former Alaskan governor and US vice presidential candidate in the 2010 election, Sarah Palin, referred to herself during the campaign as a "momma grizzly." What she instantly conveyed in this comment was the well-appreciated and powerful biological imperative about the so-called weaker sex. Females are not normally physically aggressive like their notoriously warmongering, brutish, and brawling male counterparts, but females are capable of snapping in vicious violence when necessary—especially to protect their young. Sarah Palin is an avid outdoors enthusiast and hunter, so she likely knew this animal behavior firsthand, but all women (and men) resonated with the biological truth: Don't ever get between a mother and her child.

A mother moose and her calf had wandered onto campus at the University of Alaska, Anchorage, and the two were quickly adopted as the unofficial campus pets. They were cute. Comical-looking animals with big black eyes, large furry ears, and a goatee-like beard, moose evoke the Bullwinkle cartoons of simple-minded, friendly creatures. The females, who lack the regal antlers of males, are especially amusing beasts.

Moose are enormous, though. The largest species in the deer family, moose stand more than six feet high at the shoulder when mature. Males weigh up to 1,500 pounds, females typically half that. Moose are peaceful animals. Being herbivores, they have no reason to attack other animals to feed and they are not aggressive toward humans. Despite this, more people are injured by moose than by bears and wolves combined.

As with most grazing animals, moose are preyed upon by carnivores, but moose are far from defenseless. Protected not only by their bulk, moose, unlike other members of the deer family, can kick in all directions, including sideways. Unlike with, say, a horse, there is no safe direction from which to approach a moose. Moose have very flexible joints and sharp, pointed hooves, and the 1,500-pound beasts kick powerfully with both their front and back legs.

As seen captured on video a seventy-one-year-old man, Myong Chin Ra, is walking carefully on a slick, snow-covered sidewalk bordered by three feet of shoveled snow, ambling in his gray wool cap toward the entrance of the Campus Sports Center at UA Anchorage, to meet his wife for lunch. The moose and her calf are not far from the glass doors leading into the facility. The calf is standing in the corner of the entryway and its mother stands behind. Both of them are facing the corner. The animals are off to the side of the vestibule; neither one blocks the entrance.

Suddenly, the mother moose turns and looks back at the elderly man over her left shoulder. The man briskly diverts his course to scamper away, but his escape is slow on the icy path. Suddenly the moose charges explosively in three body-length bounds at full speed toward the fleeing man. She rises up on her hind legs and crashes down on the man's shoulders from behind with her two forelegs, bringing her full weight crushing down on him as he attempts to escape. Mr. Ra crumples facedown onto the ground. The momentum of the tackle propels the moose over the victim's prostrate body but she delivers a powerful kick to the man's back with her right hind leg as she tramples over him. She pivots instantly; swinging back toward the fallen man she begins stomping him with all four legs, dancing over the helpless victim and delivering rapid, powerful blows like a flurry of invisible punches from a prizefighter, but each blow delivers the force of a jackhammer. The moose continues the vicious attack, circling again and coming back to finish him off with a blaze of vicious kicks. The curious baby moose comes over to watch the fray. The mortally wounded man lay on his right side in the fetal position trying to protect himself from further attack, but the animal kicks and stomps him about the head and upper body.

Myong Chin Ra was rushed to the hospital, but he died from the large number of severe blows he received within a period of less than three seconds.

Mike MacDonald, a game expert, explains that the moose was acting instinctively to protect her young. Even though the elderly man made no threatening actions at all, the situation from the perspective of the moose appeared to place her calf in potential danger.

"She had the building behind her and snow berms on either side of her, so she was pinned in."

Although the man had not provoked the animals in any way, witnesses later stated that students had teased the pair hours earlier. "There were people standing around throwing snowballs, yelling, whistling, shouting, trying to get their attention," said Ann Gross, a director at the university's day-care center next to the sports center.

Appreciating that this was a natural, instinctive defensive behavior, university authorities decided not to kill or capture the moose because of this freak incident. Instead, they posted signs on campus educating the college community about the need to keep a safe distance from the animals.

Five days later, the same moose attacked psychology professor Bruno Kappes in a very similar chance and unprovoked encounter.

"I think my response was a normal panic response," he told *Anchorage Daily News* reporter Sheila Toomey a few days afterward, with obvious gashes on his battered forehead.

"Fortunately I had seen the power and viciousness of this animal, and that was significant in appreciating what the animal could do to me.

"I knew when the ears went down and it stretched its head out at me, I knew it was preparing to launch toward me . . . otherwise I probably would have just said 'Nice moose,' and tried to walk around it." Instead, Kappes instantly darted away as fast as he possibly could.

Kappes was saved by the hardwired defensive threat-response circuits in his brain. Instantly he felt his body's own powerful fight-or-flight response engage, and that explosive neurophysiology saved his life.

The adrenaline rush kicked in just as it was supposed to, he said to the reporter afterward. "That's why I was able to jump probably 10 feet in the air. They said I also changed direction in midair."

"It was like the professor was just shot out of a cannon," campus police officer Richard Altman said, as he rushed toward the scene and drew his gun. "Back up! Get behind me!" Altman commanded. Kappes responded as ordered.

The psychologist, who, ironically, is an expert on PTSD, suddenly found the tables turned. "Let's say this: I look for moose in my house."

For the safety of people on campus, the decision was made to kill the moose.

This episode with the mother moose illustrates the Family trigger of rage. It is a primordial, instantaneous trigger of violence shared by many animals, large and small, weak and strong, predators and prey. The young are the weak point in the cycle of life, and nearly all animals have been programmed through evolution to immediately sacrifice all to save their offspring from danger. This is the core of what it is to be a parent.

The magnitude of the danger can be immaterial to a mother who suddenly finds her child in jeopardy. Her response is immediate. Security cameras in the London Underground system captured an example of this on July 23, 2014. A whoosh of strong wind produced by a subway train blows against a blue stroller carrying a baby as the mother is distracted tending to her luggage. The stroller rolls slowly off the platform and falls onto the steel tracks. Instantly the young mother leaps onto the tracks, snatches her baby, stroller and all, and hoists her child back onto the platform to safety. Then she springs back up onto the chest-high platform with the sudden strength of a gymnast as the rails begin to glow with the reflection of headlights from the oncoming train. She escapes seconds before the train barrels into the station. Her male companion drops to his knees to embrace the stroller. It could have been a scene from a superhero movie, except that the petite blonde with her hair piled up on her head and wearing a "girly" backpack stuffed with baby supplies, looks seriously miscast.

## Threat-Response Neurocircuitry

The circuitry for the Family trigger is tripped when a mother rat snaps violently to protect her young and attacks a male intruder. Laboratory research shows that this response involves three brain areas: amygdala (sentry for danger); hypothalamus (hormonal and autonomic regulation); and septum (triggering bodily responses). There are discrete circuits in each of these brain regions that operate in controlling maternal aggression. For example, olfactory cues from a male intruder that communicate to the MEA region of the amygdala are critical in triggering maternal aggression in rodents.

Other brain regions are also involved. The cerebral hemispheres contain a hollow, fluid-filled space called the lateral ventricle. An arc of brain tissue surrounding the lateral ventricle forms the limbic system. This brain region functions to help an individual cope with their environment and especially cope with other members of the same species in the environment. In short, the limbic system is the brain's sentry for danger, as was discussed previously in the chapter on fear. Naturally this system needs widespread connections throughout the brain to carry out this complex function of threat detection and rapid response. Many of these functions influenced by the limbic system are automated behaviors related to food, sex, and threats of various types, which are conveyed to us as feelings or emotions—fear, hunger, anger, and so on. There are interactions between the amygdala and the hippocampus to coordinate time and space with experiences, past and present, in the present threat environment. These activities recording context and experience make the hippocampus critical for forming and recalling many kinds of memory, and this connection between fear and memory is the substrate for PTSD.

The amygdala, a central part of this system, serves to monitor complex status information about the body and changing environmental factors and then grab the attention of the cerebral cortex to make us consciously aware of the most critical matters as they arise. We cannot attend to everything in our environment at once; that would lead to sensory overload and paralysis. The brain's complex and constant monitoring of our internal and external environment is conducted automatically and unconsciously, only reaching our awareness as an emotion when the amygdala has assessed an immediate threat and has called upon some action by the body or intellectual faculties to address it. (It is a pickpocket!) Thus the limbic system is the neural mechanism pivotal in all social behaviors.

The amygdala receives sensory information from higher-order areas of the cerebral cortex. Different senses enter the amygdala through different pathways; for example, auditory information reaches the amygdala through the medial geniculate body. Knowing the anatomical name is not as important as understanding the concept of how different sensory abilities feed into the amygdala through separate lines of communication. The amygdala is like the Pentagon with its hotline to the White House (cerebral cortex). Other connections link the amygdala to the frontal lobe, which is where higher executive functions are carried out—functions critical to any threatening situation. Other fibers connect to the lower brain regions to control hormones, autonomic responses, and to the brain's "relay center," the thalamus, for sending sensory information up to the cerebral cortex. Controlling this relay center is how the amygdala brings specific events in our environment and internal states to our conscious awareness and filters out others. For example, pain is suppressed in the midst of a life-or-death battle. General anesthesia removes pain by preventing the cerebral cortex from responding to the signals streaming into our brain through pain fibers. If the signals never reach the cortex, we do not feel the pain, which explains why a soldier in battle can suffer a serious wound with no pain or conscious awareness of the injury: The amygdala has signaled the thalamus to cut off pain signals to the cerebral cortex, which enables his conscious faculties to fully engage with the life-risking threat. One can also see how dysfunction in the limbic system can lead to mental illnesses that affect emotion, threat responses, and antisocial behavior—illnesses such as mania, obsessive-compulsive disorder, or psychosis.

Humans depend most heavily on their sense of vision, but to most animals the world is experienced as a complex and dynamic realm of smells. The olfactory system connects to the limbic system, but not for analysis of an odor, but rather to evoke the emotion that is associated with it. The smell of smoke provokes alarm, the smell of food provokes hunger, and the smell of rot provokes repulsion, for example. The strong feelings evoked by specific odors are what give the sense of smell such power over us. Can you eat anything with the smell of vomit in the air? Could you sleep peacefully at night with the smell of smoke seeping from your attic? In most mammals, specific odors will trigger complex social interactions, including violent aggression.

"Septum" means wall of separation, or partition. In neuroanatomy, it refers to the thin membranous tissue that separates the left and right fluid-filled ventricles of the cerebral hemisphere. There are several clusters of neurons in the septal region, called nuclei. One of these clusters, the nucleus accumbens, lies just to one side of the septal nuclei. This is part of the brain's reward system. This is the circuitry for the positive reinforcement we receive—the emotional boost, when we succeed at anything. Mood-enhancing drugs, such as cocaine, cause release of the neurotransmitter dopamine in the nucleus accumbens, delivering an artificially induced sensation of reward. Addiction to alcohol and other drugs involves changes in the nucleus accumbens reward system. As one would expect, the brain's reward system plays a pivotal role in motivating a person's behavior to seek novelty and danger or to shun them. The anxiety that grows into craving associated with drug and alcohol dependence is generated by changes in function in an addict's nucleus accumbens that decrease dopamine levels. Similarly, general anxiety feeds through this same reward pathway in an individual confronted with the choice of taking a life-risking action or not.

When sensory information reaches the cerebral cortex, it is relayed to both the hippocampus and amygdala to rapidly assess whether the experience is novel or potentially threatening. The hippocampus links new sensations with memories of the past, looking for novelty and familiar rewarding experiences. The mundane is filtered out. Memory is not a video recording of sensory experiences; memory is very selective. In fact, memory is not a record at all but rather a mental reconstruction. This is why eyewitness testimony at trial and in general is so unreliable. By far, most of what we experience is rapidly forgotten. As we all know from personal experience, the emotional aspect of an event is a critical feature that determines whether an episode will be remembered. No one ever forgets a traumatic experience or any other emotionally charged event, even if it is encountered only once. You will never forget being mugged, for example; or your first kiss. Emotionally charged events are those the system identifies as having survival value and therefore need to be remembered to direct behavior appropriately in the future. It's this interchange between the amygdala and hippocampus that filters out the noise of everyday existence, attaches emotion to meaningful events, and stamps the patterns of neural activity re-creating that experience in the mind to be retained for a lifetime if deemed important enough by this truly astonishing unconscious bit of our brain.

Now we pick up from where we first encountered blindsight, the ability of unseen visual stimuli to influence defensive behavior, which was described in the experiments on subjects who are blind in one eye but nevertheless are influenced by threatening faces shown to the sightless eye. Sensory information reaches the amygdala by at least two routes. One path is a direct line from the thalamus, because this is the shortest and fastest pathway from the body's sense organs to the brain. This "subterranean" shortcut, tunneling beneath the cerebral cortex, is unconscious. We have no awareness of sensory information as it reaches the amygdala through the thalamus, but this high-speed communication route evokes rapid emotion and instantaneous reaction to a potential threat. The amygdala initiates this threat response by activating three neural circuits: First the amygdala stimulates the appropriate autonomic and endocrine systems located in the hypothalamus and brain stem to flood the bloodstream with adrenaline and stress hormones to trigger the fight-or-flight response or to evoke other profound emotional sensations from goose pimples to vomiting.

Second, the amygdala shoots information back to the hippocampus to reaffirm the emotional significance of what has simultaneously been sent there by other inputs. The message in English would be something like, *This is important! Remember it!*

Finally, both the hippocampus and amygdala relay signals to the cerebral cortex in the sensory association regions, where memories are made and context is synthesized and deliberation is added to set the body on a course of action. Now your conscious brain is being brought into the loop and you become aware that you are engaged in a threatening situation or defensive battle. We have already seen evidence of this in the study on people blinded in one eye from damage to the visual cortex who nevertheless demonstrate a heightened response to threatening images shown to the blind eye.

The second route into the amygdala is "aboveground," through the cerebral cortex, where sensory input is relayed from sense organs and then analyzed in detail to extract intricate information and meaning, such as where the object is, what the object is, and where the object is going. Complex circuitry in the cerebral cortex makes associations among all of one's senses, and cogitates on the significance of it all. This enormous computation requires a great deal of processing time, and in a sudden life-threatening situation, such as a rattlesnake strike, this route to action would be far too slow. However, this slower route to the amygdala can provoke the same reactions as the "subterranean" route, but it does so in response to far more complex environmental situations. One can also work themselves into a state of fear by this cortical pathway, as kids do when they tell ghost stories around the campfire.

It also works the other way: Picture a world-class rock climber as he creeps precariously up an impossibly sheer granite face on dime-edge footholds. As he does so, he starts to hum the scarecrow's theme from *The Wizard of Oz*, "If I Only Had a Brain":

*doo da* doot *da do do doodo*
*da doot da doo doo doodo*
*da doot dooo do da dooooooo*

The neurocircuitry for singing resides, of course, in the cerebral cortex. The pathways from the cerebral cortex to the subcortical fear- and threat-detection circuitry can suppress fear as well as reinforce it. The fear circuits are screaming up to the conscious brain, *We're gonna die! We're gonna die!* Singing, a "positive attitude," and especially humor, puts a damper on the unconscious fear and rage circuitry to allow a person to maintain control (the proverbial whistling in the dark).

*Yes, I know it's sketchy,* the cerebral cortex tells the amygdala by singing in response to those gut-twisting emotions that the mute unconscious brain is frantically pumping as panic up to our conscious mind, *but I've got this.*

The descending input from the cerebral cortex as it regulates the threat-detection circuitry is the neuroscience behind the iconic casual understatement and gallows humor of those who must operate in the most treacherous situations. Rodeo cowboys, adventurers, and men and women in the military all use these techniques to positive effect in order to operate in the most life-threatening conditions.

In Marcus Luttrell's book *Lone Survivor,* he recounts a tale of awe-inspiring bravery in a horrendous battle in which his teammates were caught terribly outnumbered and surrounded by Taliban forces. Luttrell and the handful of SEALs decimated the enemy with skill, determination, and bravery, but the battle left only one of them alive. Luttrell's account is replete with examples of the mechanism of cortical control of fear circuitry to persevere and succeed in a situation where all but the most elite among us would cower.

After jumping off a cliff to escape being blasted by bullets and RPGs (rocket-propelled grenades), Luttrell describes how his buddy Mikey Murphy recovered from the leap of faith amid the ricocheting bullets and deafening explosions:

> He still had his rifle strapped on. Mine was resting at my feet. I grabbed it, and I heard Murphy shout through the din of explosions, "You good?"
>
> I turned to him, and his entire face was black with dust. Even his goddamned teeth were black. "You look like shit, man," I told him. "Fix yourself up!"
>
> Despite everything, Mikey laughed, and then I noticed he'd been shot during the fall. There was blood pumping out of his stomach. But just then there was a thunderous explosion from one of the grenades, too close, much too close. . . .

SEALs never give up. Never. It is cortical control of their fear and rage circuits that helps them to persevere.

Later, after most of Luttrell's team are mortally wounded but still 100 percent in the battle, firing their rifles and killing the enemy as they themselves lay fatally injured and trapped:

> Mikey worked his way alongside me and said with vintage Murphy humor, "Man, this really sucks."
>
> I turned to face him and told him, "We're gonna fucking die out here—if we're not careful."
>
> "I know," he replied.
>
> And the battle raged on.

As described in Mark Owen's book *No Easy Day*, running practical jokes, like the purple dildo that keeps cropping up in the team members' gear at unexpected times and places, and the bra strap slipped onto the author's backpack by a teammate as they escape from Osama bin Laden's compound, were commonplace. They served to take the edge off so that the team could operate under stresses and dangers that few of us can easily imagine.

An important point about the sensory input to the subcortical pathway is that while the information is transmitted to the threat-detection circuits at the highest rate through this shortcut path, this pathway can convey only the most rudimentary information—the minimum that is necessary to alert us to danger. In receiving visual input from subcortical pathways to the amygdala, we cannot perceive an image. Forming an image would require far too much information processing that must pass through layers of cerebral cortex and reach many different cortical regions, but the amygdala does not need to form an image to exploit visual input for threat detection. Like a motion detector in a burglar-alarm system, which rapidly switches on warning lights and alarms when an intruder moves into range, the amygdala will do the same when an object—be it a baseball or a person's fist—zips into your visual field and you jump to safety as you simultaneously duck and raise your hands in defense. Even though you can't see it as anything more than a blur, that thing—whatever it is—suddenly appearing in your visual field should not be there. The object has violated your perimeter of safety and  you are under threat. You will dodge and strike at it instantly, despite the fact that this is an extremely complex and highly coordinated sensory-motor response. Likewise for auditory input. In a scary movie a sudden bang causes you to jump even though you can't discern if the sound is a gunshot, a slammed door, or an innocent hammer striking a nail. That abrupt loud sound could herald impending doom. There is no time to think. There is not even time to perceive the sound clearly. React! Kick the heart into high gear! Clench muscles for maximum strength! The sensory signals take the subterranean shortcut to the amygdala while they simultaneously split off and head toward the longer pathway into the cerebral cortex. Let the cerebral cortex catch up afterward and inform you of what the potential threat was that you have just avoided.

As I mentioned in chapter 1, dramatic pioneering research by Walter Rudolf Hess using electrical stimulation of the hypothalamus showed that there are automatic circuits of rage and aggression in the brain and that these "attack areas" reside in the hypothalamus. However, stimulating brain tissue with electrodes is not very precise. An electric shock spreads broadly, exciting larger areas of brain tissue and also activating nerve fibers that pass through the area from distant regions. Stimulating fibers of passage can lead to an incorrect conclusion about where the neurons that control a certain brain response are located. Modern methods are providing a much more detailed map of brain circuitry, and one of the fascinating discoveries is that different automatic attack behaviors are controlled by specific subregions of the hypothalamus. Very discrete circuits within the hypothalamus, amygdala, and other brain regions are involved in rapid responses to threat, and they are activated by very different types of threats. That is, the LIFEMORTS triggers have a corresponding and highly tuned circuit for each trigger.

Let's have a look at the Family trigger as an example. This trigger circuitry, responsible for "momma bears" and superwoman confrontations with subway trains to protect their young, was identified by neuroscientists in 2014. Research on lab rats and mice has identified two specific subregions of the hypothalamus that are part of the maternal-aggression circuitry activated by the Family trigger. When a strange male rat is placed into a cage with a female rat tending her pups, the mother will immediately attack the male intruder. To identify the precise brain circuitry involved in maternal aggression, new and more fine-tuned methods than electrical stimulation are needed. One of these new methods allows researchers to see the activated circuits in the brain tissue with their own eyes.

Researchers use techniques that enable them to determine directly, by looking at brain tissue on a microscope slide, whether a neuron was firing rapidly just before the tissue specimen was taken for examination. In the 1990s researchers discovered that when neurons fire impulses, chemical signals reach the nucleus of the neuron to activate genes that are needed to make proteins in response to high levels of electrical activity. One of these genes is called c-*fos*. This gene makes a protein called Fos that binds to specific DNA sequences to "turn on" other genes—that is, to start the process of making mRNA from the DNA genetic code, to then make a specific protein, such as an ion channel, or cell signaling molecule, to better cope with the high level of neural activity. C-*fos* and other genes that act in this way are called "transcription factors" because they start the process of transcription (synthesis) of mRNA from the DNA template. Only genes that need to be turned on when a neuron fires rapidly have the DNA binding sites that Fos will recognize and bind to. This explains in part how the right genes out of the tens of thousands of genes in the cell's nucleus are activated by the right stimulus. So when scientists see c-*fos* protein being made in a neuron, they know that this neuron was firing rapidly just before they took the brain sample to examine on a microscope slide. It is possible to see the amount of Fos in a neuron by using a staining technique that makes neurons stain darker (or shine brighter, if a fluorescent dye is used) in proportion to how much Fos is in the cell.

Neuroscientist Simone Motta and colleagues looked at the hypothalamus of a mother rat under a microscope just after a male intruder was placed in her cage with her and her newborn pups, which caused the mother rat to attack the intruder. Motta and her colleagues saw a specific spot in the rat's hypothalamus that looked like it had been stippled with a black ink pen. These neurons here were loaded with Fos, meaning that the rat's tiny collection of neurons had been firing rapidly in response to the Family trigger activated when the male intruded into her cage and provoked her to attack to defend her young.

These neurons were seen only in a small speck inside the hypothalamus, but when this tiny spot of neurons was surgically removed, the researchers found that the mother behaved normally in every other way, but attacks on male intruders ceased. (The brain is comprised of two symmetrical halves, the left and right, so this region had to be removed on both sides of the brain to fully abolish the mother's protective attack response to a male intruder.) This spot resides in a general region of the hypothalamus called the hypothalamic attack area, which has been discussed previously in the pioneering experiments by Walter Rudolf Hess on initiating aggression in cats by electrical stimulation of this brain region. The spot itself is called the *ventral premammillary nucleus* (PMv). However, removing this tiny maternal aggression trigger spot in the hypothalamus does not affect other automated attack responses to different threats, as will be discussed. The PMv is part of the same Family trigger neurocircuitry that was likely activated in the mother moose's brain when she suddenly attacked and killed Myong Chin Ra and later attacked Professor Bruno Kappes on the campus of the University of Alaska at Anchorage. It is not difficult to imagine how mothers, with different tendencies to protect their young or neglect them, could have corresponding differences in the PMv nucleus of their brain that controls the Family trigger.

By now you understand that no single part of the brain completely controls an animal's aggressive response to a threat. The environmental triggers and the priming factors that change the threshold on when the trigger is activated are very complex, requiring threat detection and accurate response in complex social situations while the brain is engaged in countless other crucial tasks. The responses of aggression or retreat are also very complex behaviors, so many brain regions have to be engaged properly to launch an attack, freeze, or run.

Expecting that the amygdala must be involved in maternal aggression, not only the hypothalamus as just discussed, the scientists used the c-fos staining technique to discover if any part of the amygdala was also activated in the mother's brain by the male intruder. Looking under the microscope they clearly saw two spots in the amygdala that stained strongly for c-Fos in response to activating the Family trigger. Both of these stained areas of the amygdala are known to receive input from the olfactory region. (These are the posteroventral and posterodorsal parts of the medial amygdalar nucleus [MEApv and MEApd], if you are a neurosurgeon or just interested in knowing the names of the exact spots.)

It makes perfect sense that aggression in rats can be triggered by olfaction, because the sense of smell is of primary importance to rodents. Rats and mice inhabit dark places and are nocturnally active, so their vision is not that good. Their senses of smell and touch through their whiskers are the most important senses for these rodents. Behavioral scientists had already shown that the *smell* of a male intruder was the prime trigger for launching aggression in mother rats. Cut the nerves from the nose to remove the sense of smell, and mother rats do not attack the male intruder. It is interesting that the PMv region of the hypothalamus, which is the linchpin of the maternal aggression response, has neurons in it that are known to respond specifically to odors only from the opposite sex. The whiskers also help a mother rat investigate an intruder immediately before an attack to defend her young.

Another part of the amygdala, the posterior amygdalar nucleus (PA) also showed strong staining for c-Fos. Neurons in this region are known to have receptors for stress hormones (mineralocorticoid receptors). Interestingly, when these stress hormone receptors are blocked in aggressive male rats, the animals become docile. This circuitry in the amygdala must link hormonal stress to the Family trigger circuitry, explaining in part how other aspects of a given situation, such as stress or hormones, can lower the threshold for pulling a specific trigger for rage. "There are always other factors," Secret Service Agent Scott Moyer says concerning when people snap and commit a crime.

Likewise, another extension of the lateral amygdala, the BSTv area (bed nuclei of the stria terminalis), which is also activated in maternal aggression, has receptors for adrenaline and it receives input from the forebrain, which is the brain area associated with worry and deliberation. The BSTv also has connections to the hypothalamus that control autonomic responses and the release of hormones that control stress, mood, and anxiety. (Those hormones include oxytocin, corticotropin-releasing hormone, thyroid-stimulating hormone, TSH-releasing hormone, somatostatin, and dopamine). The BSTv is therefore in a position to strongly influence aggressive behavior and neuroendocrine responses involved in maternal aggression. Hormonal stimulation of this brain circuitry is the basis for the "roid rage" response in bodybuilders who take testosterone to build muscle mass and experience sudden rage as a side effect.

Looking more broadly throughout the brain, the researchers also saw other spots activated in circuits that would be necessary for the maternal aggression response. This included three other regions of the hypothalamus: (1) the medial preoptic nucleus (MPN), (2) the ventrolateral part (anatomical terminology for lower and to the side) of the hypothalamus called the lateral hypothalamic area (VMHvl), and (3) another spot to the side of the hypothalamus called the lateral hypothalamic area (LHAtu). Also a nucleus in the septal region (bed nuclei of the stria terminalis, BSTv) showed c-Fos staining. Importantly, the researchers found in rats that when the first spot that was identified in the hypothalamus was surgically removed (the PMv), the other spots were not stained for c-Fos (that is, marked as active) when a male intruder was present, which meant that the PMv is "upstream" of the other brain regions activated in maternal aggression.

It is important not to let the Latin tongue twister names get in the way. They are just geography; as is true for the Left Bank, the Eiffel Tower, and the Arc de Triomphe, until you become familiar with the geography, the names don't mean much. But the conclusion from this research is not a difficult concept. The important thing to understand is that there is a specific circuit for maternal aggression in the unconscious brain and that the behavioral response is triggered by one spot, the PMv in the hypothalamus, that is particularly sensitive to a male intruder.

## LIFEMORTS Trigger Circuits

Drawing exact parallels from rodent brains and rodent behavior to human brains and behavior is difficult, but these newly identified circuits can be roughly extrapolated into the LIFEMORTS triggers of human fear and aggression in the following way: The Life-or-limb trigger is the defensive response to injury, which is associated with pain. If something or someone inflicts pain on you, you will immediately respond aggressively to prevent further injury. Similarly, animal aggression in response to attack by a predator also encompasses the Life-or-limb trigger to defend oneself violently.

Aggression among individuals of the same species is how dominance is established in animals, from fish to primates. In humans, with our unique ability to use complex language, verbal insult achieves the same purpose: the Insult trigger, as examined in chapter 4 on the claw-hammer homicide at Carderock. The Organization trigger (social order) also involves violence within the same species to maintain social order. Not all animals are highly social, but those that are utilize violence to maintain compliance with social rules. Fear and threat circuitry activated by members of the same species (conspecifics) in animals, provokes anger and violent responses in human brains when other people do not follow social rules—if they don't stop for a red light or cut in line, for example. We have already considered the Family trigger to protect one's family in our discussion of animal research on maternal aggression.

The amygdala has several subregions, which we will refer to simply by their initials: L, BL, BM, ME, and CE. (For those interested in neuroanatomy, these correspond to lateral amygdala, basolateral amygdala, basomedial amygdala, medial amygdala, and central amygdala.)

A brief primer on the neuroanatomy of the amygdala will be helpful in understanding the fear and threat-detection circuitry in this brain region. The amygdala is an almond-shaped lump of tissue deep inside the temporal lobe of the brain. The amygdala looks different depending on how you slice through it, but in looking face-on at an MRI of the human brain at a slice through the top of the head, shoulder-to-shoulder, taken at about the level of the ears, the amygdala looks like a pair of narrowly spaced eyes, framed by the temporal lobes on the left and right sides of the brain. Each of the ingoing and outgoing connections in the amygdala forms a small knot of neurons, and these are clustered at different spots inside the amygdala. Early anatomists could clearly see these neuron clusters in their microscope slides, but they named them long before the function of any of these hubs of communication in the amygdala was known. So, unfortunately, the names (and the initials we use to refer to them) reflect little more than their location inside the amygdala. Thus we have central, lateral, and medial nuclei (clusters of neurons), as well as nuclei that sit in transitions between zones and in sub-domains within zones; for example, basolateral (lower-left), and centrome-dial nuclei (in the middle of the middle!).

By analogy, one can view the amygdala as a map of Manhattan. The different nuclei correspond to different neighborhoods, and neurons corre-spond to individual buildings. Like police tracing a phone call to a specific building in the city, neuroscientists want to trace circuits to find the specific neurons responsible for fear and threat-detection inside the amygdala. The island of Manhattan is a tongue of land bordered on the west by the Hudson River and on the east by the East River. The streets of Manhattan are laid out in a gridwork pattern, numbered from near the tip of the tongue (southern extreme) sequentially to the base of the tongue (northern extreme). Major avenues run from north to south. Fifth Avenue splits midtown Manhattan into East Side and West Side, numbering from First Avenue on the East Side to Eleventh Avenue on the West Side.

This gridwork divides Manhattan into general regions, which is helpful in communicating your desired destination to a cabdriver, for example. "Uptown" refers to property above Fifty-Ninth Street. Central Park, a rect-angular green space, is situated at the core of Manhattan, as its name sug-gests. "Downtown" refers to the tip of the tongue below Fourteenth Street. By analogy, the basolateral nuclei (BL) of the amygdala might correspond to the West Side of Manhattan, the central medial nuclei to Central Park and the Upper East Side of Manhattan; the dorsal amygdala would correspond to Northern Manhattan.

Manhattan is further divided into subregions of smaller neighborhoods, each of which has its unique identity and boundary, but the names them-selves tell you nothing about the character of the neighborhood; they are just labels. For example, Harlem is an uptown neighborhood, and Greenwich Village is one of the downtown neighborhoods. Likewise in the basolateral region of the amygdala there are smaller neighborhoods (nuclei). For exam-ple, there's the lateral nucleus (L), basal nucleus, accessory nucleus, and others, and in the centromedial region of the amygdala we have the central (CE), medial (ME), and bed nucleus of the stria terminalis (BNst).

This is a lot of terminology to swallow, but it is not a difficult con-cept. Furthermore, the terminology can be confusing because different peo-ple may refer to the same spot by different names. Just as New Yorkers sometimes argue over whether a neighborhood should be called Hell's Kitchen or Clinton, so too do neuroanatomists like to quibble sometimes over anatomical names. There are, however, many more neighborhoods in Manhattan than named nuclei in the amygdala, so it doesn't take a brain surgeon to learn their names if you want to consult a neuroanatomy book. However, for present purposes it is not necessary to learn the names of any of the various "neighborhoods" in the amygdala. The important point is that different LIFEMORTS triggers connect through different circuits passing through different communication hubs in the amygdala. From these points the circuits interconnect with other brain regions, such as the cerebral cortex and hypothalamus.

In rodents, response to predators is conveyed through connections from the olfactory system to the ME, L, and BM nuclei of the amygdala. By analogy, this roughly corresponds to Midtown and the Lower East Side (ME), the West Side (L), and somewhere around the Greenwich Village / West Village region (BM). These connections have been determined using methods like the ones that uncovered the circuitry for maternal aggression.

Fear that is learned, rather than innate fear, involves the lateral (L) region of the amygdala and causes us to associate certain cues with danger, much the way you'd learn that a certain part of your neighborhood is dangerous after dark. Damage to the L region of the amygdala prevents rats from learning that a certain part of their environment is dangerous. This is studied in the laboratory where rats learn that if they venture into certain parts of the test cage, they will receive a nasty electrical shock to their feet through the steel floor. Under normal conditioning they learn to avoid this part of the cage. The L region sends connections to the CE and to the BL. If drugs are used to block activity in the CE, rodents no longer exhibit fear in response to that area of the cage, but here's an important point: The same animals still exhibit fear that they have learned to associate with predators. This shows that different fear and threat circuits are linked to different LIFEMORTS threat triggers in the brain.

Meanwhile, if the ME portion of the amygdala is blocked, rats exhibit exactly the opposite response to fear associated with predators versus foot shock. This shows that the fear-of-pain circuit and the fear-of-predator circuits are in different modules inside the amygdala. These parallel circuits for different types of threats extend beyond the amygdala; indeed, they span throughout a large network of brain connections.

In recent studies, researchers compared a rat's brain responses to a predator—namely a cat—with its brain responses to another rat intruding on its territory. This parallels the Life-or-limb trigger for defense when the rat sees a cat, a dangerous predator, with the Environment trigger to defend one's territory when the test rat sees another rat intruding into its cage. Researchers found that introducing a strange rat into the test cage activated the dorsal medial part of the hypothalamus. Electrical stimulation of this region in humans elicits panic attacks. However, when a cat is presented, a different region is activated—the ventral lateral part of the hypothalamus. When the scientists removed one of these two regions, either the fear response to only the Environment trigger (intruder; another rat) or the Life-or-limb trigger (predator; a cat) remained intact, depending on which brain region was surgically removed. This shows that the Life-or-limb and Environment trigger circuits are separate.

"We have also some findings regarding restraint," Dr. Newton Canteras replied in response to a question I'd asked about the Stopped trigger of rage circuitry, "but they are still very preliminary." Identifying the neurocircuitry for the LIFEMORTS triggers is at the cutting edge of neuroscience research, but it is clear that different threats activate very specific and distinct circuits. It is reasonable to assume that part of the reason people find different kinds of situations frightening and are provoked to sudden rage by different triggers is that the threat-detection and fear circuits responding to different types of threats are not developed to the same extent in everyone. Thus one person could be provoked by minor insults (Insult trigger) but have a slow fuse in responding to seeing other people violate social rules (Organization trigger), and another could have the opposite tendency.

Consider the output circuits from the amygdala that drive specific behavioral actions. A rat's behavioral response to a predator and an intruder are quite different. The sight of a cat causes the rat to freeze in fear, whereas an intruding rat causes the resident rat to attack. Indeed, it's two different pathways leading out of the hypothalamus that activate the periaqueductal gray brain region in response to these two different triggers. This information travels to the conscious brain to generate the sense of fear via connections from the periaqueductal gray to the thalamus and then reaches the cerebral cortex to bring awareness of the situation.

This line of research raises interesting questions, not fully answered by experimental studies at present, about whether individual differences in a person's reaction to a threat—a mugger, for example—would depend in part on innate differences in the strength of these two circuits in different people. This would help account for how different people respond differently to an identical threatening situation—one person may freeze and the other may fight. Based on what we know about other brain functions and the circuit activity that supports them, this seems reasonable. It also seems reasonable to expect that the relative strength of the two circuits that control freezing or fighting in this experiment with rats and cats would depend in part not only on inherited genetic differences that guided the circuits' development, but also on the individual's life experiences, which could have strengthened or weakened either one. Extensive experience operating in the face of danger, such as a Navy SEAL who drills constantly under intense stress, or traumatic events in early life, might reasonably alter the balance of freezing or fighting in different individuals.

As discussed in reference to heart-rate variability, reflexive threat responses are modified by prefrontal inhibition of the amygdala, but information processing in the cerebral cortex can also activate the threat-response circuit (the "top-down control" we learned about in chapter 1). Recall what happened when terrorist Umar Farouk Abdulmutallab attempted to detonate plastic explosives concealed in his underwear on an airline traveling from Amsterdam to Detroit on Christmas Eve, December 24, 2009. Suddenly there was a loud pop and smoke billowed up from one of the passengers on Northwest Airlines Flight 253. Everyone surrounding the person fled for safety, but one passenger—Jasper Schuringa—instantly leaped over rows of seats to tackle the underwear bomber and put out the flames with his bare hands, subdue the terrorist, and thus save every person on that airplane. Why did he react this way when everyone else had the opposite response and probably the most natural reflex to a loud noise and smoke—to flee? Is it that Schuringa is preprogrammed to attack a threat, whereas others are preprogrammed to flee? In part this is likely, but a Navy SEAL that I asked points out another factor, based on his own training and experience responding immediately to countless life-risking threats. Situation analysis in the cerebral cortex to engage the defense reflex (top-down control) could have been an important factor in this incident:

"I think the guy who reacted that way understood the situation. 'Hey! I need to act offensively or something's going to happen.' Everybody else in their mind thought, 'I need to get out of here to protect myself.' They didn't understand the bigger significance of the threat. Somebody smoking [from a bomb] on an aircraft—that's not good for anybody on the aircraft."

SEAL training places great emphasis on rapid analysis (cortical activity) to activate appropriate subcortical reactions to threats.

"It is one thing if we're on the street and some guy's cooking off a bomb a hundred yards away; I'm not gonna go running over to him, I'm just gonna get everybody to back away and he can blow himself up. That's fine, but in an aircraft, I would say that person had the cognizance to think, 'If I don't act, everybody including myself is done.'

"Break that entire situation down and assess. A lot of people who might have fled might have been women and kids who assessed that 'Hey, I can't do anything if I tried. What am I going to do to this much bigger guy if I tried?' Whereas maybe that guy who dove on him had something in his past, some training—sports, or something where he understood he needed to act offensively in that moment."

Clearly the "lizard brain" simplicity of the past century has had its day. Much of this threat-assessment analysis utilizes the cerebral cortex. So does analyzing a complex social threat such as an intruder entering one's environment. The cerebral cortex is an essential part of the threat response.

## To Fight, Freeze, or Flee

A mouse is roaming around its cage curiously, sniffing and wiggling its whiskers, but there is a slender fiber-optic cable coming out of its head connected to a bank of electronic equipment. Suddenly a blue flash of light from a laser illuminates the fiber-optic cable and the mouse freezes instantly as if it had encountered a cat. Light-sensitive ion channels have been genetically inserted into specific neurons in the mouse's brain. When these channels are activated by blue light, they cause the neuron to fire electrical impulses. This method of stimulating neurons (optogenetics) is superior to using electrodes to stimulate brain tissue, because the light-sensitive channels can be inserted specifically into only the neurons that the researcher is interested in exciting without having the stimulation spread to other areas. In this case, optogenetics is being used to uncover the function of neurons in a particular spot in the amygdala. Other light-sensitive channels can be used that inhibit electrical firing of neurons. This particular mouse has had these ion channels inserted genetically into neurons in the CM region of its amygdala. With the flick of a light switch, we know precisely what behavior these neurons control. Neurons in the CM region of the amygdala cause the mouse to freeze in fear. These neurons send their signals out of the amygdala to the brain stem—in particular the periaqueductal gray area. Thus they are the action channel—or output path from the amygdala—for freezing in fright.

A subcircuit in the periaqueductal gray (PAG) called the ventral lateral nucleus is where these amygdala outputs that cause freezing send their signals. From there, signals go out to other regions of the brain, controlling movement by activating motor axons that run down the spinal cord and eventually out to the muscles. This causes the muscles to clench and the animal to freeze. A different subcircuit in the PAG is responsible for the opposite response—attack. That's what the rat will do instantly when the fiber-optic light stimulates these particular neurons located in the dorsal lateral nucleus of the PAG. Scientists have control of fight or freezing by tapping into the appropriate circuit. When they stimulate these cells optogenetically, the mouse begins to frantically run in an attempt to flee, as if seeing a ghost. Scientists can activate this region of the PAG on either the right or left side of the animal's brain, and this causes the mouse to run in circles—either clockwise or counterclockwise, depending on whether the left or right PAG flight neurons are stimulated.

It is interesting to note that the PAG is also an important structure in pain signaling. (Recall our discussion in chapter 1 that just as New York City is the financial capital of the United States, it is also the publishing capital. So too can the same brain region have more than one function.) An alternative hypothesis might have been that stimulating these neurons was causing pain rather than activating the flight behavior and instead the mouse was fleeing in an attempt to escape intense pain when researchers activated the neurons with the laser. But tests of pain threshold in these animals proved exactly the opposite. Stimulating the neurons that cause the mouse to flee in fright also caused a powerful analgesic effect, strongly *inhibiting* the sensation of pain. This was tested harmlessly by dipping the mouse's tail in hot water and measuring how quickly it flicked it away, depending on how hot the water was. The circuitry that combines threat and pain circuits explains how pain is suppressed while fleeing in fear.

The scientists also engineered a protein into neurons of interest so that they would emit light when they fired. The researchers were able to watch these neurons flash through a fiber-optic cable inserted into the brain and magnified by a microscope as the animal reacted to threats. The scientists saw these freeze-or-flee neurons in the PAG flash when the mouse was provoked to either freeze in fright or to flee.

We have now dissected the rage circuitry in the brain, explored at a cellular level how it works, and seen how essential this circuitry is to our life in a positive way, but also how these circuits releasing sudden aggression can become activated inappropriately, often with regrettable outcomes. In the next part of the book we look beyond the individual. How do the circuits in our brain that control sudden aggression and violence affect societies and other interpersonal relationships—from couples, families, and coworkers to global politics? All of these interactions are profoundly affected by the neurocircuits of snapping and aggression, both in essential ways to maintain cooperation and peace among people and in regrettable instances where these interpersonal relations become dysfunctional and result in sudden conflict. The scope of interpersonal relations will bring us now to examine closely two of the most powerful triggers of rage in the brain of our species: the Mate trigger and the Tribe trigger of aggression.

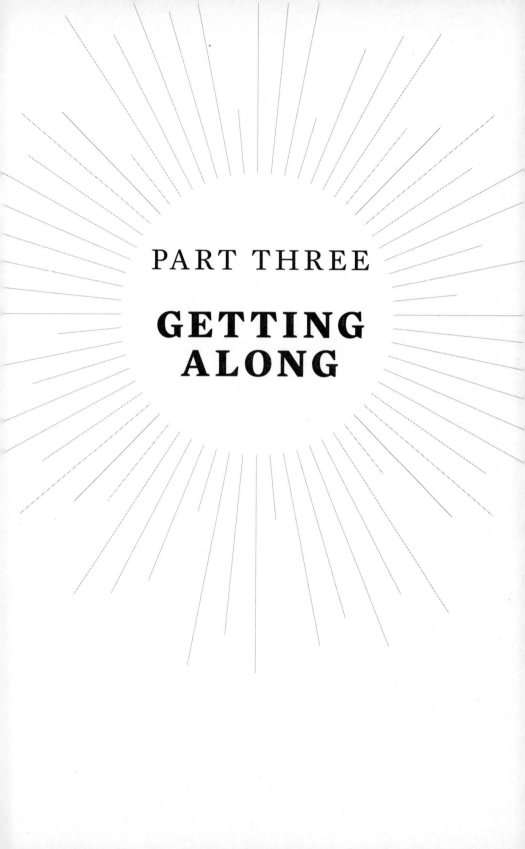

# PART THREE

# GETTING ALONG

# 10

# Sex . . . and Love

Love does not begin and end the way we seem to think it does. Love is a battle, love is a war; love is a growing up.

**James Baldwin,** *Nobody Knows My Name*

Males and females are different. This delightful fact of life makes the world go 'round. Sex differences are at the core of biology, the driving force of evolution, and in many cases fundamental to finding the best path to good health. On close inspection, nearly every part of the human body differs slightly by sex. Even the parts that would seem to have the identical function in men and women, such as hands and feet, are immediately recognizable as either a man's or a woman's. Even our bare bones can tell an anthropologist if they once supported the body of a man or a woman. It should come as no surprise that the same is true of our brain, the organ that controls all bodily functions and all behavior. So, men's and women's innate responses to danger and the ways they snap—however slightly—are different.

This chapter explores these differences, and in particular the underlying specializations in the brains of men and women that drive sex and its intimate relationship with violence. The subject raises many intriguing questions: How are sex and violence linked? What are the differences in propensity for rage between males and females, and how do the triggers for rage and the rage reaction itself differ between the sexes? Is female rage different from male rage? How does cross-gender aggression differ from aggressive rage within the same gender? Women warriors: what makes some women attracted to danger—for example, undertaking professions in the military or police? How do men and women differ in terms of conscious and unconscious brain function in fear and threat detection? Understanding how the rage circuit operates in different genders can save relationships and lives—even a great many lives.

It was an ambush. Fifty enemy fighters had launched a complex attack on the small US military convoy with a hail of heavy machine-gun fire and a cacophony of explosions from RPGs spraying shrapnel, flame, and black smoke with each concussion. The convoy was trapped in a kill zone formed by a barricade of strategically parked cars. The US soldiers were caught out in the open, taking heavy fire from the enemy dug in and protected by irrigation ditches bordering the road.

Twenty-two-year-old Sergeant Hester in the Military Police squad for the Kentucky National Guard snapped into action. Leading a small team of soldiers Sergeant Hester charged through the kill zone, maneuvered into a flanking position, and assaulted the trench line, chucking hand grenades and firing M203 grenades to cut off the enemy's escape route. (M203s are explosive grenades launched from a tube beneath the barrel of the M4 assault rifle designed for combat in close quarters.) Hester, teaming up spontaneously with Sgt. Timothy Nein, cleared two trenches. Charging into the direction of fire, Hester shot and killed three attackers. When the firefight ended and the dust began to settle, three US soldiers from the ambushed convoy lay injured, but there were no fatalities. Twenty-seven Iraqi soldiers were dead. Six were wounded, one was captured, and the remaining enemy combatants fled. Sgt. Leigh Ann Hester was awarded the Silver Star for valor—the first woman to receive the honor since WWII, and the first woman ever to receive the award for heroism in direct combat.

"Your training kicks in and the soldier kicks in," Hester says. "It's your life or theirs. . . . You've got a job to do—protecting yourself and your fellow comrades." These are the same powerful motives that drive any soldier in combat (the Life-or-limb and Tribe triggers).

To state the obvious, men and women do not compete together in Olympic weightlifting. But at the same time, there is not a separate IQ scale for the two sexes. Neither are bravery, patriotism, or determination qualities unique to only one sex. Many women have been awarded the Silver Star for valor before Sergeant Hester received that honor. They earned it for their heroic action as nurses in combat zones, serving their country bravely before women were permitted in the armed forces.

The defensive neurocircuits of rage and aggression are built into the brains of both men and women, but the circuits have been honed by evolution to best suit the biological differences between the sexes, and also to accommodate gender differences imposed by society on men and women.

In 2007 an eleven-year-old girl typed her heart out on her blog. Not so unlike others her age, except that her heart was aching for something far beyond the typical preteen dreams of other girls. She yearned for a better world. She saw injustice and could not accept it. As a child and a female, her only weapons in this battle were her courage and commitment. In the Taliban-controlled Swat Valley of Pakistan, strict sharia law denied girls education, stifled their liberties by secluding women in their homes, and imposed forced marriages on them at an early age. Malala Yousafzai's outspoken views began to attract widespread attention in the media, and this threatened the Taliban's beliefs, authority, and mission. In defiance, she openly traveled to and from the Khushal Girls School, which her father founded.

"For my brothers it was easy to think about the future. They can be anything they want. But for me it was hard, and for that reason I wanted to become educated and empower myself with knowledge," she said.

"I had a terrible dream yesterday with military helicopters and the Taliban. I have had such dreams since the launch of the military operation in Swat," she wrote in her diary on January 3, 2009.

Nine months later, on October 9, 2012, while she was riding home on the school bus from classes, two armed men stopped the bus filled with children. "Who is Malala?" one of the men asked.

"I heard the firing, then I saw lots of blood on Malala's head," one of the other students recalled. "When I saw that blood on Malala, I fell unconscious."

Malala and two of her classmates were shot. The bullet entered Malala's forehead above her left eyebrow, pierced through her cranium, penetrated through her neck, and lodged in her back. Parts of her brain controlling speech and movement of her right arm and leg had been damaged. Her facial nerve had been severed, paralyzing the left side of her face. She was in grave danger of death.

After receiving the best possible care in Birmingham, England, where her family emigrated to escape the Taliban's death threats, Malala, now with a titanium plate replacing her shattered skull, endured a slow and painful recovery. The young girl learned to cope with disfigurement and disability, but her commitment to her struggle against injustice did not waver. Two years and a day later, Malala Yousafzai, at the age of seventeen, became the youngest recipient of the Nobel Peace Prize in history. When the award was announced, Malala was in school. While the world media awaited her reaction, Malala insisted on finishing her chemistry class. Then she went on to history and English class, and only after school was over did she respond to reporters. Speaking graciously of the honor of sharing the Nobel Prize for Peace with Kailash Satyarthi, age sixty, an Indian citizen who worked to free thousands of child laborers, Malala was taken by the symbolism of the award uniting in peace two countries that have been fierce enemies for decades. "It gives a message to people of love between Pakistan and India and between different religions," she said. "One [Nobel recipient] is from Pakistan, one is from India. One believes in Hinduism, one strongly believes in Islam."

Here we see the life-risking rage of commitment against injustice is a core trait of human beings. Regardless of sex or age, this is the way our brains are wired. For Malala it was the Organization and Tribe triggers that propelled her actions. Tragically, it is not possible to simply dismiss the acts of the two gunmen who would shoot innocent children on a school bus as an aberration of mental illness. These horrible acts of violent rage were driven by the same two defensive triggers of rage in the Taliban's minds: the Organization and Tribe triggers. This is the double-edged sword of *Homo sapiens* neurobiology that can slice both ways, for good or for evil. Even in England, Malala and her family remain at risk. "Some people are silent," a businessman in the community told reporters about the neighborhood's reaction to the Nobel Prize. "They don't like her and her father, but others are quiet due to the possible threat from the militants."

It is foolhardy for a 135-pound man to get into a physical confrontation unnecessarily with a 250-pound adversary, even more senseless for a 135-pound woman to do so, given the much greater physical strength that biology has bestowed upon adult males. For this reason, the innate responses of men and women to threats differ, as do the brain circuits that are specialized to best serve men or women in danger. The marked differences in physical strength and in reproduction between men and women have guided evolution of the human brain for the greatest success and survival of each sex over eons, but threat detection, response to danger, and aggression are necessary for any individual regardless of sex.

There are two important consequences of this biological sexual dimorphism: The sex-based differences in the brains of men and women give each sex a different type of advantage in dealing with dangerous threats. Secondly, as environments change, so do the forces that guide biology and behavior, and the modern world is utterly transformed from what it was only a few generations ago. Breaking out of a cycle that encompasses every other form of life on the planet, human beings have developed the ability to control their reproduction through technology. This very recent development has enabled women to control their destiny in a way that was previously impossible. At the same time, other technological developments have made the differences in physical strength between men and women irrelevant in many situations in the world we live in today. It does not matter at all whether the person piloting the F-16 fighter jet scrambled on September 11, 2001, to save the United States Capitol had two X chromosomes or one X and one Y, just as gender made no difference to the Taliban fighter on the receiving end of Sergeant Hester's M203 round. Biology tells us that the present is the product of the past, but not a repetition of it. The future will develop from its footing in the present, but it will not sustain it. In Israel, a country where men and women have very separate roles assigned by traditional religious teachings, women serve in large numbers in the armed forces. Women comprise one-third of the active-duty Israeli Army, compared with about 14 percent in the United States. About 3.3 percent of all direct-combat roles in the Israeli Army are filled by women fighting together in squads with men. Sapir Yehudain, a twenty-five-year-old woman who served in combat in the Israel Army, says, "They [her male comrades] understood that we could do everything they could too."

Still, women in everyday life are at risk of threats and aggression from men.

# When the Police Are Called

## WISH LIST
*Donations Requested from the House of Ruth, a Women's Shelter*

Baby wipes

Pampers sizes 3, 4, 5, and 6, Pull-ups sizes 4T, 5T and 6T

Crib sheets

Infant and Toddler Sleepwear

Baby lotion, baby cornstarch powder, diapering cream Baby formula, baby food (stages 2 and 3), baby cereal baby snacks

Individually wrapped snacks and juice boxes for children coming to therapy after school

Baby bottles 8 to 9 oz., pacifiers, aspirators, thermometers, baby Tylenol and Motrin

New High chairs

Strollers

New convertible car seats only (no used).

School Supplies . . . Toys . . . Arts and craft supplies . . .

The list goes on, documenting, as if artifacts strewn about after a disaster, the shattered lives of babies, women, and children fleeing their homes to seek safe refuge in shelters like the House of Ruth to escape violent men who attack their spouses, girlfriends, and children. The clinical euphemism for this horrible rage is "domestic violence" or "domestic abuse." A study in England and Wales found that 30 percent of the female population has experienced domestic violence.

Objectively disturbing insight into domestic abuse was played out before a nation who saw a security-camera video catch the brutal reflexive violence of the star NFL running back for the Baltimore Ravens, Ray Rice, slugging his girlfriend in the face inside an elevator and then dismissively dragging her unconscious body out into the hallway after a verbal altercation. The athlete's powerful left hook to his girlfriend's jaw was so instantaneous, even in slow motion it is difficult to catch it. This brutal and inexcusable violent act was not a conscious behavior. Rice snapped. As so often happens, both the perpetrator and victim are soon shocked by the behavior and seek reconciliation. "I love my husband. I support him. . . . I want people to respect our privacy in this family matter," Janay Rice said when the incident became public after the two were married.

This security video shows why current approaches that seek to control snapping in domestic violence are often fruitless. The men who do these things often regret it immediately. Well-intentioned techniques offered to suppress the anger in domestic violence are not enough. What men (and sometimes women) must recognize is *why* they are suddenly experiencing a fuming rage of anger welling up inside them toward the person they love. If the individual can recognize that the rapidly unfolding incident is pushing on one of the deadly LIFEMORTS triggers, then two things should happen immediately: First, the cerebral cortex will register a potentially violent or life-threatening trigger, and second it will evaluate whether or not the situation is indeed a life-or-death threat that these deadly subconscious circuits of defensive rage are designed to counter with violence. If violence is not called for—for example, you are in an elevator in a screaming argument with your girlfriend—then the rage will quickly subside, because you will understand that it has to. Snapping violently doesn't help you in your relationships, doesn't work on a broader social level, and doesn't do you any good biologically. You will remain angry, but less so, because you will know why you are angry at a biological level. This will strongly activate the prefrontal cortex to inhibit the automated snap of violence programmed into the hypothalamus. At that point the opportunity to engage the conscious mind in deliberation, employ psychology, and utilize techniques of anger management can be brought to bear.

## Beyond Domestic Disputes

Some 100,000 jubilant people flooded Tahrir Square in Cairo, Egypt, in a spontaneous celebration of freedom—Egypt's dictator Hosni Mubarak had just fallen from power. CBS news reporter Lara Logan waded into the crowd with her news team to capture the historic transition. "It was unbelievable. It was like unleashing a Champagne cork on Egypt," Logan describes in her *60*

*Minutes* interview. "I've got to be there, because this is a moment in history that you don't want to miss."

Her video captures the roar of the crowd screaming in excitement, chanting revolutionary slogans, bullhorns barking, hands clapping, and automobile horns honking while flashes of smoke and fireworks, camera strobes, and improvised butane-tank flamethrowers lit the mob in celebration. As the world would soon learn, Logan, a seasoned war correspondent with blond hair and green eyes, the mother of two young children, was viciously attacked and sexually assaulted by a mob of men as she attempted to report the breaking news. The attack began the instant her TV camera battery died, plunging the reporter and her team into darkness. From the instant it began it was a sexual assault.

"I'm screaming," she said, "thinking if I scream, if they know, they are going to stop. You know, or someone's going to stop them or they will stop themselves, because this is wrong. And it was the opposite. The more I screamed, it turned them into a frenzy."

Stripped naked, brutalized, and sexually assaulted by hordes of men for half an hour, she would likely have died if she had not eventually been protected by a small group of Egyptian women and finally rescued by the Egyptian Army. She required four days of hospitalization to recover from her physical injuries.

Was this a case of politically motivated aggression and violence? Was Logan targeted because she was a reporter from the United States, or was she sexually assaulted because she was an attractive woman? Were her attackers a gang of criminal thugs infiltrating the crowd or members of Mubarak's criminal security force? Alternatively, was this a riot of ordinary Egyptian men caught up in a mob frenzy of excitement and adrenaline, resulting in gang rape? Did mob mentality reduce these men to animals, like the boys in William Golding's *Lord of the Flies*? The fundamental question about this rage attack is whether the violence was sexually motivated or instead, if sexual violence was used as a way to brutalize a victim for political or ideological reasons.

Because of her celebrity, Logan's sexual assault and battery attracted world attention, but she was not the only one.

"The atmosphere was one of jubilation, excitement, and happiness as I walked, accompanied by two male companions for safety along Kasr El Nil bridge. . . . Women, children and fathers smiled, waved, and cheered happily at the camera, calling out the widely used phrase 'Welcome to Egypt!

Welcome!'" wrote twenty-one-year-old British journalist Natasha Smith, who was in Tahrir Square to film a documentary.

"In a split second, everything changed. Men had been groping me for a while, but suddenly, something shifted. I found myself being dragged from my male friend, groped all over, with increasing force and aggression. I screamed." Her attack matched Logan's gang assault. "Men began to rip off my clothes. I was stripped naked. Their insatiable appetite to hurt me heightened. These men, hundreds of them, had turned from humans to animals."

In an article Smith published a year after her assault, she states that there were forty-six cases of sexual assault on the evening of Sunday, June 30, 2013, in Tahrir Square during mass celebrations on the anniversary of the revolution. She cites statistics by Amnesty International researcher Diana Eltahawy, but Amnesty International believes the actual number is much higher. According to Human Rights Watch, at least ninety-one women were raped in Tahrir Square over four days beginning on June 30, 2013: "Yasmine El Baramawy, a 30-year-old musician, told Human Rights Watch that she was raped and assaulted on the evening of November 23, 2012, for 90 minutes after going to a demonstration in Tahrir Square. After men knocked her to the ground, they ripped her clothes and cut her blouse and bra. As the attack continued, the group around her increased from about 15 to more than 100 men."

As El Baramawy reported, "I looked up and saw 30 individuals on a fence. All of them had smiling faces, and they were recording me with their cellphones. They saw a naked woman, covered in sewage, who was being assaulted and beaten, and I don't know what was funny about that. This is a question that I'm still thinking about. I can't stop my mind from thinking about it."

Were all of these men psychopaths? The facts suggest an uncomfortable conclusion that these attacks on women were sexually motivated by men in an environment where such animal brutality is tolerated. In Egypt, 99.3 percent of women have experienced some form of sexual violence. Sexual violence is used against women by police in Egypt and other places as a most brutal means of attack, and sexual assault is also used by men in the general population for cruel violence against women. Egypt is not exceptional in this respect, as the tragic story of a young woman gang-raped and tortured by six men, including one juvenile, for an hour on a bus in Delhi, India, on December 16, 2012, illustrates. "Burn them alive," the twenty-three-year-old paramedic trainee scribbled in a handwritten statement about her attackers while semi-comatose and fighting for her life in the hospital before she died of her severe injuries. The attack sparked international outrage and exposed Indian society as a place where sexual violence against women, often by gangs of men, occurs frequently but is not forcefully prosecuted.

So-called honor killings of women who are perceived to have brought dishonor on the family in countries like India and Pakistan still take place today; some are even committed in the United States by immigrants from countries where honor killings frequently occur. In 2014, Farzana Iqbal, a twenty-five-year-old pregnant woman, was beaten to death publicly by her father and her other family members in front of a court in India because she had married the man she loved rather than the man who had been prearranged for her by her parents. Iqbal's husband, Mohammed Iqbal, said police did nothing during the fifteen minutes of violence and murder that took place right outside the Lahore High Court.

As my daughter and I were being chased through the streets, back alleys, shops, and restaurants in Barcelona trying to elude that gang of pickpockets determined to extract revenge, I felt a powerful sensation that I had never felt before—I felt like prey. In the midst of the chase with my daughter at my side, a sudden revelation overcame me: This is a feeling that *all* women must have—the sense of being prey. Even as we ran to escape the gang members this sorry realization stirred a sense of pity in my gut for the plight of my two daughters and, I suspect, all women. The possibility of being sexually assaulted never crosses the mind of a man, but that reality is never far from a woman's mind. Quickly it became obvious to me that my daughter was very good at the role of eluding predators. Kelly would consistently spot the bad guy before I did and would draw him to my attention. During the two-hour chase I came to rely on what was clearly her superior skill at spotting the predator as my mind raced to devise tactics to evade them, and to engage them physically if that became necessary to defend ourselves. It was a powerful partnership.

Studies show that both men and women recognize an angry male face faster than an angry female face. Presumably this is because males are larger and more aggressive, and therefore men present a greater physical threat. Much the way our brain circuits are tuned to spot snakes instantly, so too has our brain's perceptual system been fine-tuned to spot the face of an angry man as quickly as possible. The emotions of anger and fear are portrayed differently in facial expression, conveying instantly and unambiguously the mental state of any other person regardless of race or language. Interestingly, both men and women detect angry faces faster than fearful faces. This subtle refinement in the brain's perceptual systems to quickly distinguish these two similar expressions makes sense. Both expressions signal impending danger, but an angry face originates from the source of an immediate threat, whereas a fearful face warns of danger in the vicinity. A quicker response to anger gives an individual an advantage in responding appropriately.

Other studies show that women are faster than men in recognizing nonthreatening facial expressions such as happiness or sadness. This makes women more attuned to socially relevant expressions. Men, however, are quicker at recognizing angry male faces than women are, which is consistent with their direct engagement in activities requiring physical aggression and defense that have been their domain throughout the course of human evolution. Our pursuers in Barcelona were not expressing outright anger on their faces; they were hunting us down, and somehow, Kelly always seemed to spot them closing in on us before I did.

Violence against women is a deplorable criminal act. It is universally condemned by all decent men and women and it is an inconceivable behavior for most men, but sexual violence is a human behavior that is committed in a state of rage, so we are forced here to confront it analytically. All behaviors are the product of the brain. Looking at this species, *Homo sapiens*, in an objective fashion the way a zoologist would observe any animal behavior, we can see this violent human behavior does occur in certain circumstances and in certain environments. There is no question that sex and violence are linked. Sex (the Mate trigger) is one of the most powerful triggers of rage.

Reflecting on his vast experience with crime, Clarence Darrow, the famed defense attorney noted for defending high school teacher John Scopes for violating Tennessee law against teaching evolution in public schools, wrote, "It seems to me to be clear that there is really no such thing as crime, as the word is generally understood. Every activity of man should come under the head of 'behavior.'" Crimes are human behaviors that are prohibited by law and enforced by sanctioned violence by societies to maintain social order or to preserve power by rulers over populations. Without this social regulation of human behavior, human societies degenerate into violent chaos.

Recent statistics show that 30 percent of all Internet traffic is pornography. Porn sites have more visitors each month than the Internet giants

Netflix, Amazon, and Twitter combined. Porn industry revenue in the United States is twice the combined revenues of ABC, CBS, and NBC—$12 billion annually, despite the growing trend for free pornography. Some 40 million adults in the United States regularly visit Internet pornography websites. As many as 10 percent of viewers report uncontrollable addiction to Internet porn.

Sex is a normal biological function, so the appeal of pornography on a neuroscience level is not surprising, but 88.2 percent of pornography contains physical aggression, and 48.7 percent of pornography contains scenes with verbal aggression. The targets of physical and verbal aggression in pornography are overwhelmingly female, 94 percent.

From the viewpoint of brain function, addiction to pornography is no different from addiction to alcohol, drugs, or gambling. All of these addictive behaviors activate the release of the neurotransmitter dopamine from the brain's reward system (in the dorsal striatum). Both sexual activity and physical aggression stimulate the same reward center to release dopamine and stimulate the brain's pleasure circuits.

Playing video games also activates these pleasure circuits and triggers the release of dopamine. By monitoring brain activity with fMRI, researchers have found that the medial forebrain, nucleus accumbens, amygdala, and orbitofrontal cortex—all part of the brain's pleasure networks—become active during video gaming. Interestingly, activation of these regions during video gaming is much stronger for men than for women. (This may reflect the design of the games in general, however, which tend to be targeted toward males.)

Anita Sarkeesian, a video gamer who writes critically of the violent sexual degradation of women in video games, was forced to flee her home after she was hounded by violent threats of rape and murder because of that criticism. In October 2014 she abruptly canceled her scheduled speaking appearance at Utah State University after someone threatened a mass shooting at the event.

"There's a toxicity within gaming culture, and also in tech culture, that drives this misogynist hatred," she says. Concerns over the harmful effect of violent video games on children and adults drives movements for new legislation to restrict violent and sexual content in video games, but these debates perpetually revolve around the eternal question of whether art motivates human behavior or reflects it.

What these facts show is that sex and violence are linked in the human brain, because both activities strongly stimulate the brain's reward and pleasure systems, especially in males. In studying aggression in mice, Craig Kennedy, professor of special education and pediatrics at Vanderbilt University, provides evidence that violent aggression can be addictive because it activates the dopamine reward system. In his experiments, an intruder male mouse was placed into the home cage of another male mouse, which immediately provoked an aggressive attack. This is the Environment trigger of violence, but here is the surprising finding: When the researchers then trained the home mouse to press a lever to permit the intruder mouse to return to the cage, after fighting with the intruder, the mouse would press the button to invite another fight, just as if it were delivering cocaine or another drug that stimulates release of dopamine in its brain. When the researchers blocked dopamine action in the mouse's brain with a drug, the mouse stopped pressing the button to seek another battle.

"Aggression occurs among virtually all vertebrates and is necessary to get and keep important resources such as mates, territory and food," Craig Kennedy said in a news report on his research. "Almost all mammals are aggressive in some way or another. . . . It serves a really useful evolutionary role probably, which is you defend territory; you defend your mate; if you're a female, you defend your offspring."

But this brain circuitry can lead the brain to seek out violence, to pick fights for no apparent reason other than to derive satisfaction of the rewarding feeling that comes from the aggressive encounter. Interestingly, the neurotransmitter serotonin is involved in both sexual gratification and violence.

The US armed forces have been scandalized by persistent news stories about sexual assaults on enlisted women. Reported sexual assaults in the US armed forces surged 50 percent in one year (2013–2014). According to the Pentagon, the Marine Corps recorded an 86 percent increase in sexual assault reports in 2013. The Army saw an increase of 51 percent, compared with a 46 percent increase in the Navy and 33 percent increase in the Air Force. In all, 5,061 people reported being sexually assaulted in the armed forces in the year 2013, and the majority of victims were younger than twenty-five.

The scandal provoked outrage among many in Congress and among the general public. But sexual abuse in the United States is not a problem that is unique to the military. The same year, fifty-five colleges and universities reported having open sexual violence investigations under way; this list included the most elite private schools in the country as well as public institutions. This led the White House to launch a task force to find ways to address the problem of sexual assault of undergraduate women on the nation's campuses. "Colleges and universities can no longer turn a blind eye or pretend rape and sexual assault doesn't occur on their campuses," Vice President Biden said in announcing the White House effort.

But the problem of sexual assault is not confined to the military or colleges. A survey published in a scientific journal in 2014 found that more than half of respondents reported that, as female scientists, they had experienced sexual harassment or sexual assault in the workplace. A surprising 64 percent of the female scientists reported being sexually harassed, and 20 percent stated that they had been victims of physical sexual assault in the workplace. Younger women are at highest risk, with 90 percent of the female scientists who were assaulted being trainees, undergraduate, graduate, or postdoctoral students. These three environments (military, universities, and laboratories) and the people who inhabit them differ dramatically. What they have in common is an abundance of young women working together with men.

## Self-Harm

"I have cut myself only once in the last twelve years." The words that come from the highly successful, intelligent, and extremely motivated petite blond woman shatter the impression that crystallizes from her outward composure.

"I started self-harming at seventeen with a pencil-sharpener blade. My home was incredibly difficult and there were young children that needed as much protection or buffering as I could give them. I would hurt myself to release the tension, so that I could go back and carry on absorbing as much as I could. I was alone and so learned that way of coping with my own emotions. I continued to self-harm, but by then it was more serious as I was using a scalpel until I was thirty.

"I can feel that it is building up—can feel it coming. It will reach a pitch where it almost cannot be ignored, like going to the toilet, like being absolutely starving. It just gets to a point where it has to be acted on. Something will happen and that will trigger something. . . . Rage can be a huge part of it.

"The really crucial part of it is that self-harm works like no drug I have ever seen. It brings people down. They are softened. They feel better. They are calm. And depending on some degree to the degree of the injury, that [relief] can persist for quite a few days."

"Does this feeling happen *after* someone responds?" I ask her, wondering if perhaps the frenzy of people responding to the medical emergency might fulfill a possible starvation for personal care and attention.

"No. Just the act of doing it, and sometimes seeing the wounds. I think it can produce a shock. A genuine biological shock—drop in blood pressure. There is something very validating, I think, about having seen [the bleeding cuts]. That degree of injury—that is self-validating: 'This is how bad it is. Look what's happened.'"

According to a recent study, 1 in 12 young people, mostly girls, engage in self-harming, such as cutting, burning, or taking life-threatening risks. Many carry this self-destructive behavior into adulthood. Self-harming is one of the strongest predictors of who will go on to commit suicide.

"It's literally like a bomb is going to go off in your head," the young woman explains. "Something will happen; it reaches a pitch where serious self-harm has become inevitable. I have contempt and rage for myself. When I hurt myself the thoughts are that I should have known better than to ever try and make it different. That I am so stupid. What did I expect? I am nothing. I should know that by now. In that space I hate myself for trying and I hate myself for hoping. It's pretty bleak. When I am self-harming I am in a cold rage. I do it. Look at it. Judge if it's enough and then carry on until I am literally sated, totally soothed. The peace I feel is brilliant. I can manage anything then.

"It would be very interesting to know what has happened in my brain! It feels absolutely dramatic. (It looks pretty dramatic too!)"

Her question could be easily answered with an fMRI brain scan, but that will never happen. It is not ethical to harm people (or to let them harm themselves) simply to study the effects. That bright line in scientific ethics cannot be crossed. But there is little doubt what an fMRI would show. As has been discussed earlier, the sight of blood does induce a mild physiological shock response. The slowed heart rate, sudden drop in blood pressure, and increased tolerance to pain would indeed act instantaneously—faster than a drug—to squelch anxiety and stress.

But this is an act of violence driven by unconscious forces in a state of cold rage. The behavior does not come as a conscious decision within the cerebral cortex; it wells up from emotional unconscious circuits deep in the brain that control behavior by issuing powerful urges. To carry out an act of violence in a sudden rage state, one of the nine LIFEMORTS triggers must be tripped. Self-harm is not rational.

In addition to identifying the trigger of violence, other important questions about self-harm arise: Why violence and not some other behavior in response to the stress? Why is self-harm so common in adolescent years? Why are girls much more likely to engage in self-harm? To what extent does a child's environment during rearing contribute to the violence directed at oneself?

People like the woman who shared her experience with me frequently describe the self-harm as being driven by an intense sense of shame or inadequacy. This falls within the Insult trigger of violence. Shame is a stressful emotion that develops from comparing yourself with others, or with your own expectations of your personal standards, and finding yourself inadequate. Shame is a personal insult. It is all about establishing your rank in society, and this explains why the behavior that is evoked is violence. Humans, as with other social mammals, are utterly dependent on society for survival. An individual's place in the social organization is one of the most important factors in one's quality of life and well-being. Most commonly, physical aggression is how rank in society is established in the animal world, especially for males.

Haley Peckham, a neuroscientist and psychiatric nurse trained in the UK, cares for children between the ages of twelve and seventeen in an adolescence public health ward. "We have a huge amount of self-harm/suicides. This is a common problem for adolescents." The remainder of her patients are kids with drug-induced psychosis and others with personality disorders or apparent psychosis.

I asked her what causes the kids in her ward to snap aggressively and lash out violently against others or to inflict harm on themselves.

"Often it can be another young person in the ward getting attention, by which I mean—we have this thing called 'code gray,' which is an aggressive unarmed threat. We have lots and lots of them where we are. So what happens is, security turns up and nurses turn up and there is a lot of effort put into talking someone down. That can very frequently send off other kids into—not usually aggressive, but quite often self-harming-type behaviors. There can be many reasons for this, but it can be through feeling not important enough to get *that* many people to be able to come to you."

"What types of people do this?" I ask.

"They are people who are very, very unlikely to become aggressive with other people. They are people who might find it very difficult to express their own needs, to be assured enough to ask for what they want, so they tend to be much more inhibited in that way. So it tends to get directed toward themselves. They wouldn't think of hurting someone else. Nevertheless, if what they did to themselves, they did to someone else, they would be in jail."

This explains why females more frequently direct their violent impulse in response to the Insult trigger toward themselves rather than toward others. A male, equipped biologically by evolution with physical strength and aggression, who experiences the same trigger of rage as the woman would more likely engage in a brawl. This brutish behavior is how a man will more frequently react to the Insult trigger of violence. This makes for male bullies and fistfights. Such men seek violence and direct it toward another man to achieve the same satisfaction and release of stress that a person experiences from self-harm. Blood, injury, and the "high" of overcoming social defeat provide the same physiological and emotional outcome.

"I've only ever had chairs thrown at me by boys," Peckham says when I ask her if she sees a difference in snapping behavior between girls and boys in her care. "I think I am quite lucky because I am small. I'm gentle, and I am a girl. So I think that keeps me out of a whole lot of violence. There is nothing to be achieved in picking on me.

"I've been around much more self-harm with girls. Head-butting [against a wall], or often you can see that they are in this state—vibrating with needing to get something out, literally vibrating but not knowing what to do— pacing or just bashing things. If you can tolerate being around someone who is doing that, banging their head or punching, punching, and they are hurting themselves, that's OK, but when you lay hands on someone, then you become fair game. They are trying very hard not to direct it to you and they are hurting themselves. Then you intrude on them hurting themselves, and then the violence can be directed at that person, or directed at the person who has triggered it or anyone who has witnessed the shame."

This also explains why self-harm peaks in adolescents. Adolescence is a period of life when an individual is seeking to establish his or her place in society, and at this age one's peer group is the most important aspect of their environment. Another critical environmental factor is their home life and upbringing.

"When you talk to these kids I hear so many tales of emotional, sexual, physical abuse. Of fractured families, of school bullying, Facebook bullying, and lack of—lack of *nurture* and good attachment," Peckham says.

All of these environmental factors relate to one's feelings of self-esteem and comparative success in family and social life. This is the period of life when the brain is being developed according to experience.

"We live in a different environment [from our evolutionary past where violence was necessary and appropriate for survival], but the pressures are still enormous; where shame is still . . . Look, *everyone's* climbing, *everyone's* competing. There is still *huge* pressure and *huge* competition and there are still people who are on a good cycle, where they have a good start and they can make their way, and there are other people who are absolutely flat-out trapped, and it really doesn't matter how much effort they make, because they are pretty much trapped."

Quite often, it is females who are inclined to feel trapped, because of social and biological factors.

## A Guy Thing

Despite the enormous number of elements that enter into criminal violence, gender is by far the most important factor. Nearly all violent crime is committed by males. Crime varies greatly in different locations, but taking my state of Maryland as an example, the 2012 crime statistics reveal rage circuit triggers in action.

Police are able to identify the offender, charge them, and take them into custody in 55 percent of all violent crimes in my state. This is termed the "clearance" rate; the prosecution rate is significantly lower. For juveniles involved in violent crime the clearance rate is 13 percent, meaning that the vast majority of violent juvenile offenders are not identified and taken into custody.

Murder accounts for 1 percent of all violent crime in Maryland. Of the 228 murders in the state of Maryland that were cleared in 2012, 93 percent of the killers were male. Five percent were juvenile—that is, under eighteen—but young people do show up in significant numbers in this data: 45 percent of murder victims are between the ages of eighteen and twenty-nine. Handguns were used in 73 percent of murders; a knife was used in 13 percent. Contrary to what one might have imagined from television dramas and news reports, drug-related murders account for only 3 percent of the total. The rate at which family members are murdered is four times higher, 12 percent of total murders. Five percent of the murders that take place in families are committed by cohabiting husbands, wives, boyfriends, or girlfriends. Eighteen percent of murders are committed by acquaintances who are not family members. Strangers account for only 17 percent of murders.

Robbery is the most common circumstance leading to murder, but heated arguments are nearly as common. In half of the murder cases under investigation there is not enough information to determine what instigated the homicide. Consider that statistic: Half the time after someone's life ends violently at the hand of another human being in murder we can find no reason for the death of the person.

We turn now to other violent crimes. Rape by force accounts for 88 percent of all rapes. Of those arrested for rape in 2012, 11 percent were juvenile, 60 percent black, 39 percent white (Hispanics included), and 1 percent Asian or American Indian. Fifty-two percent of robberies were committed on the street, while only 1 percent were bank robberies. Guns were used in 43 percent of robberies. Robberies are cleared in only 14 percent of cases involving juvenile offenders. The breakdown by sex shows the perpetrators are 89 percent male, 11 percent female, and 26 percent juvenile. Gender is again the most important factor in these crimes of violence.

Aggravated assault, in which one person causes severe or aggravated bodily injury to another person, shows similar trends. Aggravated assaults account for 58 percent of all violent crime. Fourteen percent were assaults with firearms, 27 percent were assaults with a knife. The difference between rates for these two weapons used in murders versus aggravated assaults likely reflects the greater lethality of firearms, thus putting the results of the violent attack in the murder category when a gun is used. When other weapons are used in assaults, 21 percent are with hands and feet, and 39 percent are with other weapons. The breakdown of persons arrested for aggravated assault is 76 percent male, 12 percent juvenile. Again, with respect to this chapter on sex and violence, these statistics show that most violent assaults are committed by males, and that being male is the factor, of all the factors measured, that is the most common among perpetrators.

The vast majority of domestic-violence crimes involve assault, with 16,269 assaults out of a total of 17,615 domestic-violence crimes reported in 2012. There were twenty-two murders in domestic disputes in Maryland in

2012. Most domestic assaults occur between six p.m. and one a.m., with a peak at eleven p.m. Thirty-six percent of domestic violent crimes are reported on Saturday and Sunday. Most occur on Sunday (3,360 out of 17,615). The victims are typically female—13,029 women assaulted out of 17,615 incidents of domestic violence (74 percent). The rates do not vary much by race, but most victims were between the ages of fifteen and forty-four (59 percent). The victim is typically a wife or female cohabitant living together with a male, rather than in an estranged relationship. Alcohol use was involved in 4,882 of the 17,615 cases of domestic violence in 2012—far from the majority.

In analyzing the circumstances that sparked the domestic violence, infidelity was by far the most frequent cause: the Mate trigger. Next were disputes involving offspring (Family trigger), followed by arguments over money and property (Resources trigger). Surprisingly the table of statistics shows disputes over television as a cause of domestic violence. Tussling over the telly generates more domestic disputes than gambling, sex, or violence erupting during reconciliation attempts. This can be taken as another reason to favor books over TV and may be a reason to keep the television out of the bedroom. The struggle over the TV is the Insult trigger of interpersonal dominance.

Evolutionary history and the physical differences between males and females account for the prevalence of violence committed by males. In a physical contest with much stronger males, females will likely lose. Females therefore would be wise to avoid direct physical battles with males and to use indirect methods of aggression instead, such as gossip and sabotage. This does not explain why there are so many male/male physical assaults and murders and so few female/female physical battles, unless the behavioral strategies females have had to evolve to cope with male aggression have resulted in differences in the female brain that support a different response to interpersonal threats other than violence. There is good neuroanatomical evidence for such differences in the male and female brain. Men's brains are tuned like those of a predator, and women's more like prey.

This is not to say that women are not capable of violence, as we saw in Sgt. Leigh Ann Hester's battlefield bravery. Females respond to all the LIFEMORTS triggers of rage with rising internal anger and often violence, but some triggers are more powerful instigators of snapping in violence for men than for women. Certainly women will protect their young and family with violence if necessary. They will respond to a violent attack with violence to save their own life. Domestic violence and violence related to partners are common for women. "Hell hath no fury like a woman scorned," as the saying goes.

Professor Stefan Andersson and his girlfriend Ana Trujillo spent most of the evening at a local taco restaurant where they shared a couple bottles of wine and a few tequila shots. Another man approached Trujillo at the bar and offered to buy her a drink. That sparked an argument between the couple (Mate trigger). The two left the restaurant but the argument boiled up into a fight in Andersson's apartment around two a.m. Neighbors were awakened by the noise of fighting and furniture sliding over the floor. Trujillo fatally stabbed Andersson twenty-five times in the face and head with the heel of her five-and-a-half-inch stiletto shoe, killing him. "It looked like something out of a horror movie," Assistant District Attorney Sarah Mickelson said. "There was so much blood, the police officer thought Stefan had been shot in the face."

Trujillo claimed she acted in self-defense, but she was convicted of murder and sentenced to life in prison.

Female-on-female violence occurs between women who know each other in 88 percent of cases. It is most likely to occur in outdoor locations (60 percent) and in the presence of others (91 percent). A survey found that most of the women involved in this type of violence have experienced disorder in their lives, had limited resources, and were the victims of violence themselves in multiple relationships. Sexual abuse is a common source of psychological trauma for women and girls. In the United States, 18.3 percent of women have been raped or experienced an attempted rape; of which 12.3 percent were younger than age twelve when they were first raped, and 29 percent were between eleven and seventeen. These traumatic experiences often leave long-term psychological disturbances.

"Oh, yeah, women fight," a coed told me. "Usually they fight over men." Her four female friends all nodded in agreement. Their perception is supported by surveys showing that the primary cause for female-on-female violence is in response to gossip, male partners, and personal insults. Another study showed that fighting to protect a third party is also a common cause of female-on-female violence. The presence of bystanders increases the likelihood of intra-gender violence in situations where there is a public challenge requiring a response to save face. Being hit with an object or stabbed are the most common forms of female-on-female violence; guns are rarely used.

Most incidents of female-on-female violence are within the same race (93 percent) and same age range. In a survey of females admitted to the emergency room as a result of female-on-female violence, 33 percent of the assaults were sparked by personal insult, 23 percent related to male-partner jealousy, and 22 percent related to negative gossip and rumors or in response to not being liked. Defending the reputation or physical well-being of friends and family constituted 17 percent of the cases of female-female violence. Jealousy about material goods or physical appearance also triggered females to fight each other, in 17 percent of the cases. Sixteen percent of female-on-female violence was related to using and selling drugs.

The LIFEMORTS triggers are clear in female violence even though the way the researchers tabulated the causes would lump and split some of them into somewhat different groups in calculating the percentages. The study shows that some factors that have been established for male violence, including low socioeconomic status, apply to violence committed by women as well. Aggressive posturing and aggressive action are thought to be necessary to survive in a threatening environment, which requires communicating a readiness to use violence to resolve disputes lest one become the target of victimization. That is, people reared or living in "rough neighborhoods" need to be "tough" in their behavior to avoid becoming a victim. Indirect aggression, such as rumor and gossip, are much more frequent among females than males, who rely more heavily on overt physical aggression. Low-income black women are particularly at risk for involvement in assault in the United States. This correlates with the socioeconomic factors that increase the risk of violence. According to a Centers for Disease Control and Prevention (CDC) study, 44 percent of lesbians will be physically assaulted by their partners, somewhat higher than the rate for straight women (35 percent), gay men (26 percent), and straight men (29 percent).

In June 2014, thirty-two-year-old Olympic gold medalist Hope Solo, a goalkeeper in soccer, was arrested for domestic violence for allegedly assaulting her sister and her seventeen-year-old nephew in a drunken, violent outburst. The cause of the argument was an insult. According to court documents, Solo told her nephew at the party that he was "too fat and overweight and crazy to ever be an athlete." He responded by calling her an insulting name and walked away to another part of the house. Solo followed and attacked him. When her nephew's mother tried to intervene, Solo attacked her as well. Hope's lawyer claims that she is innocent. Insult, especially under the influence of alcohol, is a cause of violent rage for both sexes.

In a less violent example of female snapping in rage, in April 2015 ESPN television reporter Britt McHenry was caught on a security camera snapping in anger and berating a towing company clerk after her car was towed and impounded. In this case, it was the Resources trigger that caused McHenry to snap. In any other circumstance, having your car taken away without your knowledge would be robbery. What is interesting from the perspective of female-female aggression is how this videotaped scene provides candid insight into female-female aggression. McHenry snapped at the company clerk by insulting the woman's status as a woman and her lowly position in society. Here are excerpts from the security-camera video:

> "That's why I have a degree and you don't," McHenry taunts. "I wouldn't work at a scumbag place like this. Makes my skin crawl even being here."
>
> Launching into a personal insult about the woman's appearance, McHenry says, "Maybe if I was missing some teeth they would hire me, huh?"
>
> The employee responds with an insulting comment about McHenry's dyed blond hair and the color of her roots.
>
> McHenry snaps back, "Oh, like yours, 'cause they look so stunning, 'cause I'm on television and you're in a [expletive] trailer, honey. Lose some weight, baby girl."

After she was suspended briefly by the television network, McHenry offered this apology via Twitter on April 16, 2015, reflecting the remorse that so often follows snapping: "In an intense and stressful moment, I allowed my emotions to get the best of me and said some insulting and regrettable things. As frustrated as I was, I should always choose to be respectful and take the high road."

Had McHenry known that her sudden rise in anger was the Resources trigger activated to engage in a fight to get back her rightful property, she might have been able to suppress her immediate reaction and use her influence to plot sweet revenge against the towing company. In fact, a month later, lawmakers in Maryland and Virginia proposed new legislation to protect consumers from predatory towing by the same company that impounded McHenry's car and others in the region.

# Sexual Violence in War

Sexual violence against women and girls in war has a long, ugly history that continues to the present day. On April 14, 2014, 276 schoolgirls in Nigeria were abducted by Muslim extremists, the Boko Haram terrorist network, for use as sex slaves. Abubakar Shekau, leader of the Boko Haram Islamist group, released a video admitting kidnapping the girls for slaves to use and sell in the "marriage market." The heinous crime of rape is a tactic of war to terrorize and defeat the enemy. Since ancient times, rape during combat has been regarded not as a crime but rather as one of the spoils of war. For insight into the ancient Mate trigger of violence in Western society, one need look no further than to the most universally respected of all sources. Consider these lesser known Bible stories:

> I will gather all the nations to Jerusalem to fight against it; the city will be captured, the houses ransacked, and the women raped.
>
> —Zechariah 14:2 (NIV)

> So the assembly sent twelve thousand fighting men with instructions to go to Jabesh Gilead and put to the sword those living there, including the women and children. "This is what you are to do," they said. "Kill every male and every woman who is not a virgin." They found among the people living in Jabesh Gilead four hundred young women who had never slept with a man, and they took them to the camp at Shiloh in Canaan.
>
> —Judges 21:10–12 (NIV)

> The people grieved for Benjamin, because the Lord had made a gap in the tribes of Israel. And the elders of the assembly said, "With the women of Benjamin destroyed, how shall we provide wives for the men who are left? The Benjamite survivors must have heirs," they said, "so that a tribe of Israel will not be wiped out."
> . . . So they instructed the Benjamites, saying, "Go and hide in the vineyards and watch. When the girls of Shiloh come out to join

in the dancing, then rush from the vineyards and each of you seize a wife from the girls of Shiloh and go to the land of Benjamin."

. . . So that is what the Benjamites did. While the girls were dancing, each man caught one and carried her off to be his wife.

—Judges 21:15–23 (NIV)

Sexual violence against women that stems from regarding females as property is recorded in the Bible outside the context of war. Concubines were female slaves who functioned as a "secondary wife." In addition, poor families could sell their daughters in dire times (Exodus 21:7–10; Judges 19:1–29). This practice was perpetuated to meet the sexual desires of males and also to cement political alliances.

Korean women were forced into prostitution by Japan during WWII to service the military men. Estimates range from 20,000 women to 200,000, including not only Koreans but women from many countries occupied by Japan during the war—China, the Philippines, and others. The women were incarcerated into several "comfort stations." Approximately three-quarters of these "comfort women" died, and most survivors were left infertile as a result of sexual trauma or sexually transmitted diseases.

"In the 'comfort station' I was systematically beaten and raped day and night. Even the Japanese doctor raped me each time he visited the brothel to examine us for venereal disease," a survivor of the forced prostitution testified.

Prepubescent girls were repeatedly raped by Japanese soldiers, those who fought back were executed, and the women served as many as twenty-five to thirty-five men a day. Rapes committed by Soviet servicemen during combat operations and during subsequent occupation in WWII are well known. Estimates of the number of German women raped by Soviet soldiers range up to 2 million. In many cases, women were victims of repeated rapes, as many as sixty to seventy times.

## Insight from Our Distant Relatives

Why are humans typically monogamous? Only 5 percent of other mammals have a single mating partner. From a Darwinian perspective, having more female partners should be an advantage to a male by spreading his genes through many more offspring. In studying DNA of a variety of mammals, researchers have recently suggested two explanations for human monogamy. "Females changed their diet to foods of higher quality that were clumped, and defended that food more aggressively," University of Cambridge zoologist Dieter Lukas says, referring to the consequences of developing agriculture. This led to large, exclusive territories becoming associated with individual females. These territories became too big for one male mammal to successfully defend more than one of them.

Infanticide is another factor, as supported by studies on primates. "Infanticide is a real problem, particularly for social species," University College London anthropologist Christopher Opie says. The large brain of humans takes years of childhood to develop, giving more opportunities for a rival male to kill the child and impregnate the female (an evolutionary advantage from a Darwinian perspective of survival of the fittest). Thus, male primates that stick with their child, and therefore with one mate, have an evolutionary advantage by warding off intruding males. From this perspective we can see the distant roots of the Mate, Environment, and Family triggers of defensive rage.

In their 1996 book *Demonic Males: Apes and the Origins of Human Violence*, researchers Dale Peterson and Richard Wrangham consider malefemale relationship violence through studies on primate behavior. Male chimpanzees fight even when in captivity where they are provided with ample care and food. Their battles establish a social dominance, and the alpha male reigns supreme. There are complex social coalitions formed among rival males to attack the alpha male. The authors describe the death of an alpha-male chimp one night by a combined attack by three male rivals. In the morning the zookeeper found several of the alpha male's toes, fingers, and both of his testicles on the floor—bitten off in the attack for dominance.

Among the aggressive behaviors of primates that have been passed down to us are aggression for social dominance (especially among males) and the ferocious defense of territory by both males and females. The sexually distinct bodily features of many male animals, such as horns and tusks, exist for combat between rival males to establish social dominance and for the aggressive acquisition of females for mating. These battles are ritualized contests where most mammals defeat the rival without battling to the death, but in chimpanzees, battles to the death do occur, just as in human conflicts.

Peterson and Wrangham conclude that there is something about the apes

that specially predisposes them to violence. Sexual violence occurs in orangutans, and male gorillas kill infants frequently. Female orangutans, who
must care for infants for eight years, prefer larger, aggressive males as partners. Some matings of orangutans are so violent they are also termed "rape"
by the British researcher John MacKinnon, who has observed them in their
native habitat. Applying the term "rape" to animal behavior is problematic,
but there is no doubt that the matings are forced by the male using extreme
aggression. "The females showed fear and tried to escape from the males, but
were pursued, caught and sometimes struck and bitten." By some accounts,
one-third to 88 percent of orangutan matings in the wild involve aggressive
attacks by males that drive the females into submission.

Other primates use violence in mating as well; for example, Jane Goodall
has categorized violently aggressive matings as "rapes" among chimpanzees.

One extraordinary account of a violent mating by an aggressive male
orangutan that Peterson and Wrangham describe in their book was unques-
tionably a rape, because the victim of the sexual assault was the camp's Indo-
nesian cook. The cook screamed for help. "[She] fought the ape with every
ounce of her strength, beat him with her fists, attempted to ram a fist down
his throat, but with no effect. . . .The cook stopped struggling. 'It's all right,'
she murmured. She lay back in my arms, with Gundul [the male orangutan]
on top of her. Gundul was very calm and deliberate. He raped the cook."

Peterson and Wrangham argue that rape cannot occur if social alliances
exist to combat it. In orangutans the females are solitary, whereas female
gorillas live in troupes that protect them from strange males, making gorillas
safe from "rape." There is a rich literature on evolutionary psychology based
on studies of primates and other social mammals that provides compelling
insight into human behavior, but this is outside the domain of this book on
the neuroscience of sudden rage and aggression. The important point is that
our brains and behaviors today evolved from our prehistoric hominid ancestors, and before that time from nonhuman primates who have brains that are
so similar to ours anatomically that the nonhuman primate brain, down to
the cellular level, is almost indistinguishable from the human brain. For this
reason, most countries have banned brain research on chimpanzees for ethical reasons.

# Tangled Neural Networks of Sex and Violence

"What I call the four F's: feeding, freezing, fighting, and . . . mating," quips neuroscientist David Anderson of Caltech, introducing his research on the neurocircuitry of sex and violence at a scientific meeting. Sex and violence are linked in culture, seen throughout the animal kingdom, and evident in the extreme as pathological dysfunction. Dating tips on the website Match .com list teasing and physical aggression as one of the sure signs to women that a guy likes you: "The boy at school who kept teasing you or punching your arm; he might have been crazy about you." The work of Anderson and others, using advanced methods that have been developed only in the last few years to study neural circuits, shows that there is a neurobiological explanation for the link between these two behaviors. The brain circuits of sex and violence are intertwined.

Feeding, freezing, fighting, and "mating" are all powerful unconscious urges controlled by neural circuits in the hypothalamus, so the answer to the intriguing question of whether there's a neurobiological basis for a relationship between sex and violence is likely to be found in this deep brain region. New research teasing apart these hypothalamic circuits reveals that different automatic behaviors, such as different responses triggered by different types of threats, are controlled by separate circuits in the hypothalamus (as we explored in chapter 9), but an exception is found in the apparent connection between violence and sex. More perplexing is that these two contradictory behaviors of love and hate are seemingly polar opposites. How could these contradictory behaviors possibly be controlled by the same set of neurons in the hypothalamus? Both powerful behaviors are activated by different stimuli in different situations and they serve different purposes. An alternative explanation could be that there are two separate circuits controlling these diametrically opposed behaviors ("love and hate") and they are somehow cross-wired in certain circumstances? The search for answers to the neurocircuitry of sex and violence is uncovering part of the neural mechanism behind the Mate trigger of violence.

Consider this important fact: From a behavioral and physiological point of view there are common features in both aggression and mating. Both behaviors evoke intense states of arousal—indeed, the most intense states of arousal possible. At some level, then, the two behaviors must activate similar neural circuits to evoke such similar extreme states of arousal. Also, both behaviors, when the outcome is successful, evoke potent feelings of reward. Finally, in the natural world, aggression and mating are interrelated behaviorally, and both are co-regulated by similar environmental influences and by the internal state of the body. Physiological states that arise during mating also promote aggression. Male animals are more aggressive at the mating time of the year, when females are ovulating. Most fatal encounters between people and moose occur during mating season, when bull moose become aggressive in seeking out mates and are engaged with other males in contests for mates. In mice it has been shown that prior sexual experience can increase aggressiveness. Seasonal differences, environmental context, and hormonal states mutually reinforce both aggressive and mating behaviors, so similar brain systems are involved in both behaviors.

It has been known for some time that regulation of mating behavior is controlled by the same part of the hypothalamus that Walter Hess identified with his electrical stimulation experiments on the cat brain in the 1920s, the hypothalamic attack area. This brain region encompasses the medial hypothalamic nuclei. But electrical currents from an electrode can also activate nerve fibers that pass through the site of stimulation, so the neurons controlling sex or violence might actually lie in different parts of the brain and only become artificially activated as they pass by the electrode. Alternatively, the neurons controlling sex or aggression might be located in the same "attack" region of the hypothalamus, but the two opposing behaviors could be controlled by different neurons mixed together like salt and pepper. These perplexing questions have puzzled scientists for decades, but very recently neuroscientists have discovered the answers.

Using the method called c-Fos staining described in chapter 9 to identify neurons that have been active just before examining the brain tissue, Anderson's research team identified individual neurons that became active after a mouse engaged in an aggressive encounter with another mouse or, instead, had just engaged in mating behavior. They found that the same set of neurons in the medial hypothalamic nucleus became activated by both behaviors. This would suggest that sex and violence are controlled by the same hypothalamic neurons. But in looking at the tissue alone, it is not possible to determine if the very same neuron that had been activated by sex was also activated by fighting, or whether two different types of neurons handle each behavior separately. The two types of neurons are clustered together in the same brain region. To make the distinction, it would be necessary to monitor the activity of a single neuron during both fighting and mating. This remarkable experiment has been done.

Electrophysiology has advanced greatly beyond the pioneering studies Hess performed with his slender wire used to stimulate the cat's brain. Today researchers use microcircuit fabrication techniques designed for integrated circuit production to make microelectrode arrays, looking something like microscopic combs. Once implanted into the brain these electrode arrays tap into a dozen or more separate neurons at once. Moreover, these precision microelectrodes can be implanted in the brain of a freely behaving mouse and studied for days. Dayu Lin, while working in Anderson's laboratory, implanted such microelectrodes into the brains of mice to record electrical signals in individual cells in the medial hypothalamic nuclei as the mice went about their normal activities. What this research found was that some hypothalamic neurons became active during fighting, and others were active during mating, but many of the very same neurons were active during both mating and fighting.

Strictly speaking, this experimental result revealing that many of the same neurons are turned on in the brain's attack region during both mating and fighting does not necessarily mean that the neurons are critical for either behavior. They just might be responding to some aspect of brain function or brain state that is present during both sex and violence, such as arousal. To test the very important question of causation, the researchers used genetic engineering to turn on or turn off these specific neurons using a drug. When these neurons of sex/violence were silenced, mice no longer engaged in aggressive behavior toward an intruder mouse placed in their home cage. This shows that these neurons are necessary for violent attacks, not just a by-product of them. However, mating activity was not at all affected in these mice when those neurons were silenced. Perhaps the fourth *F* is just more important than the third?

Follow-up research showed that there were in fact two different types of neurons in the medial hypothalamic nucleus that became activated during fighting and mating; about 40 percent of these neurons identified by c-Fos staining had a membrane receptor for the hormone estrogen. (Estrogen is present in both males and females, although females have higher concentrations.) The previous approach of turning off both sets of neurons in the brain region (with and without estrogen receptors) might have gummed the works in a way that would not happen naturally, so the researchers tried a more precise approach.

Optogenetics allows activity in individual neurons to be turned on or off by light stimulation. If a gene (for example, one making estrogen receptors) is expressed selectively in one type of neuron, scientists can use genetic engineering to insert an artificial gene into that particular type of cell. The artificial gene that is introduced makes a membrane protein that either excites or inhibits the ability of the neuron to fire electrical impulses (action potentials) when a blue laser is turned on. The light can be delivered to select spots in the brain by using slender fiber-optic cables implanted in the brain while the animal is freely engaged in its normal behavior, as was described in chapter 9 in studies on the fear circuitry in the hypothalamus and limbic system.

When the researchers switched on the blue light to activate the neurons that had the estrogen receptor, the mice began to fight. The battles ceased when the laser light was switched off or when scientists specifically inhibited firing in those neurons. This shows convincingly that the neurons with the estrogen receptor on their surface have a causal role in both initiating attack and in maintaining fighting. Again, mating behavior was not affected.

Thus far, the results would lead to the conclusion that aggression and sexual behavior are controlled by different sets of neurons mixed together in the ventral medial hypothalamus, but that different neurons in this region control either fighting or mating. But if this is so, where are the neurons driving sex? The next experiment provided a surprising answer.

If, rather than blasting the neurons with a strong laser beam, the researchers carefully increased the brightness of the laser, like turning up a dimmer switch, they found that such low-intensity stimulation of the exact same neurons that caused fighting now caused mating. So powerful was this control over the animal's behavior that weak stimulation of the neurons provoked mating behavior toward an intruder placed in the cage, rather than an attack, regardless of whether the intruder was male or female! This astonishing result means that all the various and complex sensory and emotional signals that go into deciding whether to mate with another individual, act through this common decision point in the ventral medial hypothalamus. By increasing the intensity of stimulation while the mouse was engaged in mating behavior, researchers found that they could switch the behavior from mating to fighting. Although it is dangerous to anthropomorphize from animal behavior, the videos of this experiment leave the poor female looking very perplexed as mating or fighting is interrupted or initiated by scientists increasing or decreasing the intensity of the laser beam. So the same neurons in the hypothalamic attack region control both violence and sex, but it is the level of activity in these neurons—both the number of neurons activated in the circuit and the intensity of activity in the individual neurons themselves—that determines whether the behavior will be mating or aggression. Perhaps the intense arousal at the jubilant celebration in Tahrir Square provoked a combination of violence and sexual behavior in the minds of some groups of men in the square.

This newly discovered neurocircuitry provides important new insight. Parallels between these animal studies and some normal and abnormal human sexual behavior that involves an interaction between aggression/pain and mating (from "rough sex" to deviant sexual behaviors) become apparent.

A team member in major league baseball shared a secret that some baseball players use to exploit the connection between aggression and sexual arousal. "I know players who, before their plate appearance, they'll go in and they will look at pornography right before they go to the plate. They do this in an effort to have a kind of adrenaline boost or something. Let's call it an animalistic type of approach to the 'at bat.'"

The intriguing difference between these neurons that have receptors for estrogen means that other factors in the body can influence the general level of activity in these neural circuits. The complex set of factors in our environment—season, hormonal state, motivation, arousal, inhibition—all influence both aggression and mating behavior. Thus, the split decision to fight or flee, or to mate or attack, depends on the complex set of factors influencing the general level of neural activity in these circuits. We will return to this interesting question of setting the threshold for triggering a snap response in chapter 12. In my case with the attack in Barcelona, the question of whether I would have acted the same way toward the pickpocket on a different occasion or if I would always act in the same way in the future is answered by these animal studies: no. The split-second decision to snap aggressively in defense or for any other reason is set by a complex combination of situation-specific environmental factors and internal states of stress, arousal, or physiology in the body at the time. Clearly this must be the case, because such powerful behaviors with such significant consequences must be highly regulated and matched appropriately to the circumstance and bodily state. Similarly, in cases of snapping inappropriately with violence, as in Ray Rice's case when he knocked his then girlfriend unconscious in an elevator, we can extrapolate from animal studies that the neurons in the ventral medial hypothalamus of Rice's brain had been strongly activated, most likely in part by internal and external stresses that had revved up these circuits before he entered the elevator. Like the laser beam of increasing intensity, factors in combination overstimulated these circuits to provoke a sudden aggressive response.

So the environment and internal states strongly influence sex and violence through this hypothalamic nerve center. One's environment during rearing or childhood influences these circuits as well, and also the strength of cortical circuits that regulate the neurons that drive sex and violence in the hypothalamus. This contributes to understanding at a cellular level why children raised in an environment where domestic violence occurs will be more likely to exhibit those same behaviors as adults. This is also why, as is evident in many places around the world, social environments that tolerate sexual aggression and violence against women create a vicious vortex that drives sexual brutality against females.

## His-and-Hers Brains

According to UC Irvine neuroscientist Larry Cahill, an expert on differences between male and female brains, "Sex differences exist in every brain lobe, including in many cognitive regions such as the hippocampus, amygdala and neocortex." In addition to anatomical differences between the sexes, functional studies show that the brains of men and women operate differently in several respects; that is, different brain structures are engaged in the brains of men and women for certain tasks. Performance in a given cognitive task may not differ between the sexes, but men and women are sometimes using different strategies to achieve the same result. Finally, there are numerous chemical differences in neurotransmitters between the sexes, not to mention the obvious hormonal differences. Hormonal fluctuations throughout a woman's menstrual cycle are known to have striking effects on the cellular anatomy of parts of the brain (notably in dendritic spines), on cognitive performance such as verbal and spatial memory, and emotional states. The incidence of mental and neurological illnesses differs greatly between the sexes; for example, women have a much higher incidence of multiple sclerosis, depression, and fibromyalgia, but men have a higher incidence of schizophrenia and ADHD. This is clear evidence of important underlying differences in anatomy and function of the male and female brain.

Unwholesome family life can alter development of threat-detection circuits in the brain of young girls that persists into adulthood and predisposes women to developing mood and anxiety disorders as adolescents and young adults. Boys are also negatively affected by family stresses during childhood, but the lasting effects on their brain seem to influence only one of two neural circuits controlling response to threats, anxiety, and fear.

Recent neuroimaging research in several laboratories has shown that the wiring of a child's brain is permanently altered by abuse and other stressful experiences early in life. The earliest research concerned children who were exposed to severe sexual abuse, but more recently, seemingly less harsh stresses have been found to alter the wiring of specific circuits in a child's brain, such as verbal abuse from their peers in middle school. Such children suffer greater psychological problems as adults, and those problems can be traced to the differences in brain circuitry. A recent study by Ryan Herringa and colleagues at the University of Wisconsin followed a large group of children as they grew up, and found that girls are especially vulnerable to maltreatment in childhood; moreover, the maltreatment they experienced was comparatively subtle. Stressful family life—for example, financial stress or poor parenting—alters the development of threat-detection circuits in the brains of boys and girls. Thus, the young brain is very sensitive to stresses in family life; severe criminal-level abuse of children is not required to cause these effects.

As was described in chapter 9, threat detection involves interaction between three brain regions. The amygdala, deep inside the brain, detects novelty and danger in the environment and learns quickly to respond to and avoid threats. The amygdala therefore regulates our emotions of anxiety and fear. The hippocampus, located near the temples of the skull, is crucial for mapping our environment, forming memories of events, and learning the context of when experiences, including threats and stresses, are likely to be encountered. The prefrontal cortex, beneath our forehead, is the higher-level cognitive region of the brain that can evaluate complex information to make decisions and direct our attention and behaviors appropriately. This part of the brain is the last region to develop and in humans it is not fully developed until the early twenties.

Using fMRI, Herringa's study found that boys experiencing family stresses early in life had weaker functional connections between the prefrontal cortex and the hippocampus, but girls who experienced the same stresses developed weaker connections between the prefrontal cortex and both the hippocampus and amygdala. Girls thus suffer a "double hit." These changes in brain circuitry correlated with the level of psychological problems the individuals experienced in later life. These differences were measured while the children were in a brain scanner simply engaged in their own thoughts. This technique measures the brain's "resting-state" functional connections, rather than pinpointing specific parts of the brain that become activated by a particular stimulus. The method tells researchers how the brain is wired rather than what parts of the brain are engaged in specific cognitive activities taking place at a specific point in time.

The explanation for the differences between boys and girls seems to be that different brain regions develop at slightly different times in everyone, and there are differences between the sexes in brain maturation just as in body maturation. Stress and abuse affect development of brain circuits most strongly during the period when a circuit is forming and maturing. This is why traumatic events—for example, combat experienced by military personnel—do not have the same effect on an adult as they have on a child.

Since the hippocampus provides context, this enables a person to experience fear in the appropriate environment rather than being anxious and fearful everywhere. "Females may have greater fear/anxiety responses generally following maltreatment experiences," Herringa observes, because these girls have reduced connection to the amygdala as well as to the hippocampus. One of the changes Herringa would expect in these girls is that they would be more likely to suffer PTSD after traumatic experiences as adults. "Childhood maltreatment is a significant risk factor for adult PTSD," he says, and these connections in the brain's fear circuits are known to be involved in PTSD.

The changes in brain wiring caused by maltreatment in childhood seem to be the brain's way of coping with the hostile environment. "This could very well be adaptive in a stressful or threatening environment but may also come at a cost of increasing risk for anxiety and depression [as adolescents and adults]," Herringa explained in an email.

These new insights offer possible new gender-specific treatments for anxiety and depression. "For example, a therapy involving contextual safety vs. threat learning could be helpful for both sexes," Herringa says. This treatment would allow a person to realize that their feelings of anxiety and fear are not appropriate to normal situations. Both boys and girls showed decreased connectivity to the hippocampus after maltreatment, so "extinction therapies" using context to reduce anxiety and fear could be helpful to both sexes. An example of such a therapy would be having a boy or girl who suffered a traumatic experience in a particular location return to that location repeatedly with supervision under safe conditions, which should work to rewrite the memory of the trauma associated with that particular location. But even without rewiring from hostile experiences in early life, the brains of men and women in general operate differently under stress.

"There is this pattern across five or six labs," Larry Cahill told me on a visit to my laboratory. "When you stress a male there is a tilt towards the right hemisphere in the amygdala activation." In females under stress, the left amygdala becomes activated. This was seen in fMRI scans. What these scans cannot show (because imaging takes several minutes) is that this sex difference in using either the left or right amygdala to respond to a sudden threat happens in a split second. Cahill discovered this immediate response by flashing his subjects pictures, some of which showed emotionally charged stressful images, while recording the brain-wave activity from the left and right cerebral hemisphere in both men and women.

A brief explanation of brain waves that are evoked by sensory stimulation will be helpful in understanding what Cahill found. When a signal from the eyes (or any other sense) reaches the brain, the arrival of the electrical event causes a ripple in ongoing brain-wave activity. This ripple of electrical activity in the brain is called an "evoked potential," because the brain-wave response was evoked by a sensory stimulus. But subsequent waves of brain activity ripple out after the initial tsunami of electricity reaching the brain from the senses. These "after-effect" waves are even more intriguing. These electrical signals represent the brain networks processing or "cogitating" unconsciously, in a sense, on the event that has just occurred. Most of what constantly assaults our senses flows through our mind as background noise, but any sudden unexpected feature in our environment instantly grabs our attention. This happens because we are not able to consciously perceive all of the ongoing sensory input to our brain, but our unconscious mind is in fact doing just that—constantly monitoring everything—looking for anomalous events in our environment of such significance that they warrant being elevated to a higher level of awareness or triggering an immediate reflexive reaction.

Say a gruesome picture of a car wreck suddenly flashes on a screen during a series of otherwise tranquil images. That anomalous image sends a shudder through our awareness and we immediately attend to that image, remembering it clearly, while the various other tranquil pictures fade out of consciousness and are lost from our memory. When this happens—that is, when a novel feature suddenly grabs our attention—the EEG recording of our brain waves shows a sudden characteristic rise in activity, called the P300 wave. This wave is called P300 because it is a positive (P) rise in brain-wave voltage and it peaks 300 milliseconds (only one-third of a second) after the novel stimulus. This P300 wave is not evoked by any of the tranquil images. Intriguingly, the P300 wave reflects not the simple transmission of sensory information to the brain but rather the brain processing the signifi-cance of the sensory input that has come in. That is, the P300 wave reflects the instantaneous and unconscious decision to engage our attention and ori-ent our cognitive function toward the novel stimulus. If someone were to say to you, "I take my coffee with cream and dog," a P300 wave would erupt in your brain from the unexpected input "dog" instead of "sugar." Did you "feel" the P300 wave rise in your own brain just now?

When I felt the thief's fingers delicately slip into my pocket to snatch my wallet in Barcelona, a P300 wave would most certainly have surged in my brain like a huge breaker rolling in amid tranquil surf. That anomaly in my sensory world tripped an urgent alarm within one-third of a second, fully engaging my brain and body and orienting my attention to address the threat, causing me to reflexively grab the unseen robber by the neck as he turned to dash off with my wallet.

But male and female brains operate differently under stress, and the P300 wave proves this in Cahill's experiments. "In the first 300 milliseconds of a scary thing coming into view [like a gruesome picture of a car wreck] there is a right lateralized P300 response in [the] male and a left lateralized P300 in the female." By this, Cahill means that the right cerebral cortex gets activated in males under stress and the left cortex gets activated in stressed females. Who could have predicted that the brains of men and women would cleave in such a major way under pressure? What is the possible significance of this differential lateralization of brain activity?

The answer, Cahill suspects, has to do with differences in information processing between the left and right hemisphere. "Gist versus detail," he says. "The right hemisphere of the brain is better at processing lower-frequency [less detailed], bigger-picture—seeing the forest but not the trees. The left hemisphere is geared towards processing the higher-frequency details. Males and females are doing both [when not under stress], because both males and females need gist and detail processing to survive."

Why should the two halves of our brains operate so differently? "Detail by definition gets in the way of gist, and gist by definition gets in the way of detail," he says. The elegant solution seems to be that the brain divides up the problem to permit independent analysis of detail by the left hemisphere and big-picture gist by the right hemisphere. The left brain is breaking down the problem analytically and the right brain is gathering diverse information to assemble a comprehensive picture of what is going on. This is how the human brain solves the contradictory challenge of perceiving detail and at the same time grasping the big picture. But in times of stress, brain function in men and women suddenly works differently.

"What happens when things get stressful, when things get emotional?" Cahill asks. "The male brain, for reasons that I don't understand, is busy processing the gist of the situation [using the right hemisphere], and in women the left hemisphere is busy processing details."

"During that stressful situation in Barcelona," Cahill explains, referring back to when Kelly and I were being pursued the gang of robbers, "your brain was tilting more towards a larger-scale strategy, whereas hers was tilting towards a higher-frequency detail analysis."

It was true in our case; Kelly rapidly picked out the criminals in the crowd amid the complex barrage of information we were confronted with in that foreign city, whereas I was looking at the forest, devising strategies to escape or fight.

To understand this difference in left-right/male-female brain function under stress, look quickly at the image below. What is the first thing you see?

$$
\begin{array}{ccc}
 & \$ \ \$ & \\
\$ & & \$ \\
\$ & & \\
 & \$ \ \$ & \\
 & & \$ \\
\$ & & \$ \\
 & \$ \ \$ &
\end{array}
$$

Men under stress will tend to see a large letter *S*, whereas women under stress will immediately see a dozen dollar signs. A woman's brain under stress shifts to activate the left cortex through her left amygdala, whereas a man's brain under threat or stress shifts to engage his right amygdala. Since you are presumably not under stress at this moment, both your left and right amygdala are being activated equally by the image to allow your brain to both encompass gist and discern detail, but in times of stress or sudden threat, what you see will depend on your sex. On average, the two sexes cleave into specialized left- and right-brain function under stress to take either a helicopter view of the situation or get down in the weeds and analyze the details.

It is interesting to speculate why we have this sex difference in brain function. "The best answer I have for that is that sex selection across many species in the female tends to be more detail-oriented than does the male," says Cahill. Think of female birds evaluating potential mates by subtle differences in their plumage or the performance of their courting behavior. "Two rams going *bam!* batting horns—there's nothing subtle about it. In general the face literature [fMRI studies of brain responses to different faces] indicates that the female is better at picking up the emotional details than is the male." This is only Cahill's best hypothesis, he emphasizes, and there may be other factors involved in this sex difference in brain function under stress.

"There is strong indication that sex hormones are playing a role in stress response," he says. "There are different stress effects related to the menstrual cycle. They are generally potentiated in women during the high-hormone luteal phase of the cycle. Stress responses are also clearly blunted in women who are on hormonal contraception." Factors influencing triggers of rage will be considered in chapter 12, where research will be presented showing that hormonal factors have a major influence on brain function, threat detection, and snapping. What is clear from these studies is that the anatomy and function of male and female brains differ, and the sex differences are pronounced in circuits that detect and respond to threat. This neurobiology results in a partnership between males and females that is far more powerful than either sex alone.

## What Women Want

She's checking out your online profile.

"I am a scientist who enjoys bird-watching and canoeing."

Interesting! she thinks.

Then she scrolls to the next profile—another scientist:

"I enjoy whitewater kayaking, and I study alligators in the wild."

She passes on you with your canoe, and in eager anticipation sends the kayaker an electronic "wink."

This, according to a study by psychologist John Petraitis, is what most women will do, but why?

John Petraitis limped painfully into his office with his left foot in a black knee-high Velcro cast. His right wrist was wrapped in a matching black cast to stabilize his thumb tendon, recently repaired by surgery.

"Skiing deep in the trees makes me come alive," he says enthusiastically, gazing at the gorgeous snow-covered mountains surrounding his office in the Department of Psychology at the University of Alaska at Anchorage.

That explains the snapped Achilles tendon and hand surgery. Many guys are drawn to danger. Whether it's aggressive skiing, motorcycle racing, or rock climbing, why are men and boys attracted to risky activities?

Part of the answer, according to Petraitis's research, together with coauthors Claudia Lampman, Robert Boeckmann, and Evan Falconer, is supported by an experiment analyzing responses to online profiles in a mock electronic dating service. A lady's choice for a first date may be swayed by factors extending back in time to when sharp stones, rather than Sharp computers, were the most advanced technology.

Petraitis was investigating the psychology of adult substance abuse when he was struck by the conspicuous differences in risk-taking behavior between the sexes. The highest rates of cigarette use, heavy alcohol use, binge drinking, and illicit substance use are seen in young people between the age of fifteen and twenty-five. Males have higher rates of all these risky activities, and males show up at emergency rooms in much higher numbers with traumatic injuries. They die at higher rates in outdoor accidents such as skiing and car crashes, and they are more often victims of homicide.

Partly these gender differences in risk-taking could be cultural. Boys are encouraged to display dominance and courage, accept dares, and take risks, whereas girls tend to be socialized to be cautious, social, and to show concern for others. Girls play with dolls. Boys play with "action figures." But the preference for risk-taking behavior in males is seen across all cultures, suggesting something more than socialization may be at work in drawing men and boys to risky pursuits.

The research team suspects that gender-specific behaviors that have been favored over eons of evolution in the battle for survival have left their imprints in our DNA and they are still guiding our mate choices today. As every biologist knows, evolution is about sex. When it comes to sex, females are the ones who make the decision about mates. Males audition.

Consider the garish male peacock with such ridiculously showy tail feathers that actually make it harder for them to fly and easier for predators to spot them. The male birds strut about displaying their flashy tail feathers to impress the peahens in hopes of mating with them. The females, seeing the handsome bird with such a dangerously flamboyant plumage, think, "This guy must be amazingly fit to have survived with those dazzling tail feathers." Genetic fitness, superior ability to survive in the face of dangers and handicaps, that's what females are seeking in selecting their mates. A mate that can survive great risks must be exceptionally good at avoiding predators and acquiring food.

Many modern women will object to having their mate choices reduced to the pea-brained level of a bird, but the data paint a clear, broad picture. The researchers devised a list of 101 pairs of behaviors in a mock dating service in which each question paired a higher-risk option with a lower-risk choice. For example: Do you prefer a person who enjoys canoeing or one who goes whitewater kayaking? The choices included many more subtle risks, such as whether one prefers their salsa mild or spicy. The questionnaire was given to both men and women, and what the results showed is that women greatly preferred guys who engaged in the higher-risk behaviors. Guys, in contrast, did not show any preference for women based on their risk-taking profile.

But here's the really clever part: Half of the paired questions dealt with the sort of risks that human beings would have faced thousands of years ago, and the other half dealt with modern risks, such as driving while talking on a cell phone. Neither guys nor gals cared a whit about modern risks in selecting first dates; in fact, these modern risks were likely to be viewed as unattractive and foolish.

Females couldn't care less about a guy who enjoys sticking forks in toasters. There was no electricity in the Stone Age. The risk-taking behaviors women preferred are the ones that deal with overcoming gravity, dealing with wild beasts, crossing water, being indifferent to nasty or dangerous foods, and engaging in human conflict. These are what the research team calls "hunter/gatherer risks," the kind of risks our caveman ancestors would have had to deal with. Modern risks, like playing with electricity, fooling with deadly chemicals, taking risks of identity theft, or driving without a seat belt, did not impress the ladies one bit.

Why is risky behavior so pronounced in *young* males? Again, the answer is sex.

"Female fertility is a rare commodity," Petraitis explains. Males remain fertile into old age, but not so for females. "A twenty-year-old male competes with a sixty-year-old male [for attractive women]." The two age groups use different strategies to attract younger women. "Younger males are faster, stronger; they can bounce back from injury or adversity. Older males have more resources to provide for women." So each group competes for young women in arenas in which they are more likely to win. "Young males are greater risk takers and adventurers to demonstrate their fitness," he says.

He cites statistics on the biological facts of life to make his case. Males are fertile for sixty years, or 22,000 days. Females are only fertile half as many years, and they are only fertile twenty-six days per year, whereas males are fertile every day. Do the math and in an entire lifetime, women are fertile fewer than 850 days, compared to 22,000 days for men. Also, women's investment in fertility is much greater, considering the nine months of pregnancy and years devoted to rearing a young child. Women have to be choosy.

Human behavior is complex, and one important insight, such as the hunter-gatherer risk appeal identified in this new study, cannot explain everything about male risk-taking. Petraitis suspects that males may also engage in risky activities to elevate status among other males. These new findings also do not explain why many women engage in risky activities, but he is devising experiments to investigate these questions.

For guys this research provides revealing insights into our male urge to risk life and limb in tests against gravity, water, fire, wild beasts, and dangerous food, but if you are thinking that taking risks is the way to impress women, you are missing an important point. Male fertility is cheap. If a peacock with an outrageous tail gets eaten, well . . . there are plenty of others. Likewise for the guy who gets gored and trampled by charging bulls in Pamplona, Spain.

Energetic and fit with a neatly trimmed graying beard, one might easily imagine Petraitis as the kind of guy who would eagerly attempt a 720 off the half-pipe to impress his lady (who happens to be one of the coauthors on the paper). But maybe he shouldn't.

# To Be a Man

The drug dealer, with shoulder-length dreadlocks and a handgun tucked in his waistband, was smoking synthetic marijuana and sipping Cuervo tequila with a cluster of other men on Martin Luther King Jr. Avenue in the crime-ridden neighborhood of Southeast Washington, DC, in the summer of 2014. It was four thirty a.m. and the PCP dealer, who had more than a decade of drug arrests and gun convictions, was only minutes away from becoming involved in a kidnapping and sexual assault.

A blue Toyota swerved into a nearby alley and came to a halt. This was a common spot for PCP sales. The drug dealer approached the car. Muffled sounds of a woman's screams radiated from the vehicle. "Help me! Help me! He's raping me."

The drug dealer peered through the window in the dark and saw a man on top of a woman viciously attempting to rape her. Instantly he yanked open the car door, drew his gun, and pointed it at the assailant. "Get the fuck out!" he commanded.

Lateef Sharperson jumped out of his car (and nearly his skin) and fled, running with his pants dangling around his knees. The PCP dealer ordered the woman to slide over and he got into the vehicle beside her and started it up. He drove the terrified woman home to her mother in the assailant's car and waited for the police to arrive. Sharperson was soon apprehended by police and convicted of kidnapping and first-degree sexual assault of the nineteen-year-old Trinity Washington University student. The drug dealer, who was due to appear in court himself on charges of selling PCP, suddenly became a Good Samaritan.

"I got nieces, a little sister. I got a mother. That ain't cool," the drug dealer testified before the jury of eight women and four men.

"A Good Samaritan doesn't mean Boy Scout," Assistant US Attorney Kenya Davis told the jury, in an effort to prepare them for the witness's lawless background. In hope that the men and women would accept the testimony of a criminal, Kelly Higashi, head of the sex offense and domestic violence unit of the US Attorney's Office, said, "He happened to be the one who rescued the woman who was being raped and who thought she was going to be killed."

The young woman calls the nameless man "my angel." Cooperating with police is dangerous in the drug dealer's world, so the newspapers have kept his identity secret.

Before police arrived at the mother's home, the drug-dealing Good Samaritan stashed his gun, he admitted to the jury. Sharperson's attorney, Leonard Long Jr., latched onto that stashed weapon. Confronting the man on the witness stand, he asked why, if he was a Good Samaritan, did he not admit to police that night that he had a gun?

"I'm not that much on the good side yet," the man in dreadlocks responded from the stand, setting the female jurors giggling.

Sharperson was sentenced to thirteen years in prison for abduction and sexual assault on the basis of testimony from the only witness to the attack: the Good Samaritan drug dealer.

The Organization trigger is the spark of immediate rage and violence and a peculiarly human trait that binds us into a cooperative society. While the triggers of rage and violence in men are the cause of much abuse against women, countless women have been protected or rescued from sexual assault by another man who, on an instantaneous impulse, snapped to action and risked his own life to engage in a violent battle with an individual threatening or physically harming the woman. "I got nieces. . . . That ain't cool."

This same snap reflex of a man coming to the aid of another person is seen in response to all the other LIFEMORTS triggers. "I do not expect to stimulate or create heroism by this fund," Andrew Carnegie said when he established the Carnegie Hero Fund in 1904, which gives cash awards to ordinary people who save the lives of others. "That heroic action is impulsive." Since 1904, 80,000 cases of extreme heroism have been recognized by the fund. Nine out of ten of the awards for heroism have been given to men. This is the same gender difference seen in statistics of violent crime. Ninety percent of violent criminals in prison are men, but men, in far greater numbers than women, will also instantly risk their life for a woman, child, or stranger in danger. Men do this reflexively despite the highest stakes. Nearly one in four Carnegie Hero medals is bestowed on a dead man. In each case, a man instantly surrendered his life for someone else.

On Saturday, April 26, 1986, the Chernobyl nuclear power plant sixty-five miles north of Kiev in Ukraine overheated and exploded, exposing the hot nuclear core to the outside world. Fires raged for three days, sending almost fifty tons of highly reactive fallout into the atmosphere. Once considered the breadbasket of Europe, the land surrounding the plant was poisoned by the radioactive fallout, which will continue to poison the area for 25,000 years. Some 336,000 people in 434 towns and villages in Belarus were evacuated from their homes and relocated outside the contaminated zone. As devastating as the Chernobyl disaster was, it could have been much worse.

Ten days after the explosion, an impending disaster was developing that would cause immeasurably greater devastation than the initial explosive meltdown. Water that firefighters had poured on the burning power plant for days had collected into a highly contaminated pool beneath the reactor core. The reactor core itself had melted down into a searing, intensely reactive lava oozing its way unstoppably through the reactor floor. The instant this radioactive ooze broke through and reached the water below, it would set off a massive thermal explosion that would send a deadly plume of radioactive steam across most of Europe.

Three men—Valeri Bezpalov, Alexei Ananenko, and Boris Baranov—volunteered to dive into the radioactive pool and swim using scuba gear to find the safety valves to drain the water away. Boris's lamp failed shortly after submerging, but the three divers groped through the darkness until they found the safety valves, opened them, and drained the pool to avert the disaster. The molten lava did soon burn through the reactor floor, but by then the water was gone.

The action of these three men saved the lives of hundreds of thousands of people throughout Europe, but all three men died after two weeks of suffering with lethal radiation sickness. There are men who will risk death to save another person, but in this case there was no risk, only absolute certainty that each man would die in the act of saving others. Ananenko was an engineer who knew where the safety valves were located. Bezpalov was a soldier and engineer at the plant. Baranov was an "ordinary" worker at the plant who offered to hold the lamp while the other two men turned the safety valves. Reactor number 4 is now covered by a concrete sarcophagus, and in the distance, amid radiation levels still five times above normal, stands a memorial to these three lost workers of Chernobyl.

So it is important in discussing sex and violence not to lose perspective, as illustrated by the case of Johnny McGirr. The thirty-three-year-old firefighter was seated next to two young boys on a Virgin Australia flight in 2012 when a female flight attendant ordered him to trade seats with a woman. "You can't sit next to two unaccompanied minors," she said. Appealing to a female passenger, she demanded, "Can you please sit in this seat, because he is not allowed to sit next to minors."

Johnny McGirr is not a pedophile. But four airlines—British Airways,

Qantas, Air New Zealand, and Virgin Australia—prohibit adult male passengers from sitting next to unaccompanied children out of "fear of sexual assaults." McGirr and many men like him who suffer similar discrimination on airlines with this policy feel deeply humiliated by being pegged as a criminal sexual predator for no reason other than being a male.

This discrimination against all men, because some men commit sexual crimes, highlights an important point. In examining the broad scope of human behavior on the neuroscience of gender and aggression, and the predominance of males involved in violent crime, it is vital to remember that the vast majority of men are not violent. They are not aggressive or abusers of women in any way. All members of society, men and women, with the very few exceptions of a criminal minority, are appalled by the attacks on women in Tahrir Square and by other crimes against women. These acts of animal brutality are repugnant and intolerable.

On the contrary, the neurocircuits of threat detection and sudden aggression, and the complementary differences in brain function in men and women, are what bind humans together into peaceful and cooperative groups. Countries, tribes, coworkers, families, and couples are all united by the neurocircuitry at the core of the human brain that exists for mutual protection. The reflexive concern for others, at a neuroscience level, derives from this brain circuitry in our hypothalamus.

It is the strong bond between man and woman that forms, in the biological sense, the foundation for all levels of social organization built upon it, from family to country. Males prevent violent crime. Interpol crime data show that in countries with fewer males in the population, murders, rapes, assaults, and other violent crimes increase. This may seem counterintuitive, but this fact should not be ignored. The exceptional cases of male criminal violence should not distract from the obvious: the biological role of males as protectors. This is why, after all, nature has given men bigger muscles—to provide for and protect their loved ones.

## The Ties That Don't Snap

To fall in love. Of all the verbs that could be applied, "fall" is the one we always use. We "make" friends, "get" into situations, "enter" into an agreement, "undertake" a challenge, "have" fun, but in love—we *fall*.

This essential aspect of the human emotion called falling in love transcends the English language. In French it is *tomber amoureux*; in Spanish, *caer enamorado*; in Russian it is *vljubit'sja*, and in the Thai language the expression translates into "fall into a hole or trap."

Of all the verbs that we might have used, we say "fall" because love is a sudden unconscious brain process that operates beyond our control. We are powerless to initiate it willfully or to control it once we have fallen in love. The sudden kiss between man and woman is the stereotype plot point of romantic movies. Falling in love is indeed a snapping behavior. Sometimes love is sparked by nothing more than a silent locking of eyes between two people in a crowd. It is the same unconscious brain process and mental telepathy that Secret Service Agent Scott Moyer described, albeit in the opposite context of identifying a potential foe. When love strikes, you know it. It overpowers and consumes us.

We have examined the nine LIFEMORTS triggers of rage and the unconscious, automated brain circuits that generate sudden snap reactions to danger, and in this chapter we have seen how the Mate trigger can be the source of violence between the sexes and of violence within the sexes in competition for mates. But the powerful biological attraction between the sexes, and the compassion that the neurocircuitry of love generates, is a potent antidote to violence between the sexes, between individuals of the same sex, and among groups of people.

As with any human emotion, love can be examined and understood as a brain function that drives behavior. As with all emotions, love is the result of unconscious brain circuits, and the neurochemicals involved in this neurocircuitry of love are an important component of the LIFEMORTS system of defensive triggers. The neurocircuit of love energizes the Mate and Family triggers, which is what makes these two circuits of rage such powerful motivators of vigorous or even violent reaction to protect our children or our mates, or to evoke ruthless behavior in competition for mates. Love also fortifies the Environment, Organization, and Tribe circuits of snapping in rage by compelling us to provide for and protect our family and children and the social structures that support them. Remarkably, the same neurocircuit that was designed by evolution to bring males and females together to generate new life and to provide the necessary prolonged care for offspring by forming intense emotional bonds between mother and child, man and wife, father and family, also resulted in the emotions of compassion and altruism that bind humans into larger groups and societies.

Love is a singular and powerful emotion combining feelings of deep affection, delight, pleasure, and cherishing another person. All of the emotional components that form the core sensation of love are the result of specific neurotransmitters operating in specific brain circuits. The neurotransmitters serotonin and dopamine generate the sensation of pleasure, reward, excitement, and addiction. The neurohormone oxytocin generates the powerful feeling of attachment to another human being rooted at a deeper level in the mind than conscious reasoning, and with a grip on behavior that is more powerful than any other, including our self-interest if our loved one is in peril. This emotional response is called heroism or altruism when applied to self-sacrifice for another human being.

One of the most remarkable aspects of the human brain is that this bodily organ develops after birth, and it is not fully formed for nearly two decades spanning from infancy through childhood and adolescence. This prolonged postnatal development of the human brain requires that human offspring must be cared for, fed, taught, and protected for the incredible period of nearly twenty years. This requires strong pair bonding between man and woman, family, and a supportive social structure to care for and teach children. None of this would be possible without the powerful brain circuits of attachment that bind men and women together as lifelong partners. This circuitry of love anchors a mother's behaviors to the care of her infants and children, with compassion, care, and self-sacrifice, and it motivates the father's behavior toward providing for and protecting his family. (Here we are tracing the biological roots of the human brain and behavior. The modern world is complex and different from the environment of our ancestors in the human line. The modern world embraces other types of relationships and gender roles, but these roots of our past anchor our brain and behavior in all the variation of relationships that we enjoy today.)

The neurocircuitry of love that binds family and society, when applied to a higher spiritual power, is the same circuit that sustains religion. Love, compassion, altruism, and care for others are universal to all religious and spiritual beliefs. Objects of spiritual worship uncovered with bones are irrefutable archeological evidence that the bone fragments that may have lain buried for thousands of years are human. The paucity of artifacts of worship and ritualistic burial practices of Neanderthals is convincing evidence that these forerunners of *Homo sapiens* lacked something that is essentially human. This brain circuit of *Homo sapiens* caused a biological revolution that brought members of our species together into families, tribes, cooperative groups, and complex societies—all of it, fundamentally, to bring forth the next generation. This cooperative bonding enabled division of labor, specialization of individual expertise, and the development of technology. Spiritualism and religion have always had a profound role in guiding human interactions and in controlling human behavior, notably moderating behaviors that stem from misfires of the LIFEMORTS triggers. All of these complex social structures and interactions of human beings were predicated on the need to care for human offspring for the two decades it takes for the brain of *Homo sapiens* to develop to adulthood, and this all depends on the neurocircuit that generates deep emotional attachment between the sexes and their offspring, which we call love.

Neuroscientist James Swain and his colleagues conclude that interpersonal relationships constitute the foundation upon which human society is based, and that a specific neurobiological circuit and the secretion of oxytocin are responsible for this caregiving system.

Pair bonding, parent-infant interactions, compassion, and altruistic behavior toward others derive from a neural circuit that extends from the medial preoptic area of the hypothalamus and the ventral part of the bed nucleus of the stria terminalis, and the lateral septum. Much of the cortical-to-hypothalamic circuitry involved has already been discussed in the context of snapping, and it includes the amygdala (alarm), striatum/nucleus accumbens (motivation and reward), and cingulate cortex (decision-making). These same brain structures keep appearing and reappearing in this book not because the brain is so simplistic that there are only a few critical components to bother naming but because the snapping response in the diverse contexts being examined here engages these core brain functions. The inferior frontal gyrus, orbitofrontal cortex, insula, periaqueductal gray, and dorsomedial prefrontal cortex, which have already been described earlier in this book, regulate complex social-cognitive functions that are selectively engaged when the maternal brain responds to infant distress, whether it is hearing a baby's cries or seeing a baby's picture.

Empathy can be seen by brain-imaging studies on mothers while they observe and imitate facial expressions of their own child. Mothers respond to emotional expressions of their own child with activation of the insula and other cortical regions involved in imitation (mirror neuron system), in addition to the amygdala. Experience influences how these circuits develop.

Mothers who report higher maternal care during their own childhood have more gray matter in cortical regions involved in maternal empathy and they have higher levels of activation in these brain regions in response to hearing their own infant's cries compared with those of other infants.

The neurohormone oxytocin is critical in pair bonding and in parenting. Women administered oxytocin nasally show increased neural responses in the insula and other cortical regions in response to the cries of an unrelated baby. Oxytocin administration also decreases neural responses in the right amygdala. Heightened activity in the right amygdala is linked to anxiety and aversion; thus, reduced activity under the influence of oxytocin promotes positive motivations and caregiving responses—even nasty ones like changing a dirty diaper, for example. The levels of oxytocin are higher in women who give birth vaginally compared with those who had their children by cesarean delivery. Mothers who had vaginal deliveries for their children also show greater brain responses in the insula, striatum, and cingulate cortex in response to hearing baby cries. Regardless of the type of delivery, mothers show higher oxytocin levels during breastfeeding. It has been shown that mothers who formula feed have weaker response to their own infant's cries in the insula, striatum, amygdala, and superior frontal gyrus. As will be discussed, these same circuits of empathy that are the basis for maternal-infant bonding are activated in non-caregiving situations that involve interactions among other people in society.

The neurocircuitry of love between man and woman is also engaged in spiritual belief, with very similar consequences promoting peaceful coexistence among people. Religious people highly activate a pathway from the inferolateral to dorsomedial prefrontal cortex when they contemplate the actions of supernatural and spiritual beings and agents in their life. Engaging these pathways associated with spiritual belief also regulates fear and pain circuits, and circuits responsible for theory of mind.

Theory of mind is the ability to perceive the emotions, desires, and intentions of other people—that is, to attribute specific mental states to other individuals, such as their beliefs, intents, desires, knowledge, honesty, and so on, often on the basis of complex subtle cues, and to do so very rapidly. In a word, theory of mind is empathy: the ability to put yourself in someone else's shoes. This ability is highly developed in human beings, and it is essential to our survival as a species because strict social structure is crucial for survival of all humans. These circuits are the neurobiological basis for altruism; that is, self-sacrifice for the benefit of someone else. Religious or spiritual belief engages these brain circuits, involving the medial prefrontal cortex (mPFC), precuneus, and connections between the amygdala and prefrontal cortex. These are all brain regions and circuits that have been encountered earlier in this book in the context of fear and threat detection. Dysfunction in these circuits is associated with mental illnesses, autism, schizophrenia, ADHD, alcoholism, and other antisocial traits.

Structural differences in some brain regions are associated with religiosity; for example, the hippocampus, critical for declarative memory, is smaller in people reporting life-changing religious experiences and in born-again Protestants, but a decrease in hippocampal size is also associated with other factors, such as depression, that could interact with religious factors. The size of the orbitofrontal cortex (OFC) is linked with religious or spiritual activity. In measuring the size of the left OFC over a period of two to eight years, researchers observed less atrophy in participants who reported life-changing religious or spiritual experiences during the course of the study and in Protestant subjects who reported being born-again while entering the study.

Studies on Tibetan Buddhists show that meditation activates the precuneus network, enhancing the awareness of perceptual and cognitive states that lie beneath the awareness of conceptual thought. During silent mantra meditation, the hippocampus and prefrontal cortex become activated, but there is no activation of the cingulate cortex. This suggests a role for meditation in memory consolidation. Expert meditators control cognitive engagement in the conscious processing of sensory-related thought and emotion by strong self-regulation of the fronto-parietal and insular areas of the cortex in the left hemisphere during the meditative state. The anterior cingulate and dorsolateral prefrontal cortex play antagonistic roles in the executive control of attention, which is consistent with meditation requiring both "mind wandering" and focused attention to achieve the meditative state.

These circuits of spiritual belief affect interpersonal actions through empathy and theory of mind networks, but they also regulate internal states, including fear, pain, and stress. Stress has powerful lifesaving benefits to the brain and body, but chronic stress has adverse effects on health. Hyperactive amygdala function is often observed during stress conditions. In an MRI study of people undergoing mindfulness-based stress reduction intervention over an eight-week period, the reduction in perceived stress level was correlated with the decrease in right basolateral amygdala gray matter volume.

Using fMRI to compare brain activity in practicing Catholics and atheists during painful stimulation, researchers found that analgesia (loss of pain sensation) is triggered by an image with religious content in believers, but not in the nonbelievers. Activation of these networks of religiosity enabled believers to detach themselves from the experience of pain by engaging the right ventral lateral prefrontal cortex. This indicates that religious belief enables individuals to engage pain-regulating brain processes. Brain imaging in Carmelite nuns, for example, while they were subjectively in states of union with God, showed widespread activation throughout many brain regions, but many of the regions involved in regulating fear, pain, and social interactions—including the orbitofrontal cortex, anterior cingulate cortex, and left brain stem—were engaged.

The power of this spirituality-induced analgesia is striking when one reads the horrifying accounts of religious martyrs burned at the stake— people who often endured the unimaginable pain and torture with supernatural resistance to the pain and horror—or when one sees monks in protest set themselves aflame with gasoline and remain sitting peacefully in the lotus position inside the flames as they burn to death.

Recall Todd Beamer on the hijacked flight UA 93 reciting the Lord's Prayer together with the 911 operator before announcing, "Let's roll."

This analysis of the neuroanatomical and physiological correlates of religious belief should not be misconstrued as challenging religious faith. In talking with Terrence Reynolds, professor and chairman of the Department of Theology at Georgetown University, about the intersection, and often the conflict, between science and religion, he said, "Religious believers ought not be in any way unnerved by the findings of science. At the same time, one has to be careful not to engage in reductionist thinking about religion.

"Religion tends to focus on questions of meaning and value, which may not be available through analytical verification processes. If one assumes that all rationality is tied to what we know directly through the five senses, that limits our understanding of 'meaning' questions. . . . By definition God is a being that transcends the senses."

Spiritual belief is a uniquely human capability. That human beings have the capacity for religious faith when animals do not means that there is something unique about the human brain that provides this ability. These neurocircuits enabling humans to have religious faith are what we are tracing here.

The neural circuitry of empathy and understanding brings people together into cooperative groups. The self-awareness these circuits convey to human beings instills a sense of awe and wonder upon the realization of the transience of an individual life in a vast and infinitely mysterious universe.

The neurocircuitry of religion, spiritualism, empathy, and theory of mind is included here in a chapter on sex and violence because it derives from the bond between men and women for sex and nonviolence, but it expands into a much larger context. Religion and the neurocircuitry engaged in spiritualism bind people into peaceful, cooperative groups and control individual and group behavior for the peace and order of society.

For centuries and throughout all cultures, religion has been one of the strongest mechanisms of self-control and social organization. "Indeed, there is a case for arguing that *Homo sapiens* is also *Homo religious*," former Roman Catholic nun Karen Armstrong says in her book, *A History of God.* "Men and women started to worship gods as soon as they became recognizably human; they created religions at the same time as they created works of art."

As our understanding of nature and the human body advances, religion and spiritualism are increasingly being supplanted by science. "Indeed, our current secularism is an entirely new experiment, unprecedented in human history. We have yet to see how it will work," Armstrong observes.

Conversely, history and current events show vividly that religion is and always has been a source of violent aggression between individuals and among groups of people. Armstrong acknowledges the enormous conflict and destruction carried out in the name of religion, and she perceives the watershed moment that mankind appears to be approaching: "Like any other human activity, religion can be abused, but it seems to have been something that we [*Homo sapiens*] have always done."

Fundamentally, the neural pathways in the human brain that allow our species to have spiritual experiences and beliefs are instrumental in peaceful coexistence as well as self-control, but when these religious beliefs become perceived as dividing people into different groups, the Tribe circuit of rage and aggression can overcome the theory of mind circuitry, giving us a sense of empathy for others. The LIFEMORTS triggers reinforce one another, but they can also come into conflict.

# 11

# A World of Trouble

My first wish is to see this plague to mankind banished from off the earth.

**George Washington, on war, in a statement of 1785**

Step back and view our species objectively from the outside, the way a zoologist would carefully observe any other animal, or see us the way every other creature perceives human beings. The brutal reality could not be more evident or more horrifying. We are the most relentless yet oblivious killers on Earth.

Our violence operates far outside the bounds of any other species. Human beings kill anything. Slaughter is a defining behavior of our species. We kill all other creatures, and we kill our own. Read today's paper. Read yesterday's, or read tomorrow's. The enormous industry of print and broadcast journalism serves predominantly to document our killing. Violence exists in the animal world, of course, but on a far different scale. Carnivores kill for food; we kill our family members, our children, our parents, our spouses, our brothers and sisters, our cousins and in-laws. We kill strangers. We kill people who are different from us, in appearance, beliefs, race, and social status. We kill ourselves in suicide. We kill for advantage and for revenge, and we kill for entertainment: the Roman Coliseum, drive-by shootings, bullfights, hunting and fishing, animal roadkill in an instantaneous reflex for sport. We kill friends, rivals, coworkers, and classmates. Children kill children, in school and on the playground. Grandparents, parents, fathers, mothers—all kill and all of them are the targets of killing. We devise the cruelest means possible to kill other people: crucifixion, burning at the stake, beheadings, napalm, and Nazi gas chambers. We devise unthinkable means of systematic cruelty designed to produce the most extreme pain and agony possible, which we call torture. There is no animal on land, in the sea, or in the air that we do not kill. Many species that have inhabited the Earth for millions of years have been slaughtered to extinction by human beings. We kill whales, dolphins, seals, otters, sharks, and every other fish that swims or shellfish that hides in the sea. On land we kill herbivores and we kill carnivores: deer, cattle, squirrels, elephants, bears, and lions. Since 1970, human beings have killed off *half* of the entire world's animal population (vertebrate animals), according to a World Wildlife Fund report in 2014. If a virus had been the cause of this global scourge, there would be widespread panic.

And because of our incomparable brainpower, we are by far the most ingenious killers ever to have evolved. We kill with stones, clubs, knives, rope, spears, arrows, slingshot, gun, cannon. We kill with gas, violent chemical reactions, electricity, atomic fission and fusion, bacteria and viruses, fire, water, and with our bare hands. We kill with every means of transportation we have invented by exploiting their power and speed: horses, motorbikes, cars, boats, balloons, blimps, airplanes, satellites, and rockets. We modify our inventions of transport specifically and for no other purpose than to kill other humans: tanks, warships, fighter jets, bombers, submarines, unmanned drones, and missiles. We kill with passenger airplanes. We kill with fire, gasoline, poisons, cleaning products. We keep weapons for home defense and fortify our dwellings with alarms triggering an armed response from professionals trained to kill other humans. We kill with our tools: icepicks, screwdrivers, hammers, power tools, machetes, and box-cutters. In the passion of battle or in premeditated murderous cunning we can turn any object into a lethal weapon, from a high-heel shoe to a suicide vest. We kill with the Internet, using it to prey upon and trap victims or to indoctrinate and train others in killing, and to broadcast the videotaped horror of this killing to terrorize others. We kill with radio, cell phone, and radar. We kill with rhetoric and with books to instruct and incite violence. We kill as individuals, as friends, and in small teams, as gangs, as armies, and in massive global alliances among countries to wipe other countries off the face of the earth. We do this continually and we have done so for eons. Once we cohabited the world with our sibling species, the Neanderthals, and bred with them. They had bigger brains and larger bodies than *Homo sapiens*, but they are gone.

Other species—fish, birds, and creatures of the forest—intermingle except when encountering predators or competitors for a limited resource. Violence among animals is engineered by evolutionary design to give each animal a precise role to maintain the intricate food web and sustain the delicate ecological balance of nature, but all creatures in the sea, air, or on land scatter when a human being is present. The sound, smell, and image of a human being is universally feared among Earth's wild creatures, and for good reason. All animals have been programmed through evolution and experience to know that human beings are killers. Even the lowly lizard with its feeble brain is hardwired to dart away from us on sight.

Our society is structured by violence. Large sectors are devoted to conducting violence or to maintaining control of human violence. In the United States 7.4 people in 1,000 are in prison. The United States houses 25 percent of the world's prisoners but only about 5 percent of the world's population. According to the US Bureau of Justice Statistics, blacks account for 39.4 percent of the total prison and jail population; Hispanics comprise 20.6 percent. Ninety-three percent of all prisoners in the United States are male. One-fifth of our national wealth in the United States is spent on killing in defense and for advantage, and $718 billion of the US federal budget is spent on defense (2011 statistics). This figure does not include the expense of benefits to veterans, another 3.5 percent of the federal budget. Only 2 percent of the national budget is spent on education or on science and medical research. The United States spends more on its military than the next thirteen nations combined.

Genocide in WWII by the Germans, and Japanese atrocities against the Chinese and war prisoners have been reduced to vacuous numbers: 300,000 Chinese killed and 20,000 women raped by Japan in the Nanjing Massacre; 11 million people killed by the Nazi genocidal policy against Jews, gypsies, Poles, and others deemed inferior. The atrocities of war are more easily dismissed when the criminal acts are committed by the enemy, but this mass human violence against other humans is not exclusive to any nation or group of people. Secretary of State John Kerry, when he testified as a Vietnam veteran before the US Senate in 1971, stated that many soldiers had told him of atrocities committed by US servicemen. "They told the stories of times that they had personally raped, cut off ears, cut off heads, taped wires from portable telephones to human genitals and turned up the power, cut off limbs, blown up bodies, randomly shot at civilians, razed villages in a fashion reminiscent of Genghis Khan, shot cattle and dogs for fun, poisoned food stocks, and generally ravaged the countryside of South Vietnam in addition to the normal ravage of war and the normal and very particular ravaging which is done by the applied bombing power of this country." Nick Turse, in his 2013 book *Kill Anything That Moves*, documents many examples of US atrocities on the scale of the My Lai massacre and routine violent crimes against civilians and the enemy by individuals and teams of US soldiers in Vietnam. Vietnam is not an exception; it is cited here because the time that's passed since this war raged provides a vantage point for objectivity. WWII is too remote and the current Middle East wars are still too raw. Clearly, the majority of servicemen and -women could not commit such war crimes; they risk their lives and engage in violence for the most noble of causes, but such criminal violence in war is an ugly and undeniable aspect of human nature. *Homo sapiens* do such things to other members of their own species.

Even within the bounds of accepted practice in war, some actions can seem in retrospect unjustifiably brutal. The busiest airport in history is not Chicago O'Hare, or London Heathrow; it was a secret CIA-built airbase in Laos. The small country of Laos, posing no military threat to the United States, was the most heavily bombed country in the history of warfare. More bombs were dropped by the US Air Force on Laos in the 1960s and '70s than on Germany and Japan combined. The target was not a massive military industrial complex or military base. The Ho Chi Minh Trail, a meandering supply route for Viet Cong fighters into South Vietnam, ran through this region of civilian villages and farmland. From 1964 to 1973 the United States dropped more than 2.5 million tons of bombs on Laos in nearly 600,000 bombing missions—equal to a planeload of bombs dropped every eight minutes, twenty-four hours a day, for nine years. US Air Force Chief of Staff Curtis LeMay gloated, "We're going to bomb them back into the Stone Ages." Homeless and defenseless, the peasants, villagers, and refugees of Laos spent months or years hiding in holes or trenches dug into the foothills. The terrain remains pockmarked by craters today, and more than 20,000 people have been killed or injured in Laos by unexploded ordnance since the bombing ceased. Twenty million gallons of Agent Orange and other herbicides that were sprayed by the US Air Force over Vietnam decimated the environment and caused widespread birth defects and illness. Forty-two percent of all the herbicide was sprayed over food crops in an effort to starve the civilian population.

Why do we snap? Fundamentally, this is why. We are drawn to carnage, not repelled by it. Freeways are shut down by rubberneckers. Murder mysteries and horror movies are extremely popular forms of entertainment that have wide appeal. Television programs routinely feature guns and crimes of violence. Violence in sport is conducted in a highly structured manner, as in boxing and football, or it is the focal point of its appeal, as in "pro" wrestling and the fistfights accepted and expected in ice hockey. Riots break out after soccer matches, sometimes deadly. Through evolution the parts of our brain that produce violent behavior have become fortified and embellished to the extreme. One wonders if our violent brain circuits may be equivalent to the evolutionary extravagance of the peacock's tail feathers—pushed by evolution to the brink of dysfunction. These circuits of violence are energized by circuits of cunning and ingenuity into a potent, deadly mix of horrifying cruelty and savage attacks. The unhealthy alliance of these violence circuits, originally essential for survival, seems to have outgrown the circuits that keep them in check. Rather than being activated under circumstances that are strictly necessary and lifesaving, our evolutionary path expanding our ingenuity has transformed our environment such that these triggers of rage are now constantly exposed to release violence.

The prolific author, biologist, historian, and futurist H. G. Wells, who conceived of time travel and foresaw space travel and nuclear weapons, concluded that our species was doomed by its aggressive excesses. Viewed objectively, *Homo sapiens,* he concluded, must, through forces of evolution that maintain balance in nature, be replaced very soon by a superior species. In his last work, *Mind at the End of Its Tether,* published in 1945, Wells argues that the human brain has evolved to a grotesque level of violent dysfunction that has already doomed our species: "The bird is a creature of the air, the fish is a creature of the water, man is a creature of the mind. . . . A series of events has forced upon the intelligent observer the realization that the human story has already come to an end and that *Homo sapiens*, as he has been pleased to call himself, is in his present form played out."

This dark view of humanity is simple biology, as Wells sees it: "During all this period [of evolution] there has been a constant succession of forms, dominating the scene. Each has dominated, and each in its turn has been thrust aside and superseded by some form better adapted to the changing circumstances of life. Each has obeyed certain inescapable laws that seemed to be in the very nature of things. First of these laws was the imperative to aggression."

Wells traces the evolution of *Homo sapiens* from early primates, and attributes our species's rise to domination to our peculiar unbridled childlike aggression being infused by our creative minds and constantly innovating ever more potent means of killing. "Their semi-erect attitude enabled them to rear up and beat at their antagonists with sticks and stones, an unheard-of enhancement of tooth and claw. But presently their sociability diminished . . . These cursive ground apes [that is, primates who evolved from ancestors like chimps living in trees] were the *Hominidae*, a hungry and ferocious animal series."

Our arboreal ancestors "acquired quickness of eye and muscular adjustment among the branches. They were sociable and flourished wider," but when they evolved to a ground species "the usual increase in size, weight and strength occurred, they descended perforce to ground level, big enough now to outface, fight and outwit the larger carnivores of the forest world." This led to interpersonal conflict, tribalism, and interspecies combat that annihilated all other lines of *Homo* except for the most violent *Homo sapiens*. "Families and tribes may have warred against each other and the victors have obliterated their distinctiveness by mating with captive women. . . . primordial adult *Homo* [Neanderthals and others], for all effective purposes, faded out, leaving as his successor the childlike *Homo sapiens*, who is, at his best, curious, teachable and experimental from the cradle to the grave." (Wells would be amazed by modern methods to track human ancestry through the genetic imprints left in our DNA, as well as the fossil record encased in stone. Most descendants of northern Europeans, including me, as I learned from an analysis of my own DNA, retain up to 4 percent Neanderthal DNA in our genome, indelible fingerprints of the children born to parents whose matings long before recorded history united—either in love or rape—these two early species of man.)

Many dismiss Wells's pessimistic final book as the dreary musings of a dying old man suffering in sickness and grappling with the end of his life. He died the following year. Wells anticipated this reaction. In this final work Wells concludes, "Ordinary man is at the end of his tether. Only a small, highly adaptable minority of the species can possibly survive. The rest will not trouble about it, finding such opiates and consolations as they have a mind for."

Look at *Homo sapiens* through the eyes of any other species in the wild and you see one thing—*death*.

## A Rage Murder Autopsy

For insight into the roots of rage killings, consider this true murder mystery.

The body was found facedown at 1:20 p.m., on September 19, 1991, partly wedged in a rocky crevice. At first the hikers thought it was a pile of trash, but the shocking truth quickly became obvious. They had stumbled upon the corpse of a forty-six-year-old man. He was well armed and carrying a backpack stuffed with valuables. The cause of death was determined at autopsy—homicide.

The man had been shot in the upper back. The entrance wound pierced the victim's shoulder blade and severed a major artery. The projectile stopped just centimeters from his lung. Since the nerves energizing that limb were severed, the victim's arm would have been paralyzed, and he would have bled out quickly from the spurting blood pulsating from major arteries exiting the heart. There were defensive wounds on the palm of his right hand, sliced to the bone by a knife blade. The base of his skull was fractured and blood had pooled inside the brain. This cranial injury could have been the result of a fatal fall after being mortally wounded, but more likely it was the murderer's coup de grâce as he stood over his fallen victim.

Piecing together the forensic evidence, authorities arrived at their conclusion of murder because the perpetrator had taken careful steps to cover up his crime. All traces of the murder weapon had been removed. The man had been shot by an arrow, the tip of which was still lodged in the now-frozen body, but the arrow shaft itself had been pulled out by someone standing behind the fallen body. It was never recovered. That arrow could have been matched to others in the killer's quiver, providing incriminating evidence. The victim was of high status and he came from a small village, so the victim most likely knew the person who suddenly ended his life of forty-six years. Stomach-content analysis led investigators to conclude that the man had no reason to fear for his life in the time leading up to his death. He had not been fleeing the killer; he had died suddenly. The forensic analysis showed that the dead man had just consumed a very large meal—a generous barbecued slab of meat. This feasting is not the behavior of someone running for their life or engaged in mortal combat. Moreover, the man was very wealthy, but none of his possessions were taken, including precious metal that he was carrying. It was not a robbery. The authorities speculate that the murderer left the victim's valuable personal effects with the body because they would have identified the killer had he taken them back to the small community where he lived.

The autopsy was performed in July 2001 by Dr. Eduard Egarter Vigl and Dr. Paul Gostner, but the murder had been committed more than 5,350 years earlier, on the cusp of man's transition from the Stone Age into the Copper Age. The victim, now called Ötzi the Iceman of the Alps, had been interred in an icy glacier at an elevation of 3,210 meters in the Austrian–Italian Alps until one day, centuries after *Homo sapiens* had passed from the Copper Age into the Electronic Age, after having ventured beyond Europe, across the vast Atlantic, and inhabited every corner of the globe and even reached into outer space to walk on the surface of the moon, the sun's rays finally melted away the ancient ice and opened a time capsule on an archaic homicide that authorities concluded was committed in a sudden fit of rage for reasons that are lost to time.

The key clue to the victim's high status was the copper ax he was carrying and which the murderer left alone. Copper was the product of the most advanced technology of the day and only the most elite members of society at the time would have had such a modern tool fashioned from the most precious metal of the era. The murderer dared not steal it and risk becoming identified by the booty. Had the man died in a robbery, or in the course of battle, the copper ax would have been snatched as rightful plunder. The fatal arrow would still have been lodged in his flesh.

It may seem a remarkable coincidence that not only could the mummified remains of a human being from BCE 3000 be discovered intact in the twentieth century, but that the death of this man had been the result of homicide. Perhaps, but the archeological facts suggest that death by homicide was not at all rare for ancient man. The nineteenth and twentieth centuries have seen the mass deaths of millions of people in horrendous wars, one after the other, through generations. Since 1914 the world has endured horrific world wars, genocides, mass exterminations, "ethnic cleansings," riots, murders, and vicious civil wars. A staggering 100 million to 200 million *Homo sapiens* have been killed in mass battles with other members of the same species. Hitler, Stalin, Mao . . . dictators and warlords the world over have slaughtered millions of people, but these warmongers are only the most recent in a long line of ruthless warriors that can be traced back through Napoleon to Genghis Kahn to the brutal Roman Empire. But measuring the 100 to 200 million people killed in wars of the twentieth century against the 10 billion lives lived in the same period yields a homicide rate from war of 1 to 2 percent. This an astonishing figure, but compare that with the tiny populations that humans inhabited in prehistoric times: Professor Ian Morris of Stanford University calculates that prior to recorded history, human beings died at the hands of other people at an appalling rate of 10 to 20 percent.

When we speak of the LIFEMORTS triggers of rage being tripped by the radical changes in human existence in the modern world, it is important to put it in the proper perspective. Human ancestors developed from other primates in Africa about 5 million years ago. The chimp brain shares the fundamental structure of the modern human brain. "Lucy," *Australopithecus afarensis*, lived 3 million years ago. About 2.3 million years ago the first humans (hominids of the genus *Homo*), with their very large brains, arose in Africa. Some 1.5 million years ago early humans migrated out of Africa into Asia and Europe. And 100,000 years ago *Homo sapiens,* with the even larger brains that we have today, thrived in Africa, while our sister species *Homo erectus* and *Homo neanderthalensis* lived in different parts of the Old World. Humans lived at this time and for tens of thousands of years later in family groups and small social groups much like nonhuman primates. The first evidence of human culture—cave paintings, body adornment, and elaborate burials—appears only 50,000 years ago. At this time humans were forming larger, more organized societies and developing technology that would rapidly change the world. By 25,000 years ago, all other *Homo* species except *Homo sapiens* were extinct. Humans continued to develop advanced technology and large complex social groups, and they spread throughout the Old World. Even the Iceman of BCE 3000, with the same brain we have today—a brain that had developed 100,000 years earlier—is a recent development.

The environment of *Homo sapiens* is changing radically, at an explosive rate that far outstrips the pace of evolution. Through generations, evolution has been the enduring mechanism by which organisms were matched to their environment. The exponential transformation of our environment by technology in the last few thousand years has made our world utterly alien to the environment that forged the human brain in ancient Africa. Many people alive today lived in a time before computers, space flight, atomic energy, or cell phones. A little over a century ago our species relied on animals for transportation. Knowledge of germs, electricity, and the cellular and molecular basis of life are recent events in human existence. These technological developments that have transformed our environment are so recent they cannot even be plotted on the time line of *Homo sapiens* ancestry on this planet. It is clear that our species has changed the environment radically, faster than biology can change our bodies to adapt, and that we are an extraordinarily ruthless animal, uniquely separated from the natural world. Since 1980 a staggering 421 million birds have vanished from Europe, according to a 2014 census; a 20 percent population decrease. The deaths are hardly limited to vulnerable exotic species; those were the first to go, but since 1980, 90 percent of the decline is in the thirty-six most common species of European birds. The losses are attributed to humans, primarily by human-caused environmental degradation. Nearly one-third of the world's amphibian species are threatened with extinction. Amphibians have flourished on the Earth for more than 300 million years, but in just the last twenty years 168 species of amphibians have gone extinct due to environmental changes caused by humans: habitat loss, pollution, and climate change.

Human beings have changed the environment too rapidly for these animals to adapt, but are these mass extinctions "canaries in the coal mine," indicators of environmental dangers threatening our species? *Homo erectus* flourished on Earth for about 1.8 million years. *Homo sapiens* has only been on the planet for 100,000 years. To put this in perspective, if *Homo erectus*'s time on Earth were one day, *Homo sapiens* have been in existence for one hour and twenty minutes. From a biological perspective, there is little information or precedence to predict how long this latest offshoot of the *Homo* genus will survive.

Human beings are violent creatures, with an odd penchant for killing their own. What is clear is that human brains are uniquely equipped for violence and for defense against attack by other members of their own species. In part this is because we are all descendants of the warriors who survived in violent confrontations to pass on their genes. Recent DNA analysis shows that of 3,700 Austrians volunteering DNA samples, 19 percent showed the indelible traces of ancestry leading back to the Iceman who was murdered on that mountain thousands of years ago. The Iceman carried a mutation in a gene, G-L91, which resides on the Y chromosome passed on from father to son over generations after that Copper Age man was murdered suddenly by someone he knew.

History shows that there is little logic to the horrifying mass homicides in war. War is most often called "senseless," but also inevitable and necessary. Brothers killed brothers in the US Civil War. More Americans died in the US Civil War than in all other American wars combined. Peoples that Americans fought to annihilate in World War II, the Japanese and the Germans, became our closest allies within ten years of the war's end, whereas our vital WWII ally, the Russians, became our most feared enemy in a Cold War that brought the world to the brink of nuclear holocaust.

## Triggering War

In 2014, a US Navy P-8 Poseidon patrol aircraft was flying off the coast of China's Hainan Island. The large P-8 US surveillance warplane is built on the Boeing 737 platform, and equipped with a bomb bay to drop bombs, torpedoes, and depth charges. Air-to-surface missiles are clutched as if by talons beneath its wings. As the aircraft skirted the Asian coastline, suddenly a Chinese J-11 fighter jet came screaming toward it in a deadly game of chicken. At the last instant, the Chinese fighter jet flashed past the Navy airplane, barrel rolling to miss it by only twenty feet.

Air Force general Herbert Carlisle, head of US Pacific Air Forces, says that China's naval and air forces are "very much continuing to push," and they are becoming more active in international waters and airspace in Asia. "They still talk about the century of humiliation," he said, referring to China's decline as a world power in the twentieth century and the country's determination to rise to a great world power in the twenty-first century. In recent years China has conducted more military exercises farther from its shores and pressed into disputed territorial waters, some regions coveted by Chinese oil-drilling companies. "The opportunity for something to go wrong," General Carlisle warns, increases with the rising numbers of close encounters between the United States and its allied countries and the Chinese Navy and Air Force in border patrols.

Meanwhile, deteriorating relations between the United States and Russia have fueled a resurgence of confrontations not seen since the dangerous cat-and-mouse encounters of the Cold War that brought the world so close to nuclear annihilation. Russia is now conducting long-range reconnaissance and bomber missions encroaching on US territory. On September 17, 2014, two Russian fighter jets, two long-range bombers, and two refueling tankers were skirting the coast of Alaska at the edge of international airspace when US fighter jets intercepted the half-dozen Russian attack force approaching the US border. As with China, General Carlisle attributes the Russian provocations to President Vladimir Putin's intention "to reassert Russia into what he thinks its rightful place in the international order is, and part of that is continuing to push into the Pacific." In contrast to the recent incidents with China, where the Asian country's actions can be characterized as defending (rightly or wrongly) what it perceives to be its territory, the Russian military probes along the Alaskan coast are undertaken to test the borders of US territory. These are aggressive rather than defensive actions. Over the past year there has also been a sharp rise in Japanese intercepts of Russian military aircraft along Japanese territorial borders, and Russian submarines have been lurking off the coast of Sweden, provoking swift reaction by NATO fighter jets. In August 2015, Finland fired depth charges on a suspected Russian submarine in its territorial waters off Helsinki.

One must resist oversimplification of such dangerous military challenges as contests between "good and evil." Such characterizations are justified in many cases, but the Tribe trigger neurocircuitry of violence in our brain to defend one's own people predisposes human beings to form herds with their own and to reflexively adopt an "us versus them" perspective in conflicts instinctively. The United States also tests Russian territorial defenses in much the same way that Russia tests ours. In April 2014, a Russian Sukhoi Su-27 fighter jet flew within 100 feet of a US Air Force RC-135U aircraft operating on a surveillance mission over the Sea of Okhotsk. The RC-135U is a large, four-engine jet aircraft with a 135-foot wingspan, designed to collect electronic intelligence on adversary radar systems.

There is reason for concern about possible Russian territorial expansion. In April 2014, Russian president Vladimir Putin closed the entire Sea of Okhotsk—52,000 square kilometers of ocean adjacent to Russia and Japan. Prior to this the Okhotsk Sea had been open to all other countries for fishing and deep-water exploration and was a major fishing area for Japan and China.

Here we see the same LIFEMORTS triggers of reflexive violence wired into the human brain of every individual played out on a massive scale in international conflict. (Stone Age brains with Space Age weapons.) Defense of one's territory, or the Environment trigger, is the fundamental purpose of any military, and territorial intrusions are very often the cause of war. The Russian expansion into Ukraine is a recent example, but throughout history, wars constantly dissolved and redrew the territorial borders of countries. Perceived insult (the Insult trigger) is reflected in China's and Russia's current aggressive reassertion of military power. Fights over resources, such as oil deposits, are often the trigger of war—the Resources trigger. Blockades are an act of war. This trips the Stopped trigger or rage. The Tribe trigger of international conflict is so obvious it needs no further discussion. Populations of people with different religions, races, sects, social status, and political beliefs coalesce around those with similar characteristics and amass into warring factions in wars throughout the ages, as is starkly evident in current events in the Middle East, Africa, and Europe. Christian, Muslim, Jew; different sects within the Islamic faith; genocide in Africa and the Balkans; and World War II are examples of brutal conflicts and wars in anger over "tribe." If not the cause of conflict within and between societies, the Tribe trigger is always at the surface in any military action. The enemy is differentiated and alienated, often with derogatory names that relegate them to an inferior status: gooks, Japs, Krauts, and ragheads.

When any of these triggers or a combination of these triggers results in a violent action, the Life-or-limb trigger to defend oneself is tripped. When this happens, the Life-or-limb trigger provokes the ultimate and irresistible outrage of anger and commitment to war to punish or annihilate the attacker. This is what General Carlisle wisely perceives as the potential spark of a war that could sweep around the globe from a mishap in airspace between aircraft pushing the limits of territorial borders. As in an individual's brain when the Life-or-limb trigger to fight in defense is tripped, a violent response in populations is assured to erupt to defend against an attack on one's country, as we saw after Pearl Harbor and after the attack on the World Trade Center and the Pentagon on September 11, 2001. The Japanese sneak attack on Pearl Harbor was provoked by the US economic blockade of resources to Japan (Stopped and Resources triggers). The attack on September 11 was provoked by what was perceived by Islamic extremists as US intrusion into the Middle East and a clash of religious and cultural values (Tribe trigger).

Politicians can exploit the LIFEMORTS triggers unscrupulously to provoke the public into support for war. The Vietnam War was sparked by an attack on US Navy Destroyer USS *Maddox* and USS *Turner Joy* along North Vietnam's coast in the Gulf of Tonkin on August 2 and August 4, 1964. The August 4 attack prompted Congress to approve the Gulf of Tonkin Resolution on August 7, 1964, giving President Johnson power to conduct military operations in Southeast Asia without a declaration of war. The president assured America that he would not commit the country to a war in Vietnam. "We are not about to send American boys nine or ten thousand miles away from home to do what Asian boys ought to be doing for themselves," he said in his 1964 campaign for president.

History would show that pronouncement to be misguided or disingenuous. The prolonged war eventually took the lives of 58,220 American soldiers and wounded 150,000 more. Approximately 830,000 Vietnam veterans returned home suffering from post-traumatic stress disorder. The president and his supporters had invoked the powerful Life-or-limb trigger in the minds of American citizens on a false pretext. In 2005 and 2006 the National Security Agency declassified top-secret documents related to the Gulf of Tonkin incident and, together with transcripts released from the Lyndon B. Johnson Library, the two hundred documents reveal that there was in fact no attack on US forces by the Vietnamese on August 4, 1964. This is something many had questioned from the outset of the Tonkin Resolution debate, but the suspicion could not be proven for more than forty years after the events. Quoting from an article by Lt. Cmdr. Pat Paterson, "The Truth About Tonkin," published in the US Naval Institute *Naval History* magazine, "These new documents and tapes reveal what historians could not prove: There was not a second attack on U.S. Navy ships in the Tonkin Gulf in early August 1964. Furthermore, the evidence suggests a disturbing and deliberate attempt by Secretary of Defense McNamara to distort the evidence and mislead Congress."

The Vietnam War failed to unite the citizens of the United States. Instead, the war ripped the country internally into embattled factions, igniting violent protests and riots at home. Some 125,000 Americans abandoned their homeland for Canada to avoid the Vietnam draft. Approximately 50,000 American servicemen deserted. Increasingly, the public, and especially the young men who were conscripted to fight the war in Southeast Asia, questioned the legitimacy of the war and refused to support it. From a neuroscience perspective, none of the LIFEMORTS triggers were engaged in the minds of the individuals called upon to support the war or who were conscripted to risk their life fighting it. Even the powerful Life-or-limb trigger of defensive violence that initially provoked support for the war in the Tonkin Resolution was lost. Insult, the defense of honor, and Tribe, to stop the spread of communism, were the only remaining causes that could be marshaled to motivate US citizens to engage in the violence of the Vietnam War, but the arguments pressing on these defensive responses of the human brain were insufficient for an increasingly large proportion of the population. Eventually even the impartiality of journalism could not tolerate the injustice, and the highly respected news reporter and former WWII correspondent Walter Cronkite argued on air on February 27, 1968, that the war was a lost cause. A month later, President Johnson announced that he would not seek reelection. The country spent the next several years struggling to extract itself from the conflict, ending on April 20, 1975, with the last US Marines desperately escaping by helicopter from the roof of the US embassy in Saigon. North Vietnamese troops poured into Saigon within a few hours of the Marines' departure. The war was over. The conflict between North and South Vietnam simply failed to provoke the requisite sense of anger in individuals in the United States who would have to carry out the violence. Appeal to rational argument for war was insufficient because the LIFEMORTS triggers had not been activated to provoke sufficient anger and commitment to fighting in the minds of individuals. In 1977, President Jimmy Carter granted a full pardon to all Vietnam draft dodgers.

Some members of Congress and the public found it implausible that the sneak attack on Pearl Harbor on December 7, 1941, could have occurred without US intelligence being aware of the large attack force approaching the Hawaiian Islands, and they suspected President Roosevelt allowed Japan to strike the first blow to rally public support for declaring war on Japan. The Joint Committee on the Investigation of the Pearl Harbor Attack, headed by Sen. Alben Barkley, which was formed on June 20, 1946, concluded that "The ultimate responsibility for the attack and its results rests upon Japan," but two of the ten senators dissented even from this delicately worded statement. The purpose here is not to enter into that debate, but rather to demonstrate that the Life-or-limb trigger of defense is universally recognized as a powerful justification for engaging in violence in war, and that political and military leaders can exploit this unconscious provocation of anger unscrupulously.

History would show that World War II united the country in battle, in contrast with the Vietnam War, because many LIFEMORTS triggers were activated by the Axis powers, and that devastating war has come to be cited as an example of a "just war." The war was perceived as an act of self-defense (Life-or-limb trigger). Halting German aggression and genocide put the United States in alliance with others who shared the principles of defending one's homeland (Environment trigger) and defending against attacks on groups of people based on ethnic or religious beliefs, and in joining with others who Americans felt an affinity with (Tribe trigger). The accepted rules of international conduct and diplomacy had been repeatedly violated by Hitler's aggression (the Organization trigger).

To better understand the causes of war, to help avoid them and seek alternatives to violence, and to guard against being defrauded into war unscrupulously by political leaders, it is necessary to understand the LIFE-MORTS triggers of violence in the human brain. Regardless of politics, war results from anger in the minds of individuals, which is provoked by the brain pathways that have evolved to promote an individual's survival in the face of a dangerous threat and to ensure the survival of groups of people organized into societies.

Many who know John Steinbeck from such vivid classics as *The Grapes of Wrath* and *Cannery Row* may be surprised to learn that in the final year of his life he served as a war correspondent in Vietnam reporting for a column in *Newsday*. It may be that Steinbeck's extraordinary ability to bring to life, with such clarity and enchantment, the likes of Lee Chong in *Cannery Row* provokes revulsion when applied to the brutality of Charlie in Vietnam.

A man suspected of communicating, only suspected, is taken to a village center. His neighbors are forced to look on while he is taken apart little by little, starting with fingers and toes but carefully so that bleeding will not give him quick release, and when they have finished, he is a ghastly mound of butcher's meat. You don't believe it? I could show you photographs but no American paper would dare print them . . .

Two weeks later, January 21, 1967, Steinbeck reports:

At about 10 o'clock in the evening two strolling young men paused in front of a crowded restaurant and suddenly threw two grenades in at the wide-open door. One was a dud. The other exploded and tore up the people and their children. There were no soldiers in the restaurant either American or Vietnamese. There was no possible military advantage to be gained. An American captain ran in and carried out a little girl of 7.

Steinbeck's reports from nearly fifty years ago evoke scenes fresh from the Middle East. The incessant cycle of us-versus-them human violence evokes visceral despair in seeing over and over how human beings seem powerless to escape the ancient tribalistic imperative for brutal violence and aggression wired into our brain for survival in the hostile environment of the distant past.

> Last week in a remote village of a northern province of Thailand, a schoolteacher was taken by a band of infiltrating terrorists. He was killed and his head cut off and put upright in the middle of a table. . . .

Steinbeck and his wife were friends with President Lyndon Johnson and the first lady, and he embraced the president's "Great Society" programs to eliminate poverty and racial injustice. This friendship predisposed Steinbeck to strongly support the Vietnam War from the outset. Both of his sons served in the army. His son John was drafted and he was serving in Vietnam while Steinbeck was there reporting on the war. Steinbeck's support for the Vietnam War put him at odds with young Americans who denounced the war with raging dissent. They protested daily in the streets and branded the author of such compassionate classics as *Of Mice and Men* and *The Grapes of Wrath* a warmonger. Steinbeck bristled at the charge but rebuffed the angry critics as hypocrite hippies: "Couldn't some of the energy that goes into carrying placards be diverted to emptying bedpans or cleaning infected wounds?" he asked. "This would be a real protest against war."

Over the few months that he reported from the battlefield, seeing it with his own eyes, his strong support for the war began to crumble. "John changed his mind totally about Vietnam while there," Steinbeck's wife said, "and he came home to write it and spent the rest of the time dying."

Steinbeck returned home to New York and died the next year on December 20, 1968.

Steinbeck had grasped what many soldiers on the ground and military commanders had deduced, long before neuroscience illuminated the circuits of aggression and the LIFEMORTS triggers of rage. "We're the redcoats in this one, lieutenant," a full colonel quietly advised Thomas Barden, editor of the book *Steinbeck in Vietnam* when he first arrived in Tây Ninh to take command of a battery in 1970. "Don't be John Wayne. Take care of your men and get everybody home. We won't be here much longer."

The Selective Service proves that the capacity for violence exists in nearly everyone. Nearly anyone can be plucked from their routine and plunged into battle if the appropriate circuits of rage are tripped in our brain. This capacity for rage and violence is a biological imperative for survival. The Vietnam conflict was doomed because it failed to evoke the triggers of rage in the human brain. This contrasts starkly with the rage and commitment that erupted universally among Americans when more than 3,000 citizens were suddenly killed by terrorists on September 11, 2001. If a person could not go to work in the morning in an office building in Manhattan to contribute to society and support their family, then nothing else mattered. The motivation for the resulting fierce rage has little to do with politics; it is biology.

## Us Versus Them

Anyone holding the opinion that torture ended when civilization emerged from the Middle Ages is ignoring the facts. An international survey of 21,000 people in twenty-one countries on every continent compiled in 2014 revealed that one-third of respondents (36 percent) believed that torture is justified in some cases to protect the public. In China and India the large majority (74 percent) held this view. In the United States, 45 percent felt that torture was sometimes justified, well above Russians (25 percent), Argentinians (15 percent), and Greeks (12 percent). Torture is in fact a crime under federal law in the United States and in 151 other countries. Nevertheless, nearly half (44 percent) of the people surveyed around the world feared being tortured if taken into custody.

The statistics show that there is good reason for fear of torture in custody. Amnesty International reported torture in 141 countries in the last five years, in many cases carried out by government security authorities against others simply for being perceived as different. Consider this news item:

Pakistani officer kills man over alleged blasphemy: Pakistani
police said an ax-wielding officer killed a Shiite man in police custody, claiming he had committed blasphemy.

Torture and violence are often inflicted on others for taking part in anti-
government demonstrations. In November 2014 the body of journalist Par
Gyi, who was killed while in custody by the Burmese military, was exhumed
from a shallow, six-inch-deep grave with his wife present to identify the
body. Par Gyi was one of the first National League for Democracy Youth
members, but he had been driven into exile in Thailand, where he began
working as a freelance journalist under the pseudonym Aung Naing. His
body showed clear signs of torture, a broken skull, and a broken jaw from
beatings he received before he was shot five times fatally by his military
captors.

Mass demonstrations dividing people into "us versus them" often incite
raging brutal violence or torture, as in China's Tiananmen Square in 1989 or
recent pro-democracy protests in Hong Kong. Hakan Yaman, father of two,
was one of thousands of victims of police violence during the 2013–2014 Gezi
protests for democratic freedom in Turkey. Yaman was not a protestor. He
was simply returning home from work when he was mistaken for a protestor
as he passed a peaceful demonstration. Yaman was attacked by a group of
police, who beat him with clubs. They dragged him on top of a street fire to
burn him alive, but before leaving him to burn one police officer gouged one
of his eyes out.

"According to the forensic medicine report Hakan Yaman sustained
serious injuries to his head and face. His nose, his cheek bone, and the bones
of his forehead and his chin were broken. He lost one eye completely and has
lost 80 percent of his sight in the other eye. His skull was fractured from the
top of his head all the way down to his jaw and his back sustained second
degree burns from being thrown on the fire. He lost consciousness during
the attack."

In Mexico, torture and violence by police is epidemic, because of
entrenched widespread corruption among the police, government officials,
and drug gangs. Torture and violence in Mexico increased by 600 percent in
2013, compared with 2003, according to the National Human Rights Commission. Between 2010 and 2013 the National Human Rights Commission
received more than 7,000 complaints of torture and other ill treatment by

Mexican authorities. Victims who do not survive are of course not able to file a complaint. In August 2014, the Mexican federal government acknowledged that more than 22,000 people are currently missing or "disappeared" in Mexico. Bodily remains, when found, often display evidence of torture. On September 26, 2014, the mayor of Iguala, Mexico, José Luis Abarca, ordered that forty-three students at a teaching school, who were traveling in four vehicles, be intercepted before reaching town because he feared they would disrupt a speech by his wife, María de los Ángeles Pineda. Six students were killed on the spot and the others were kidnapped and reportedly turned over to the drug gang Guerreros Unidos. The students were taken to a garbage dump, forced to their knees, and each one was shot. The killers stacked the forty-three bodies into a pile together with wood and tires, doused them with diesel and gasoline, and set them ablaze. The fire burned for fifteen hours. The remaining ashes, teeth, and bone fragments were stuffed into sacks and thrown into the Rio San Juan River.

Sixty-four percent of Mexicans in the Amnesty International survey said they fear being tortured if taken into police custody. In Brazil the fear of torture in police custody was even higher—80 percent—but even in the United States, 32 percent of those polled feared torture in custody, much higher than people living in the UK, Australia, or Canada.

On June 28, 2010, a federal jury in Chicago convicted former Chicago Police Department commander Jon Burge on perjury and obstruction charges related to his denials that he participated in the torture of suspects in police custody in the 1980s. The jury found that Burge lied and obstructed justice in November 2003 when he provided false statements in a civil lawsuit that alleged that he and others tortured and abused people in their custody. On January 21, 2011, Burge was sentenced to fifty-four months in prison.

In July 2009, the state of Michigan agreed to pay a $100 million settlement in a class-action suit brought by more than five hundred female prisoners who alleged that they had been sexually assaulted by prison guards. According to a Department of Justice report on May 16, 2013, between 2011 and 2012, 80,600 inmates in United States jails and prisons experienced sexual victimization by another inmate or facility staff member. This represents 4 percent of the state and federal prison population. More sexual assaults of inmates were committed by staff, as opposed other inmates (2.4 percent compared to 2.0 percent). In 0.4 percent of the cases, the assaults occurred by staff and another inmate operating together. The rates of sexual assault among juvenile inmates in the United States are even higher. In 2013, a report found that 1,720 juveniles in state-owned or -operated juvenile detention facilities experienced sexual victimization (9.5 percent); 1,390 of them (7.7 of the total) were assaulted by staff members.

All of this brutality results from a categorization of "us versus them" made by perceptual circuits in people's brain. Second, to inflict such horrific pain and brutality in torture or terrorism, the brain must lose all empathy, and in some way cease to regard another human being who is from an "out-group" as themselves being human. John Steinbeck, seeing terrorism first-hand, could not fathom the mind of such people, but neuroscientists are beginning to uncover how this brutality works at the level of automated brain circuits.

# Tribe

"I was too young to really understand what was really going on," actor George Takei, who played Sulu, the helmsman of the USS *Enterprise* on the television series *Star Trek* said, recalling an unforgettable moment in his life when he was only four years old. "But I still do remember that day when armed soldiers—soldiers with guns, bayonets on them—came to our home to order us out. I remember that as a *very* scary day."

American-born Takei was forced with his family from their Southern California home into internment in 1942 to live in horse stables at the Santa Anita Park. After a few months, the family was sent to Rohwer War Relocation Center in Rohwer, Arkansas, and then later transferred to the Tule Lake War Relocation Center in California.

"When the camp was built we were put on a train and taken all the way across the southwestern desert to the swamps of Arkansas to a camp called Rohwer and I grew up there," Takei remembers. The Tule Lake camp was even harsher: "There were three levels of barbed wire fences and tanks patrolling the perimeter."

This dark chapter of recent American history can be viewed as a neces-sary evil to protect the greater population of the United States at a time of war when German U-boats off the East Coast threatened the homeland, except for one question: Where were the German and Italian American internment camps? Innocent German and Italian Americans were not ripped from their homes and put into prison camps; only Japanese

Americans. The reason is racial prejudice, a trait shared by all with a sense of racial identity.

"Immigrants coming to the United States could all aspire to someday becoming naturalized American citizens, except one group of immigrants: immigrants from Asia." Chinese and Japanese immigrants to the United States could not be naturalized under immigration policy at the time, Takei explains.

"When the war broke out young Japanese American men and women rushed to their local recruitment board to volunteer to serve in the US Army. But because we looked like the enemy—simply for that—we were rejected from service . . . and we were all incarcerated."

Racial prejudice works both ways: "It would be perfect for Obama to live with a group of monkeys in the world's largest African natural zoo and lick the breadcrumbs thrown by spectators," someone said in a North Korean government–controlled media broadcast from Pyongyang in 2014. "Obama still has the figure of a monkey while the human race has evolved through millions of years," raved the North Korean media in a recent and particularly ugly racist attack.

Many in Japan during World War II held a supremely racist view of the racial purity and superiority of the Japanese, and North Korea holds a similar belief in a pure Korean race and condemns contamination of its purity. North Korean women who escape to China for work and are later forcibly repatriated to North Korea are subject to forced abortions or infanticide if they are pregnant or return with children. Nazis, with their "social hygiene" programs of mass extermination, as well as colonial Caucasian Americans building a new country on slavery, regarded their own race as superior. Even the most enlightened and learned among us—for example, Thomas Jefferson—believed that Africans were an inferior race and he viewed and treated these human beings as property to be bought and sold.

Racial prejudice is wired into the human brain. Stereotyping of people as members of outcast groups is also wired into the human brain, but by somewhat separate circuits. Prejudice is our emotional response toward another group of people based on preconceptions. Stereotypes are conceptual categorizations of people that we group in our mind according to superficial characteristics. The human brain instantly sorts people into different groups along racial and ethnic lines. This may be difficult to accept, but the latest neuroimaging evidence supports this surprising conclusion. Similarly, our brain instantly classifies people into different groups according to ideo-logical beliefs and other more arbitrary categories. This brain processing happens so quickly it is unconscious, but the instantaneous categorization of different people fuels two of the most powerful drives of our species: to affil-iate or to compete with others. Psychologist David Amodio of New York University, who has studied human brain waves and used functional brain imaging to identify how our brain forms prejudices, concludes that the "us versus them" categorization of other people happens nearly as instanta-neously and unconsciously as we attach attributes to any other object. "Although prejudice stems from a mechanism of survival, built on cognitive systems that 'structure' the physical world, its function in modern society is complex and often deleterious."

John Steinbeck was rejected from military service in World War II for fears that he was a subversive and a Communist. The FBI kept a report of Steinbeck's activities, writings, and personal connections: "In view of sub-stantial doubt as to the subject's loyalty and discretion, it is recommended that subject not be considered favorably for a commission in the Army of the United States." Steinbeck, whom President Franklin Delano Roosevelt con-sulted for advice, was anything but a Communist. The Communist witch hunts of the McCarthy period are legendary, as were the actual witch hunts in the colonial period. Bias can operate implicitly, without conscious aware-ness and countervailing against reason or even an accurate perception of the facts. This instantaneous prejudicial categorization by the brain can lead to dehumanizing others to the extent that the human ability to empathize with another human being is lost. The loss of empathy, which is now understood to involve circuits in the insular cortex, is essential for unprovoked violence and torture that take place between individuals and groups of people.

Prejudice and stereotyping appear to alter early events in the brain involved in face processing. fMRI data show that there is greater activity in the fusiform cortex (part of the temporal lobe) in response to seeing the face of a person from one's own racial group. The difference in response of this brain region underlies the psychological phenomenon that we have better recognition of in-group faces than out-group faces. Brain-wave recordings show that only 170 milliseconds (0.17 seconds) after seeing a face, the brain-wave response is greater for in-group than for out-group faces, even when the definition of groups is defined arbitrarily. Studies show that out-group mem-bers are often viewed as threatening, and they elicit our vigilant attention.

Prejudice is not *located* in the fusiform cortex; prejudice evokes strong emotions of love, pride, fear, disgust, or hatred, so parts of the brain that process these emotions are also involved, including the amygdala, insula, striatum, and the orbital and ventromedial frontal cortex. These brain structures form a core network for the experience and the expression of prejudice. The amygdala receives input from all of our senses through thirteen different nuclei (clusters of neurons). The amygdala receives sensory input from all senses into its lateral nucleus, enabling it to respond very rapidly to immediate threats in advance of the more elaborate processing in the cerebral cortex. The central nucleus of the amygdala is critical for classical fear conditioning— learning to quickly react to a negative stimulus. The outputs of the amygdala, as we have already seen in threat-detection and fear responses, are also sent to the hypothalamus and brain stem structures in response to face recognition circuitry. This causes automatic arousal, attention, freezing, and preparation for fight or flight. Output from the basal nucleus of the amygdala drives actions intended to achieve a desired outcome.

The insula is a large region of cortex adjacent to the frontal cortex that functions to represent somatosensory (internal bodily) states, including visceral responses and emotions such as disgust. Activity in this brain region has been associated in Caucasian test subjects with implicit negative attitude toward black people. It is also associated with positive emotions, including empathy. Another study observed insula activity only when members of liked, but not disliked, out-groups were being harmed. Empathy depends on the victim's social affiliation, and the insula is where this is evaluated. Interestingly, the insula is also activated when a disliked out-group member is rewarded, suggesting a biological basis for the visceral response of jealousy.

The men who assaulted the women in Tahrir Square, or terrorists who toss grenades into restaurants or brutally torture and mutilate others, can do so because of the insula's categorization of people into "us" and "them." Empathy is strongly modified by race and tribe. This is how we coalesce or separate into groups with others. When Chinese and Caucasians view images of people being exposed to a painful or non-painful stimulus (either a needle penetrating the cheek or a Q-tip brushing against it), the median prefrontal cortex (mPFC) and anterior cingulate cortex (ACC) are activated, but only in response to seeing the person of the *same* race experiencing pain. Men reporting sexist attitudes exhibited lower mPFC activity when viewing sexualized images of women than men with less sexist views. This differential activation of the mPFC is consistent with dehumanizing a person into a sexual object. The absence of strong mPFC activity reflects a lack of empathy toward and dehumanization regarding other races, as well as disliked or disrespected out-group members. There is more activity in the mPFC when viewing pictures of esteemed groups than when viewing low-status groups (for example, homeless persons). This circuitry makes sense, because the mPFC is particularly important in processing social information, and it has prominent interconnections with the ACC, the insula, the orbital frontal cortex, and the dorsolateral prefrontal cortex—all regions we've already encountered in tracing the brain's fear circuitry. mPFC activity is strongly associated with forming impressions about other people, especially impressions that require mentalizing—that is, considering a person's unique perspective and motives (also known as theory of mind). mPFC is activated during our judgments about other people, but not in making judgments about inanimate objects.

Because these responses are unconscious and involve implicit rather than declarative learning, they are very difficult to change. According to Amodio, "Most forms of implicit learning are resistant to extinction [wearing down a bias by familiarity]. Implicit racial biases are particularly difficult to change in a cultural milieu that constantly reinforces racial prejudices and stereotypes."

## Pressure Cooker

The eight-year-old boy arrived in New Jersey on a spring day in 2002 with his parents. Together they were fleeing halfway around the world to escape a war-torn land of death, terror, and criminal corruption. Here the parents would start a new life and raise their children in a wholesome environment— a land of promise and opportunity. The family had grasped the golden ring— the American dream. In the footsteps of so many immigrants to this country before them, the young family abandoned their homeland, choosing to embrace a foreign country where everything was strange. The language, culture, food, people, and geography were all alien, but the new immigrants were infused with enthusiasm and driven by dreams of a better life for their children. The three—mother, father, and son—would establish a temporary home so that the rest of the family, an older brother and two younger sisters, could soon follow, and they did.

"Anzor had a job within three days of arriving in the US!" Ruslan said, praising his brother's talents and hard work ethic in support of his family. The family of six occupied a cramped eight-hundred-square-foot apartment on Norfolk Street in Cambridge, Massachusetts. Anzor built a business as a car mechanic, fixing cars on the street for less money than shops with expensive overhead costs. Through hard work and skill Anzor built a loyal clientele of customers and supported his family in their new home.

The boy was enrolled in grade school, but placed in a lower grade than his age group because of his lack of English-language skills. A quick learner, he advanced rapidly, skipping from third to the fifth grade. "He had a heart of gold," one of his teachers, Katie Charner-Laird, recalled. In high school he became an honor student, taking advanced-placement courses on an academic track to college. His destiny, only a few years away, was to become the first in his family to receive a university diploma.

Matching his academic strengths, he was also a natural athlete. He came from a long family line of strongmen, both in body and bravery. These attributes of machismo were embossed on the men of his family through inheritance and harsh environment. Many of the men in the extended family, going back three generations, were amateur boxers. His older brother was to become a Golden Gloves champion with dreams of competing in the Olympics. Their family had been through much, surviving hostility and ethnic wars in which tens of thousands of their people had been killed in brutal fighting and "ethnic cleansing"—a repulsive euphemism. Territory, politics, religion, and ethnic divisions fueled decades of horrors—both outright wars and prolonged guerilla conflicts, kidnappings, and terrorist attacks. But now the family had escaped all of that and started fresh in America.

In his junior and senior year in high school the boy was elected captain of the wrestling team by his teammates. He was awarded the MVP award in a ceremony at the end of his senior year, and he received an academic scholarship to attend college in the fall of 2011. On September 11 of that same year, the young immigrant to America became a US citizen.

Seven months later, two bombs filled with shrapnel exploded, killing three and wounding 260 innocent men, women, and children. Many were grievously injured, their lives forever derailed by a chance encounter with this boy, now twenty years old, who lay curled up in the belly of a boat on a trailer riddled with bullets from automatic assault rifles fired by swarms of police who had shut the entire city down to capture or kill him and his brother.

"Why the Boston Marathon?" Ruslan asked, reflecting on the horror that had engulfed his family in an international terrorist attack. "Why not a nightclub or a police station?"

Ruslan Tsarni's gaze focused into the distance as if to resolve some faint image in the vanishing point that might explain the incomprehensible scene. Ruslan, the uncle of the Boston bombers, an immigrant from Chechnya who had bounced the boy on his lap as a two-year-old, loved him as if a son, and who'd helped him and his brother and sisters grow up in America, was still grappling with the devastation of the nightmarish failure that had brought his family into the world's spotlight as a supposed nest of terrorist bombers.

## On Boylston Streets

The last time I was on Boylston Street, it was to give a lecture in November 2012 at a scientific meeting in the Westin Hotel. Today, Sunday, I am here again in the spring of 2012 looking out onto an empty street, barricaded; an eerie modern-day ghost town festooned with yellow police tape rippling in the cold Boston wind. I look across an enormous pile of fresh-cut flowers, teddy bears, helium balloons, baseball caps, candles, and handwritten notes. American flags sprout from the mound like brilliant poppies. Grief, still raw, is slipping away, drifting as if carried helplessly on a current, and transforming into something else; defiance, but shaken with bewilderment. Tourists gather now, out of sorrow and the need to understand. "Do we have anything to give?" a woman asks her companion desperately. "Do we have anything to give?" she repeats, and turns back to survey the makeshift shrine empty-handed. It must have begun with a single bouquet. Now it has grown into a mountain.

Many tendrils of this tragedy penetrate neuroscience, but underlying everything we all ask the same question. It is the question uttered by President Obama in disbelief: "Why did young men who grew up and studied here as part of our communities and our country resort to such violence?" How does a teenager riding a skateboard and attending college classes suddenly become a violent mass murderer of innocents, committing violence toward fellow citizens, innocent children, families, and others who intersected his sphere only momentarily through fleeting chance?

People often share or differ in political views and goals, but few could conceive or accept this kind of violence. Politics change, but such violence seems eternal. Look instead to neuroscience for insight. All behavior is the product of the brain. It is the challenge of neuroscience to understand the human brain and how it develops in every person to make each one of us unique and develop into a productive member of society—or an outcast.

No one imagines their newborn infant growing up to become a violent gang member, but in certain environments that draw becomes overwhelming. Many adolescents and young adults are unable to resist despite all parental and societal efforts to prevent it. Regardless of the violence and almost certain tragic outcome of gang association, many join gangs and become criminals at a young age. The families of these two brothers implicated in the Boston bombing had acted selflessly to protect their children from growing up in a hostile environment, uprooting themselves from their homeland and fleeing halfway around the world to seek a better life for their family. But their efforts failed horribly. We need to understand what went wrong.

Perversion of the wholesome biological process of forming allegiances and personal identities during one's late teens and early twenties is the core of the problem. The bombings and robberies committed by the Weather Underground and similar radical antiwar groups in the 1970s, for example, are fundamentally no different, except for the political veneer and the less deadly potency of their weapons of terror. Many of the young members of the Weather Underground who committed terrorist bombings and other acts of violence in the early 1970s went on to live normal and productive lives after being released from incarceration; some were only apprehended after living lives as law-abiding fugitives for decades. What is different now is that gangs have grown from local neighborhoods and pockets of radicals to become worldwide in scope, drawing the most vulnerable and compliant into the domain of the most evil among us. In the past the nucleus of a gang would have been the meanest person on the block; today it is the worst criminals on the entire globe.

The intentions of those who are drawn to violence against society have not changed, but the global torrent of instant electronic information has multiplied the capability for destruction and terror, making their criminal acts far more dangerous. The radicals of the 1970s were amateurs, but today video instructions on bomb-making are available to anyone over the

Internet. Ease of international travel enables a disgruntled young person to receive firsthand instruction on making and deploying horrific devices of mass destruction for the price of a plane ticket.

But there have been other changes. In the 1970s we had little hope of understanding the neurobiology gone awry in adolescents who become gang members and criminals. Today there is more than hope; there is data. New information and new methods of brain imaging are nurturing a new field of social neuroscience, which seeks to understand how the brain controls social interactions and, conversely, how these interactions affect the brain. In the past, such questions in brain science could only be tentatively approached through animal studies, a feeble approximation of complex human nature and the unparalleled capacity of the human brain. Today we can see inside a person's brain at work. We can see the malformations in brain structure that make it difficult for some people with certain developmental disorders to interact socially. We can see how environmental experience in early life augments or undermines normal development of brain circuits that control social interactions, emotion, aggression, propensity to violence, and empathy.

The brain systems that motivate humans to form emotional bonds are being discovered and probed. Many of these circuits of social bonding inter-act with motivational systems in the brain. Circuits involved in fear, novelty seeking, and modifying behavior based on negative events are influenced by experiences while the brain is forming—a period now understood to con-tinue actively through the first twenty years of life. Altered development of these circuits can lead to increased aggressiveness, diminished fear, and heightened anxiety. In the absence of adequate rewarding interpersonal rela-tionships and bonding to societal and cultural values, alternative means of stimulating reward pathways in the brain are often substituted through sex, aggression, drugs, and by the verbal or physical intimidation of others. Sub-stance abuse during these critical years when the emotional and social brain networks are forming can have lasting effects that increase one's risk of men-tal illness as an adult. Cannabis use in adolescence is correlated with devel-oping schizophrenia as an adult, and the molecular and cellular defects can be reproduced in experiments on animals.

Peer rejection in adolescence can lead to depression and leave marks on the brain that can be seen by brain imaging. For a teenager, peer groups are among the most powerful environmental influences. Verbal abuse in the middle school years marks the brain by decreasing connections between the left and right brain (the corpus callosum), leading to psychological problems as adults, including increased anxiety, depression, anger and hostility, and drug abuse. Neural correlates of impaired emotional processing and the neural basis of moral evaluation can be seen at work inside the human brain (including the medial prefrontal cortex, precuneus, and insula—brain regions implicated in introspective processes, as well as brain regions involved in emotion, notably the amygdala). Empathy activates the same brain circuits that process physical pain. "Unmet need for social bonding and acceptance early in life might increase emotional allure of groups (gangs, sects) with violent and authoritarian values and leadership," concludes psychologist Cort Pedersen, who analyzes the biological aspects of social bonding and violence.

We are beginning to uncover the neural basis of human behavior, violence, social integration, and how experience forms the brain. Psychiatrists today are like the heroic surgeons of the Civil War, desperately working to save lives with a crude and woefully inadequate understanding of the biology involved, but developmental neuroscience is converging with psychology and leading us to biological understanding. None of this scientific insight can excuse horrible acts of violence—many lives will never be the same, and the Boston bombers must face responsibility and justice for what they have done, but this research may help prevent such acts in the future.

# A Brotherhood of Terrorists

Cherry trees lining the streets of this Maryland suburb were ablaze with pink blossoms when the pressure-cooker bombs exploded in April. Today, January 2014, the cold, bare, black limbs are encrusted with snow. My eyes, protected by sunglasses, squint from the brilliant sunshine blazing in a cold blue sky and reflecting off sparkling snow as the wheels of my car crunch to a halt in front of Ruslan Tsarni's attractive two-story home in a quiet upper-middle-class suburb that looks very much like my own.

Amid the panic and nonstop frenzied hysteria of the media during the Boston bombing, it was Tsarni who suddenly brought clarity for me. The country was gripped in terror by the bombings. People feared that this was only the first salvo in an organized assault that would bring horrible terror and death to the country again. What terrorist organization was behind it, and what was their next move?

The nation had just seen the security-camera photos of the two suspected bombers in the crowd at the Boston Marathon minutes before the two presumed Islamic commandos planted their lethal bombs. Two fit young men, determined warriors it appeared, were members of a secret army of terrorists intent on killing civilians and bringing the country to its knees. al-Qaida was mounting its counterattack.

But the endlessly looping news scenes were suddenly interrupted by a live video feed. A man standing alone before reporters outside his home in a middle-class neighborhood was denouncing the two men in the security-camera pictures. "He put a shame—he put a shame on our family, the Tsarni family! He put a shame on the entire Chechen ethnicity!"

The uncle's passionate recrimination was a stunning turning point. These two mass murderers were *not* the tip of the spear of a new invasion of al-Qaida terrorists. These were two dysfunctional angry young men, Americans.

"Dzhokhar, if you are alive, turn yourself in and ask for forgiveness," Tsarni ordered. The older brother, Tamerlan, had just been killed in a police shootout.

"We share with them their grief," Tsarni said of the families of the bombing victims through the throng of reporters clamoring outside his home. "I'm ready just to meet with them. I'm ready just to bend in front of them, to kneel in front of them seeking their forgiveness."

Six months later I met Ruslan Tsarni at his home. We were both seeking the same thing—understanding.

His wife answered the door and I was introduced to his two polite teenage daughters. Another young boy was sitting on the floor of the living room playing a video game on a large-screen TV. The walls of the living room were encrusted with family photos of many smiling children. Ruslan, middle-aged with an athletic body, neatly cropped hair, wearing brown corduroys and a polo shirt with broad horizontal stripes, took my coat and offered me a seat on the black sofa. His gracious wife brought two clear-glass mugs of hot tea and a tray of treats—dried apricots, figs, crackers, and chocolate—setting the steaming glasses down on white paper napkins.

I had done my research and come prepared with a long list of carefully orchestrated questions, but I never needed them. As we talked freely, Ruslan unfolded his rich, complex story, beginning back in Chechnya in his father's home before his brother, Anzor, married Zubeidat, the woman who would one day become known as the mother of the Boston bombers. It is an epic story of American immigrants, driven to these shores of promise and freedom to make a better life for themselves, but it is also a tragedy. It is the story of two brothers fleeing corruption and violence in a war-torn corner of the world to seek a better life for themselves and their children in America; one of them, Ruslan, would prosper, and the other, Anzor, would fail horribly. He would fail despite his skill and hard work, and despite the earnest struggles of the extended family to help one another establish a toehold in American soil and climb to the summit of the American dream.

Ruslan never touched his tea or any of the snacks. The tea went cold as he spoke for three hours and forty-five minutes. Tracing from his family's roots in Chechnya to building a new home in America to helping his brother immigrate with his young family to the horrible events at the Boston Marathon, he carefully described all the family members and how they grew up and struggled together. He spoke as the afternoon sunbeams slanted low through the living-room window of the years of seeing his brother's family in trouble, watching that trouble ever worsening, fearing for the future of his two young nephews, and actively trying to intervene repeatedly to prevent an inevitable catastrophe.

The room dimmed as darkness fell and the sounds of his wife preparing dinner in the kitchen had long since ended. Ruslan relived the minute-by-minute events of the violent police chase that shut down the entire city of Boston, and the ferocious shootout and death of his nephew, Tamerlan. He spoke of the capture and imprisonment of his younger nephew, "Johar" (Dzhokhar), of the despair he felt for the victims, and his frustration and anger with the older nephew, Tamerlan, for carrying out such a cruel and senseless crime, as well as bitterness toward him for drawing his younger brother into such a wicked path. He ended with the torment of burying his nephew in secrecy and loneliness, because no cemetery would accept the body of the terrorist. The boy's parents had fled the country, leaving Ruslan to take responsibility for a decent burial.

"The funeral was dear to me. This is all I get."

This was an entirely different picture from the angry Muslim, religious-crazed terrorist portrayed in the media. This tragedy had nothing to do with politics. This had nothing to do with religion. It had everything to do with a dysfunctional family, and with the failure of two young men to find success in life.

"When you first saw the pictures on the news of Tam and Johar as the bombing suspects, did you find it surprising or possible that they could be the bombers?" I asked.

"My feeling was at first that it was not impossible," Ruslan said sadly.

"Why did you go on TV and make that announcement?"

"The most important thing first is to deal with the dead."

Ruslan's instinctive concern was to care for the victims who had been grievously harmed, even though his nephews were family and he cared for them dearly. The innocent victims needed help, and in a way that no outsider might conceive, I believe Ruslan felt deeply that they too were his responsibility. There is a strict hierarchy of inheritance and responsibility among brothers in Chechnya. By ancient tradition the firstborn male inherits everything, and he then assumes full responsibility to head and care for the extended family. Ruslan, after his older brother died, became the oldest surviving brother, and he took on this responsibility earnestly.

"We loved them, the children," he said about Tamerlan and his brother when they were born in the war-ravaged Chechnya region that borders Russia. "Tamerlan used to kid a lot. He was good. You should have heard him play the piano."

Recalling his brother Anzor's surprise marriage in Chechnya after his service in the army, Ruslan said his father was angry that Anzor had married Zubeidat, a woman from Dagestan. The harsh ethnic divisions inside Chechnya are difficult for Americans to appreciate, but it seems as if the marriage of Anzor and Zubeidat rocked the family the way a mixed-race marriage between a black man and a white woman might have done in Alabama in the 1950s. "Chechens and Dagestanis hate each other. They hate us worse than the Russians do. We were better off in Moscow than in Dagestan."

Ruslan and Anzor's father felt that by marrying Zubeidat, a woman of despised ethnicity, Anzor had brought shame on the family. But blood is stronger than bigotry. The birth of their first child, Tamerlan, shattered the ethnic hatred.

Nevertheless, the young family struggled in Chechnya. Zubeidat was regarded by many as an unfit mother. "She never cared for them," Ruslan said of Zubeidat. "She was total evil."

Immigrating to the United States failed to heal the difficult marriage. Ruslan's hope was that by helping his younger brother move out of the hostile environment in Chechnya to join him in America, he could rescue the troubled family. Zubeidat's alleged destructive gossiping would be undermined in America by her inability to speak English. Her reported inclinations toward thievery would be squelched. "There are cameras in shopping centers so she can't steal. Even if she goes to jail, here in America you are still a human being," Ruslan said, contrasting prisons in the United States with jails in Chechnya, where family members must quickly pay bribes to get their loved ones out before they contract tuberculosis or worse.

"All my illusions that America would change her were gone." Tamerlan and Johar, despite their talents and potential to succeed, were failing. Both young men began associating with unsavory characters. "Johar was into it—marijuana—too," Ruslan said. Both young men were struggling to enter the mainstream of American life and they were failing at the critical precipice of life to grasp an independent and successful hold as productive members of society. Johar was failing in college. Tamerlan could not find suitable employment. Their sisters were also troubled and in difficult relationships or marriages that were failing. They were having their own trouble with the police.

"These children must be safeguarded against their mother," Ruslan said of the children when they were young. "She hated us. We despised her. She was a bad mother. She has never been a wife. Not a homemaker. She never worked," he said.

Although we are hearing only one side of a complex story, no one disputes that Anzor and Zubeidat's family was struggling. Ruslan took custody of Tamerlan temporarily when the boy was thirteen years old, and Tamerlan stayed with Ruslan in his home for a time, but later he returned to Boston. "He was the nicest kid of all my nephews."

In the years leading up to the bombing, Johar and Tamerlan, despite their early successes in school and superficial appearance of prospering, were failing miserably. Ruslan said he told Johar after he was having difficulty, "Anything you want, I'll take care of it." But an uncle's ability to intercede in a troubled family is limited; indeed, such efforts are likely to be resented.

Zubeidat, according to Ruslan, was radicalizing her sons, encouraging them to embrace the community of Islamic extremists who were fighting against the decadent society of the United States of America and its corrupt militaristic government. She had never been outwardly religious before; indeed, she'd dressed fashionably for years in Western clothing, but when

Ruslan saw her in 2009 in full Islamic dress, he said, "I was shocked. Why did she need to flaunt that religious garb here? What was the point?" The family was completely familiar with Islamic dress from Chechnya, but in America the garb seemed out of place. "It is a way to demand respect and attention. Do you respect God? I am a Muslim," is what Zubeidat's new Islamic dress meant here in America, Ruslan concluded. Bella, Dzhokhar and Tamerlan's sister, soon adopted the hijab dress as well.

"You don't feel comfortable in college?" Ruslan recalls asking Tamerlan a few years later by phone after he could find neither suitable work nor success in college. "You come here then," he said. "Just get out of that house. Come here to Maryland." Ruslan sent him $3,000 to purchase a car to drive to Maryland. Anzor squelched the plan, however, resenting his brother's actions as meddlesome.

"He was an entirely different person now from the last time I saw him at nineteen years of age," Ruslan said of Tamerlan when he greeted him at JFK Airport in 2009. "Now he was a twenty-two-year-old man, but he was entirely different. It was a stranger's face.

"Let's go to Maryland," Ruslan urged him. Tamerlan made no eye contact. They met for only five or ten minutes. Tamerlan said he would come, but he never showed up.

Ruslan said, "2009 was the last time we spoke face-to-face. We spoke a couple of times on the phone." One phone call took place right after his father was attacked in Boston and beaten on the head with a pipe. Anzor was in the hospital with traumatic brain injury.

"Anzor was about to die," Ruslan told his nephew on the phone. Ruslan told Tamerlan to report the vicious beating to the police, but Tamerlan resisted.

"Do you know he is dying?" Ruslan asked.

"Speak with me properly!" Tamerlan shot back.

In an earlier telephone conversation, Ruslan told Tamerlan, "We have to do jihad," Tamerlan told his uncle.

Interpreting the word "jihad" in the same way as one would understand "being a good Christian"—that is, taking it to mean working hard and improving oneself—Ruslan pressed his nephew to reveal his intentions.

"He was trying to impress me with his Islam rubbish," Ruslan said, despising Zubeidat for filling his nephew's head with the radical Islamic

Jihad thinking. "There are many names for God, including Nature and Allah," Ruslan observed, and he went on to list the many names for God in different languages. "They all mean the same thing—something spiritual, not something religious." It was clear to Ruslan, though, that the boilerplate rhetoric he was hearing in his nephew's voice over the phone was the verbiage of hatred espoused by Islamic radicals, and it signaled a threatening turn toward violence.

Reliving the horrible bombing and his nephews' murderous flight as the two were pursued by the entire Boston police force and federal officials throughout a terrorized city completely shut down, with all residents hiding in their homes, he said, "Johar saw his brother killed." Evoking that shootout with police on a dark street he jabbed his chest with his fingers over and over as if hit by bullets. Ruslan enacted Tamerlan's last actions on Earth, stretching his arm up and pointing an empty gun to the police so he could die a martyr. "No ammunition in his gun."

"Both were very brave men, but why did Tamerlan do that to his brother?" he agonized.

"If I put myself back to nineteen [years old], I would have done anything for my brother," he reflected (The Family trigger). "He just pulled the youngest one down with him. He should have said 'run away and I'll deal with it.' That is what an older brother should have done. Why did he do it to his brother?

"You can be a murderer without feeling blame, but why sacrifice someone else? I don't know what that is.

"In a family, brothers take care of each other. In religious fanatics, they call each other 'brothers.'" It is the same in gangs and in churches, he says. "It is how they get people to do these acts, acting together as 'brothers' in arms." The horrible sight of seeing your brother shot down would traumatize anyone and turn one against the police and toward a gangsterlike affinity with violence, Ruslan imagines.

"I wish he had just killed him," he said of Tamerlan's treatment of Johar. "Better than involving him in this horror."

Ruslan struggles with questions: "Why bomb the marathon?" he asked. "Runners—they are the happiest people. I am a runner!" They (jihadist terrorists and his nephews) hated the marathon runners for precisely that reason, he concludes, because the cheering spectators at the race were happy and successful families. "Why not bomb a nightclub or police station?" he asks, if the bombing were truly motivated by political or religious conflict. Instead, they murdered happy families, something they did not have.

Ruslan recalled the time when he first came to this country and searched for an apartment. He was denied apartment rentals everywhere he went because even though he had excellent credit, the landlords would not allow children. He was perplexed and outraged, "This is life!" He still recalls that when he saw people walking their dogs in front of the apartments that he could not live in, he felt anger rising in his chest. That is how he thinks Tamerlan felt.

The Tribe trigger divides people into battling factions and every single individual must grapple with this powerful imperative. Outcasts coalesce into gangs. If an individual fails to assimilate or succeed in society, he or she will find affinity with another group that will embrace him or her, for whatever reason. If a person, especially a foreigner or racial minority, is constantly regarded as an outsider, that person will naturally be inclined to adopt that persona. It is a hardwired, necessary, and automated function of our brain to spot outsiders quickly, fear them, and dissociate from them. Muslim immigrants today are at greatest risk of being regarded first as outsiders and possible Islamic terrorists. That constant stereotyping can drive the individual to shape his identity accordingly if he is struggling but failing to assimilate and succeed. This germ of human need growing inside us is what infects the feared "lone wolves" who are exploited by extremists to commit terrorist acts or join foreign wars in solidarity with a brotherhood that will embrace them. The faults that can crumble the foundation of successful families are many, and even as they may be evident and growing, they can be difficult or impossible for others to stop.

Ruslan remains tormented by it all. "I will just have to live with this now," he said with a sigh as he walked me to my car. His words vaporized as his warm breath puffed against the cold night air like steam.

Just as light defines dark, night defines day, love defines hate, kinship defines the alien, the LIFEMORTS triggers that protect and bind us together necessarily threaten and separate people from one another. For a mother's brain to show a heightened response to her own baby's cries means that her brain shows less response to the cries of others. The triggers that provoke rage in one person's brain are multiplied when groups of people come together.

Republicans/Democrats, police/demonstrators, Muslims/Christians, rich/poor, gangs/families . . . as our brain partitions our environment it necessarily creates divisions, and the borders become battle lines for conflict.

Political, social, and economic analysis of mass conflict is important, but a fresh understanding of terrorism, war, and civil unrest is provided by looking at these age-old horrors through a modern neuroscience perspective. The Insult trigger in the human brain is played out on a societal level as riots, terrorism, and war. Recent examples include the seventeen murders in Paris in reaction to a cartoon of the Prophet Muhammad in the magazine *Charlie Hebdo*, or the cyberattack on Sony Movie Studio by North Korea that was provoked by a comedy movie with a farcical plot to assassinate the North Korean leader. All of this violence over such slight provocation seems senseless unless one can appreciate the fundamental roots of this behavior in the brain. These acts were perceived as insulting. Regardless of how misinterpreted or inadvertent an affront to dignity may be, insult will trigger human violence and it can do so on a mass scale. The violent riots after the shooting of Michael Brown, an eighteen-year-old black man, by a policeman on the Ferguson, Missouri, police force; the violent reaction to the choking death of petty criminal Eric Garner by the New York Police; and the cold-blooded murders of two New York police officers in retribution by a suicidal fanatic black man are recent examples of the Tribe trigger of violence being tripped in the human brain: the us-versus-them rage. Street riots in Ferguson broke out long before the forensic evidence with the facts about the shooting was made public. The protests in response to Eric Garner's choking death were organized weeks before the results of that investigation were released. The specific details of alleged injustice are secondary to the underlying racial and social divisions between the groups that are in conflict. This is the powerful human herding instinct—the Tribe trigger.

Tribalism arises from human herding behavior, which in cognitive neuroscience and psychology describes the alignment of thoughts and behaviors of individuals within a group through local interactions rather than by being imposed upon individuals by central coordination. Human herding behavior is an automatic, unconscious neurobiological process, which evolved in the human brain to enable us to form complex social structures. Herding of individuals into groups is the glue that binds people into social structures according to national identity, religious affiliation, and political parties. In lesser matters human herding causes us to embrace fads and fashions—why else should men fuss over the width of their tie? At its worst, human herding generates mass hysteria, spawns gangs, and ignites mob violence.

An essential element of herding in human society is that individuals often converge by modeling the behaviors and beliefs of the larger group in which they are embedded. A bolo tie and a cowboy hat are fine in Santa Fe, New Mexico, but you could be asking for trouble if you dress that way in certain neighborhoods in Chicago. This same neuroscience driving fashion, however, is what drove German citizens to embrace Nazism, and it drives different religious and political groups in the Middle East in violence today.

The Tribe trigger of rage suddenly unleashed deadly violence between rival motorcycle clubs at a biker bar in Waco, Texas, in May 2015, which resulted in nine deaths from stabbings, brutal beatings, and shootings. Eighteen people were hospitalized and at least 170 were arrested. "In thirty-four years of law enforcement, this is the worst crime scene, the most violent crime scene I have ever been involved in," said Waco Police Sgt. W. Patrick Swanton.

Although many in the Waco motorcycle gangs were likely violent criminals, the power of tribalism to spark violence between groups of otherwise peaceful people cannot be overlooked. Consider the assault on Alexian Lien that took place on September 29, 2013, on New York City's Henry Hudson Parkway by motorcyclists, which included New York City Police Department officers. Five off-duty officers were among the motorcyclists, and one ten-year veteran and undercover detective, Wojciech Braszczok, was arrested. "Not only did he fail to protect and serve, he cast his lot in with the assailants," Assistant District Attorney Joshua Steinglass said at the officer's trial.

The violence began when Alexian Lien's wife reportedly tossed a half-eaten plum and water bottle at the bikers, who were holding up traffic, popping wheelies, and slapping the tops of cars as they drove past. (The Stopped and Organization triggers snapped in Lien's wife's brain, and in reaction to her insulting provocation, the Tribe trigger separated all parties into warring factions.) One biker responded by shattering the driver's-side mirror of Lien's SUV. Another motorcyclist cut in front of Lien and applied his brakes, resulting in a minor collision with Lien's vehicle. Under attack by the motorcyclists, Lien panicked and sped away, driving over one of the bikers, Edwin Mieses, who was left paralyzed.

Much of the melee was captured on film, and it shows bikers bashing out the windows of the SUV with their helmets and dragging Lien from the vehicle to severely beat and stomp on him. Bystanders eventually stepped in and put an end to the violence.

The Tribe trigger of rage is often the spark of violent street riots. On April 19, 2015, riots broke out in Baltimore, Maryland, after Freddie Gray, a twenty-five-year-old African American, died from a severed spinal cord. Gray's mortal injury was inflicted while he was restrained by handcuffs and leg irons inside a police wagon in the custody of six police officers. At least twenty police officers were injured in the ensuing riots, 250 people were arrested, and thousands of police and the Maryland Army National Guard were marshaled to quell the riots. A CVS drugstore looted and burned in West Baltimore became the iconic epicenter of the mayhem, but many other businesses and automobiles were burned and looted. The city was placed under a state of emergency for nearly a week.

Television footage showed hordes of angry black men armed with clubs and stones, facing off against a phalanx of police in black riot gear, wearing modern armor, helmets, and shields that hark back to medieval battles between knights of the kingdom and oppressed peasants. This angry scene seems to repeat thousands of times through thousands of years of human history. As the city of Baltimore burned, many who remembered the horror of the summer of violence that plunged the United States into chaos in 1968 were sickened.

"Can we all get along?" Rodney King pleaded during riots in Los Angeles in 1992. King, a black taxi driver, was brutally beaten by Los Angeles police officers after a high-speed chase in 1991. That beating by police was video-taped by a citizen who was appalled by the brutality erupting on the street beneath his balcony. After a trial that acquitted the officers of serious charges, Los Angeles was consumed by riots in which fifty-three people were killed, two thousand were injured, and neighborhoods were looted and burned. The military was dispatched to restore order, but many neighborhoods never fully recovered and the violence spread to other cities.

Rodney King's plea echoes the bewilderment of everyone. Unfortunately, the answer could not be more obvious or more disheartening. Such turmoil and brutality are a deadly consequence of the human mind that within milliseconds of observing another person categorizes the individual into either "us" or "them." It happens as quickly and as automatically as the brain attaches the color red or green to an apple.

I went to Baltimore during the week of riots, to see for myself the scene at the burned-out CVS drugstore. The heavy thumping of helicopter blades circling overhead resonated in the air, the smell of charred wood, sirens squealing from every direction, echoing hysterically through the alleyways and streets. It was impossible to tell where they originated. In a flash a fire truck, police car, or ambulance would streak past, ablaze with flashing red lights, racing toward the violence or away from it to hospitals or police stations.

Stepping into the neighborhood surrounding the CVS drugstore triggers screeching alarms in your brain that raise hair on the back of your neck. All day, groups of men loiter on street corners, drinking oversize cans of malt liquor from rumpled paper bags and smoking. Others pass the day sitting on the stoops of redbrick row houses as if discarded. The windows of buildings are boarded with plywood weathered into a furry gray, warped and peeling. The homes and businesses have been abandoned for ages. It is a neighborhood of pawn shops, discount liquor stores, mom-and-pop corner markets with bars on the doors and windows, of bail bonds and check-cashing establishments. Faded tent cities rot under overpasses, cluttered with shopping carts and scavenged junk. It is a perilous place of danger, crime, and drugs. Twenty-five percent of the men are unemployed. They have nothing to do. Nowhere to go. Trapped, they have no way out. Children grow up in squalor and poverty.

All eyes followed me. They were the eyes of black men. I am white. There is not a thing in the world that either of us can do about that. Ours is the biological legacy of genetics, mine following a line of descent from northern Europe, theirs from Africa. It shouldn't make much difference, but it does.

My impression was that the violence by Baltimore police against Freddie Gray and the riots that followed his death were both launched by the Tribe trigger, but it was not exactly the result of racism. It was the result of tribalism, the brain dividing the world into us versus them. This impression was verified when it was eventually announced that three of the six police officers charged with assaulting Freddie Gray are black. The driver of the police wagon, who was charged with murder, is a black officer. An unfortunate result of tribalism can be festering pockets of poverty, neglect, hopelessness, divisions between the haves and the have-nots, good guys versus bad guys, and instantaneous violence unleashed by brain circuits designed for herding, defense, and mutual cooperation in groups.

But during the riots a video camera captured a stunning example of snapping in violence to protect "us" from "them." The candid video clip showed vividly why this rapid violent reaction can be lifesaving. In this case the Family trigger was tripped. When Toya Graham, a single mother of six children, spotted her sixteen-year-old African American son with a rock in his hand among looters who were confronting police, she snapped. Graham, in her frilly bright-yellow blouse and vest, stormed into the riot in a fury. She made a beeline toward her son, who was wearing a sinister black hoodie and stocking mask. She charged after her son with the same rage and tunnel vision that Kevin Ward Jr. displayed stomping against racecars to confront NASCAR champion Tony Stewart, oblivious to the personal danger. The same neural circuits of rage had taken over in her brain. The teenager stands a head higher than his diminutive mom, but she reached up and clenched her son's hoodie with both fists. She yanked it and the mask off his face and repeatedly smacked him about the head and face, spitting out salty language that would intimidate a hardened street thug.

"She knows her son and picked him out. Even with the mask on, she knew," his sister Tameka Brown said.

The crowd of looters parted in deference to a mother's rage, and the teenager cowered and retreated like money changers in the Bible story whipped out of the temple by the wrath of Jesus.

"That's my only son, and at the end of the day I don't want him to be a Freddie Gray," she explained afterward to reporters. This mother had reacted selflessly, instantly, and violently to save her son's life.

The Tribe trigger can fuel extreme violence because when assembled into groups, people can become unimaginably ruthless. Anonymity and the reduced empathy and compassion of individuals massing into herds drive cruel behaviors in groups, in which individuals participate in deviant behaviors that they could never engage in as an individual. It is not necessary to recount such atrocities as mob lynchings, gang murders, and war; we can see this played out even in interactions between political parties. For example, originally the shutdown of the United States federal government in 2013 was implemented by a faction within the US Congress that was dissatisfied with the Affordable Care Act that had been put in place by a rival party (Democrats). But the shutdown led by the Republican Party soon degenerated into an inexplicable conundrum, because the disruptive action could have no significant influence on the Affordable Care Act itself, which had been implemented through the democratic process. Marc Thiessen at *The Washington Post* and others began calling it the "Seinfeld shutdown"—a shutdown about nothing, as absurd as the premise for the Seinfeld television sitcom. Thiessen's bewilderment came from viewing the perplexing government shutdown from a political perspective rather than recognizing that what we saw play out on the national stage was the neuroscience of human herding behavior (tribalism). Many were harmed by the government shutdown in 2013, but would a single congressman or congresswoman personally restrain his or her neighbor from going to work in the morning, a neighbor who was only trying to support his or her family and contribute productively to society in their own way? Would any individual member of Congress personally stand in front of the hospital door to block a mother from bringing her child with cancer to the National Institutes of Health (NIH) to receive a potentially lifesaving treatment? Any of these actions on an individual level would be unacceptable or illegal deviant behavior, except for the psychology of herd mentality.

The capacity to understand implicitly other people's behaviors, intentions, social beliefs, and personality traits is essential for us to form complex societies. Specific neural circuits in the human brain promote these interactions between individuals in a group. Mirror neurons, for example, are nerve cells in our cerebral cortex that fire when we watch another person carry out a similar action to ours. "Mind reading" (from chapter 7), in the form of body language to divine what others are thinking and feeling, is not much different from reading text. Both are highly automated cognitive functions that derive meaning from interpreting visual signs and patterns, but reading literacy is a relatively new development in human history. Using nonverbal tests, such as tracking an infant's eye movements toward an expected location, scientists have found that mind reading, unlike print reading, develops very early in life. The important point with respect to this chapter on tribe and herding is that both reading text and reading body language are culturally inherited. It is our automatic, irresistible compulsion to mimic the actions and behaviors of others in our group that gives us the ability to communicate through body language. All of this supremely valuable human ability depends on neural circuits in the medial prefrontal cortex, connecting with other cortical regions (temporoparietal and precuneus).

Emotional contagion is also subcortical—beyond the rational mind. Our individual emotions rise and fall with the emotions of others in our group—be it raucous jubilation in a crowd at a rock concert, love among a gathering of people at a wedding, grief at a funeral, or a child's tantrum infectiously souring the mood of all around him. Laughter, yawning, and the urge to vomit are all individual behaviors that can be triggered automatically by following the herd. Each of these group-induced behaviors has survival value for the individual, but the behavior cannot be comprehended by looking at the individual in isolation. This is the neuroscience of cows and people, animals in which herding is the essence of their being.

Studies of the neuroscience of human cooperation show that people are willing to incur personal costs to punish others who violate what they perceive as social norms. Using a combination of behavioral, pharmacological, and neuroimaging techniques, researchers have shown that manipulating the neurotransmitter serotonin in the brain alters costly punishment decisions by modulating responses to fairness and retaliation. Serotonin enhances fairness, inhibits retaliation, raises the threshold for reactive aggression, facilitates harmonious social interactions, and promotes cooperative social exchange by modulating the computation of social value, the authors conclude in their paper. Oxytocin differentially modulates compromise to antagonism from within one's own group versus rivaling groups, rendering people relatively more benevolent and less competitive toward those seen as belonging to their own group.

Another study published in 2013 showed that transcranial magnetic stimulation to activate circuits in the right lateral prefrontal cortex would manipulate a person's behavior in following social norms. Another study found that testosterone administration modulates moral judgments that involve the interplay of emotions and social interactions. The study concerned testosterone exposure of males and females while in the uterus, and found that people who showed an increase in utilitarian judgments following testosterone administration had experienced higher testosterone levels in utero, while subjects whose judgments were more related to duty, obligation, or rules had experienced lower testosterone as a fetus.

An objective neuroscience perspective reveals a striking paradox, showing that this automated neural circuit is a double-edged sword of the human brain. The circuitry enabled our species to coalesce into groups for mutual protection and common purpose and to do so through violence, but these circuits can also make people snap in anger and violence as individuals and with compounded brutality in groups. There can be no patriotism without a foreign adversary, no maternal bonding without seeing other babies differently. Ironically, the same trigger to form tribes, the human herding instinct, while the cause of so much mass violence, is also the reason for human coexistence and progress. This is what drives the Internet; for example, someone somewhere desires to contribute what they know through blogs and asks nothing in return except kinship. We join en masse with others of like mind and interests in online social media.

Paradoxically, peace and cooperation in social and international conflict comes ultimately from the same neurocircuits of defensive rage that launch conflict—but with a realization that the different groups are nevertheless united in a larger context. All citizens of the United States are Americans regardless of political party, for example. At that point of realization, these same circuits unite rather than divide the groups in conflict. This is what makes the seemingly senseless cycles of mass violence so frustrating, because history shows that with the passage of time, people divided by conflict often begin to see themselves as united members of a larger group, despite their differences. Finding this common ground is often difficult, but from the perspective of neuroscience, there is no other solution.

# 12

# Beyond the Circuit

This is my simple religion. There is no need for temples; no need for complicated philosophy. Our own brain, our own heart is our temple; the philosophy is kindness.

**Dalai Lama, from *The Dalai Lama: A Policy of Kindness***

The neuroscience we have been investigating is at play every day of our lives. It is at the core of human nature. This network of neurons has enabled our species to survive, hunt, cooperate, and explore despite danger, and to defend itself since our evolutionary ancestors left trees behind and ventured upright on land.

## Risky and Rusty

Brilliant, menacing flashes illuminate thunderheads that tower on all points of the dark horizon. Their open-hulled Boston Whaler, adrift in the darkness, is the focal point of the swirling angry squall. With gill net set, drifting with the ebbing current on the vast Potomac River, their options are limited. The fiberglass boat, the only elevated object on the open expanse of the river at midnight, is a lightning rod. They know it. Static electricity energizes their eyebrows and raises goose bumps with each jolt of lightning accompanied simultaneously by an explosion of thunder. Rain streams off slick yellow rubber jackets and chest waders, and drains from the edge of soggy baseball caps in glowing streaks sliced by headlamp beams. Indifferent to the pelting storm, the two men ponder their options, calculating deftly like poker players scanning a newly dealt hand.

Leave the nets and escape to the shoreline for shelter under the trees, or stick it out just a bit longer until it is time to pull the nets? Pulling the nets now would leave them empty, a useless night's effort. And there wouldn't be so many spring nights when the tide ebbs under the pull of a full moon and so many shad are running up the river.

The net drifts off into the distance, visible only as a wavy line of cork floats that stitch a crimped seam in the inky river as it trails off in the darkness into the vanishing point. At one end a battered plastic milk jug bobs, glowing dimly in the blackness from a green glow stick dropped inside. This dim speck of firefly green will guide them back again to find their net drifting free in the dark. The boat rocks and bucks. The fishermen keep their balance on the slick and cluttered deck. They peer into the distance, bracing themselves against the roaring gusts and showering spray. They stare at their net.

Pulling it would take time. Extracting each thorny silvery shad, beautiful striped rockfish, and spiny-finned catfish, which can reach seventy-five pounds, from the net is a slippery struggle of scales, slime, and spines. Working in the tunnel vision of a headlamp, the men must free each ensnarled fish from the net as the fish struggle in their laps as if spun inside a tough, billowing cocoon of nylon webbing. As they work the nylon threads off each captured creature, the men ignore the inevitable piercing bloody pricks and slashes to their cold, wet, and swollen fingers. The sweet smell of live fish is layered above the underlying subtle stench of dried fish slime permeating the weary vessel. But do they dare pull the net now? If lightning strikes while the three-hundred-foot net is retrieved only partway into the boat, the fishermen will be snagged in place, trapped by their own net at the epicenter of the lightning strikes.

Brad is a local fisherman whose family has worked the river for generations, and Jim, a wildlife biologist for the Potomac River Commission, has fished this river every spring for decades, when silvery shad swim upriver from the open Atlantic Ocean to spawn in freshwater streams where their life began. Jim began years earlier, working with his sage partner Harley at the wheel. Harley was a fit, keen-minded skipper with vast wisdom harvested from fishing the river for a lifetime. Although Harley was lost to pancreatic cancer a few years ago, his spirit is always on board. A photo of Harley at the helm is taped onto the aluminum cover of Jim's clipboard, where he records biological data from every catch with a wet, disobedient pencil. After his death, Harley's boats, nets, and solitary nights on the river passed down to his sons, Brad and Michael.

Fleeing the storm for safety now would leave the nets fishing too long—maybe all night, ensnarling all manner of fish and bottom snags, and killing the shad. Their partners, Rusty and Michael, are somewhere out there too, tossing about the same options in their matching boat with its quirky, obstreperous, rusty outboard motor that protests like a tired old man awakened from sleep when called upon to start. It always requires much coaxing and colorful verbal threats, but eventually the worn motor will sputter to life.

Such men close to nature see the cycle of life and death every day, and they accept it. But risks are for the foolhardy. A lightning strike could end everything in an instant—with some certain but unknown probability, in the very next instant. If it happens despite their best efforts to protect themselves, so be it. But to act foolishly and take risks for greed would make these men no better than the lowly, cold-blooded creatures they hook with artificial lures on spinning reels cast from the rocking boat while they wait for their drift nets to do their work. Time to head back in to shore.

Brad beaches the boat on the dark, sandy shoreline beneath towering oaks waving, creaking, howling in the storm, and spraying down wet leaves that stick to everything. Soon Rusty and Mike appear out of the darkness. They cut their motor, and with the hollow sound of fiberglass scraping beach stone, they come ashore beside the other boat.

The fishermen quench their thirst from plastic water bottles and share smoked shad on saltine crackers, critiquing each sample as they wait out the storm. They compare the taste of flesh cured with applewood versus hickory smoke; savor the subtle differences by the sex and age of the particular fish; relish the delicate variations of different brines concocted of honey, soy sauce, salt, whiskey, and sugar that marinate the slabs of bony flesh before they are smoked. Details, brilliantly outlined distinctions, lost to all but fishermen who know every intimate facet from water to fire that makes smoked shad and its delicious rich red roe foods that men have sought from these waters since colonial times, and extending back through epochs of prehistory.

In addition to his expert knowledge of fish biology, Jim is a storehouse of knowledge about anthropology, literature, and history. He can cite annual catch statistics for shad going back to George Washington's bountiful harvests when shad were so thick in the Potomac, the descriptions seem a fantasy. Like salmon, which feast in the open ocean and return upstream to spawn in the same waters where they were hatched, shad grow to a foot and a half long in the Atlantic, then enter the Chesapeake Bay in spring and return up the Potomac River to where their life began. There they spawn the next generation to turn the cycle of life one more revolution. But as overfishing depleted the stock and the river was remodeled and dammed to suit people's needs, the fish that had been here for thousands of years dwindled. Those few that survived often could not return to their birthplace to spawn. So in successive generations the shad stopped migrating upstream much beyond the mouth of the Potomac at the edge of Chesapeake Bay.

Jim began a program to use local fishermen to help catch the fertile shad each spring and then squeeze out the abundant pink eggs to be fertilized by sperm milked from the "bucks." The fertilized eggs are then grown by the millions in a hatchery, where they are protected from predators that would consume most of the brood in the wild. Weeks later he and his colleagues release the young fish upstream, patiently coaxing future generations of shad to return and amplify the cycle. He enlists schoolteachers throughout the region to grow some of the fertilized eggs in their classrooms so that children can learn and watch the miracle of development. This is a strategy of equal patience to his scheme to save the shad, providing a way for city kids to learn that nature is scarred by man, but also nurtured by those who understand and respect the natural world. When the larvae have grown big enough to survive in the wild, Jim takes the schoolchildren and their teachers to the river to release their classroom-reared fish into the Potomac. The tiny fish wriggle away from plastic buckets into the mighty river to the cheers of children clapping in celebration. Often schoolchildren are given the chance to join the fishermen and help set the nets on nights when the weather is pleasant and the tides ebb early enough in the evening to get them home by bedtime on a school night. Many of those youngsters are now adults.

Over the years Jim's effort and patience have been rewarded. The shad are returning. Shad are rising in abundance now in this river like no other on the East Coast, and there is new legislation that would once again permit sport fishing on the Potomac for shad.

The lightning passes, leaving a drizzling rain hissing and now-silent orange flashes on the horizon far to the north. The men relaunch their boats and they are rewarded that night. They return to find their nets filled with females ripe with pink-red eggs and bucks oozing milt to fertilize them. Punctuating the gill net among the snared shad there are plenty of rockfish and catfish to make it a successful night for Jim, Brad, Mike, and Rusty.

Shad fishing is limited to springtime, so for Rusty, most of the year is spent about as far from the solitary nights on the Potomac as one could get. Rusty tends bar in Chicago. Amid smoke, raucous noise, and rowdy drunkenness, Rusty's off-season environment is also at the fringes of the mainstream of life. Brick-red and olive-green tattoos are inked onto his biceps and shoulders. A jagged green-black tribal tattoo stretches across his back between shoulders and peeks above his faded blue wifebeater shirt. Close-cropped salt-and-pepper whiskers encrust his chin and upper lip. Blue metallic sunglasses perch on top of his head as he puffs an ever-present cigarette. He is wiry and of average build or smaller. Rusty mixes in easily with the motorcycle gangs and bar-scene crowd of rebels that flock to the Full Throttle Saloon every August in Sturgis, South Dakota, where he works one month a year. Rusty was featured on several episodes of the reality TV series *Full Throttle Saloon*, which captures the wild chaos of that biker bar. In one episode, his on-camera marriage inside the saloon to his wife, whom he met on a blind date, is the centerpiece of the episode, but they are separated now.

Every August, Sturgis, South Dakota, becomes a gathering point for motorcycle enthusiasts and motorcycle gangs to live out the raucous debauchery of the "Wild One" outlaw persona of the rebel biker. The streets are filled with hot chrome machines and the summer air roars with revved-up Harleys. Tattooed, leather-skinned bikers wearing sweat-soiled bandannas and ever-present dark glasses ride into town with their "chicks" on the backs of their bikes to flood the bars and booze it up. Gang insignias and motorcycle brand names mark their T-shirts and biking jackets, and chrome skull-and-crossbones motifs ornament bikes and clothes. The booze flows inside the noisy bars, where scantily clad women gyrate onstage.

"I imagine you've seen a few barroom brawls," I say to Rusty.

"Yeah."

"What starts them? What are the brawls usually about?"

"Women," he says. "It is always about women," he emphasized.

"It's the alcohol," he went on to say.

"Yeah, I can see that. Anything else?" I ask, easily identifying the Mate trigger of rage, its threshold for snapping lowered by alcohol.

"A lot of fights over one-upmanship—like, 'My dick is bigger than your dick.'"

"Insults," I rephrase.

"Yeah—and gangs. Guys fighting other guys in gangs."

The Insult and Tribe triggers, I think to myself.

"Have you ever reacted instantly—without even thinking about it—instantly and aggressively responded to deal with this kind of thing?" I ask.

"Yeah. Tons of times," he says. "One time I jumped on this guy's head. He was huge! Put him in a headlock with my legs . . ."

His suddenly stilled eyes focus on the dangerous scene replaying in his mind.

"It was a lucky outcome. If this other guy hadn't grabbed him around the chest and pinned his arms, he would have killed me. He was huge."

"Did you think before you did that?" I ask.

"No."

"It was just an instant reflex?"

"Yeah."

"Why did you do it? The guy could have killed you."

"It's my job. I don't think about it."

Organization, Tribe, and Life-or-limb—I realize these are the brain circuits of rage that drove Rusty to snap to action selflessly to end a barroom brawl. Add a robbery, get in the way of a biker or mess with his ride, and the Resources, Stopped, and Environment triggers are tripped. Little wonder barrooms are the stage for so much snapping in sudden aggression.

## Stress and the Snapping Threshold

Now with a good understanding of the brain's threat-detection and -response circuitry, and after learning the LIFEMORTS triggers of snapping, we can examine one of the most important factors: adjusting the threshold on the trigger.

I told the Navy SEAL about how I found myself reacting instantly to the pickpocket in Barcelona and how surprised I was by my reaction. As if merely an observer, I threw the robber to the ground without any conscious thought and physically incapacitated him using a potentially lethal hold. Afterward I was frightened that I had instinctively known how to kill someone. Of course, that was never my intention but still, I found that I knew instinctively how to do it if I had to. The hip throw and headlock were moves dredged up from my boyhood in junior high school wrestling, but no one ever taught us choke holds. Choke holds are strictly forbidden in wrestling, for obvious reasons. I later looked up what I had done to the robber in a Marine Corps training manual and found that I had applied what is called a "rear choke," which cuts off blood flow through the carotid arteries. While fighting on the ground I found myself powerfully resisting a strong urge to apply what seemed to me at the time might be an even more effective hold if things went from bad to worse. I was sensing how my pectoral muscles would suddenly come into play in addition to my biceps if I slipped my hand behind his head and executed this variation, and those chest muscles were twitching in ready anticipation as I resisted doing it. I was resisting because I was concerned that it could be lethal, and all I wanted was my wallet back. But I fully realized that he probably had a knife in his fanny pack, which had slid up within his reach near his neck. If the situation turned into a life-or-death battle for me or my daughter, I suddenly knew what to do. I looked that hold up afterward too and found that the hold I was contemplating was indeed lethal. I have no idea how I knew how to do these things. Maybe I had seen them in movies. Maybe people in combat instinctively know how to fight with their hands just as a dog knows instinctively how to fight with its jaws.

Who better to help me understand this rapid unconscious reflex to fight than a member of one of the most elite combat forces who *must* be the quickest draw in real gun battles and do it successfully over and over?

"I've seen people in different states of that," he said after I told him about my reaction to the robber. "You don't process it. You were in a lot of stress and your survival instinct kicked in. It had nothing to do with you processing it. You instantly went into survival mode, which was to fight back and to fight for your life."

Something had set my trigger threshold low.

In the moments before the attempted robbery, my daughter and I had been having a wonderful time, but we had been struggling under enormous chronic stress for several days. Travel is often tiring and stressful, especially in countries where you don't speak the language, but our travel stress was greatly compounded. We had just been robbed in Paris before coming to Barcelona.

It happened while we were on the Métro on our way to the airport to leave Paris. I had just given a lecture at a Parisian university and we were

heading to Spain, where I was scheduled to speak at a neuroscience conference. I had deviated from my normal practice of keeping my cash and credit
cards split up and stored in different places when I travel, because we only
had a quick trip to the airport. I slipped up. I revealed where my wallet was,
stuffed with all our cash and credit cards, by using it to purchase a Métro
ticket. I later remembered how someone was crowding behind me at the
ticket machine while we tried to puzzle out how to turn euros into a ticket
that would take us to the right destination using the strange machine. In
retrospect, it was odd to have someone hovering over us while there were
many other ticket machines available. I was distracted by my surroundings,
focused on figuring out how to operate the ticket machine, and we were in a
hurry.

Everything seemed fine on the Métro ride until a stop midway through
the trip, when the crowded car emptied suddenly. Kelly and I were left alone
as if in a vacuum, with one very sympathetic-looking Frenchwoman with a
knowing look on her face. The pickpockets had scored and fled.

A couple of days later, on our way to see the Gaudi cathedral, we were
feeling that we were finally starting to get back on our feet. My brother back
in the United States had wired us cash so that we could find food and shelter
to survive a few more days and then return home. The difficult and stressful
process of dealing with the robbery in Paris and the loss of identification
was compounded by searching for a police station where we could file a
crime report, and then frantic international phone calls to credit card companies from the backseat of a cab as we raced to the airport, late for our
departure. We were struggling to survive with no money, far from home in
a foreign country. We were pressed by a very tight travel itinerary that had
me giving a lecture in Barcelona in two days. The stress of the robbery, the
stress of dealing with our resulting predicament, and the stress of uncertainty: Can we fly from Paris to Spain when we have no money or identification? If we get to Spain, how will we check into a hotel room with no cash,
credit card, or ID?

That morning in Barcelona had offered up challenges. My brother back
home organized a way to wire some cash to us, but getting ourselves to the
location to pick up the wire transfer had been an ordeal. We picked a nearby
company randomly from a listing on the Internet that would accept international money transfers, but it proved to be far away from the tourist area

where we were located. The cab drove us to the outskirts of the city into a very sketchy part of town. When the cabbie announced that we had arrived, I was bewildered because there was no bank in sight. The cabdriver pulled to the curb in front of a cramped storefront Internet café in a trash-strewn, run-down section of town. I asked the driver to wait for us and to keep the meter running because I would return with some cash, but more important, I knew that if the cab left us in this dismal slum, there would not be another one coming along to take us back.

Kelly and I walked into the sparse, dimly lit room containing nothing but outdated computer monitors that illuminated the faces of shady-looking men. It looked like a Middle East terrorist cell. So this is where they transfer the funds to terrorists in the United States, I thought. The scruffy-looking men leered at us over their computer terminals, marking every movement we made. I handed the store manager a fax validating the wire transfer of funds. He perused it, then reached into his pocket and pulled out a large wad of bills. As he peeled off a thick stack of bills and handed them to me, I felt all eyes in the room follow the cash into my empty front pocket. Kelly and I were certain that we were going to be robbed. The cabdriver probably gets a cut, I imagined.

So as we ascended the Barcelona Metro steps that morning we were feeling relieved and happy, but we had been dealing with chronic stress. All of our instincts had been focused for the last few days on survival.

"What's interesting to me about that is, that's *not* where I want to operate [the rapid unconscious reflex mode of aggression]. That's not where SEALs want to operate. That's not where we want to go at all, because when you get to that level where it is kind of an unconscious decision and you just spaz out to some degree, that's not necessarily the most intelligent decision. Now that you look back at it you're like, *Wow, that was pretty stupid.* I probably should have just handed over my money and went on my way."

No doubt about that. My daughter and I had been lucky, especially after what had seemed to be a fair fight with the pickpocket turned into being chased through a foreign city by a gang bent on revenge. I didn't see that coming.

"So we, SEALs or special-operations-type guys, we want to operate in the area where we are never going to get into that route. We are always going to try and process the decision-making the best we can at the next level down. That takes time and repeated training to that level where, when you're stressed—and I mean *heavily* stressed—you need to perform. Prioritize it and make rational decisions."

He cited the example of the heroic battle retold in the *Lone Survivor* book and movie, where a team of SEALs is about to be overrun by an overwhelming force of Taliban soldiers, and the men leap off a cliff to escape. That seemingly desperate act is understandable, he says. It is the same thing that happens in burning buildings and sadly what provoked many to jump to their death in the World Trade Center attack. "What are my options? Jump off a cliff or stay here and die. You are going to jump off a cliff. Understandable from a survival instinct, but it wasn't a tactically smart decision to do. It didn't make any sense. You never want to go to the low ground," but the men had run out of options. "OK, we're just going to jump. Boom!"

The SEAL goes on to say that from his experience in battle, anyone will snap violently to save their life. "I think when you push an individual to a point where that individual thinks, *I* am about to *die* unless I act, then most people will act. A lot of people say, 'Oh, if I were ever in combat I could never shoot anybody—I could *never* do that.' Trust me. You put two people in a gunfight where the pacifist says, 'OK, I would never do that,' hold a gun to their head unless they act? *Trust* me, they are going to act. There is absolutely nobody who's going to stand there and be executed when they have the ability to shoot the person back. Nobody. I don't care if they are a pacifist or not, if they have seen other people die. They have a gun. They will act."

Flash back to 1974, inside Hibernia Bank, San Francisco:

"I got my own carbine out into the open and pointed it at the assistant bank manager at the rear desk as well as at two women at nearby desks," the bank robber described after being captured by the FBI.

Holding them at gunpoint, the robber heard her partner shout, "This is a holdup! The first motherfucker who don't lay down on the floor gets shot in the head."

Moments later, the bank robber said, "I heard the rapid shots of a submachine gun and I caught sight of an elderly man stumbling out of the doorway, his back to me. I actually saw his jacket rip open as the bullets struck him."

The bedlam exploded moments after the bank robber shouted, "This is Tania . . . Patricia Hearst . . ."

The story of Patricia Hearst, the University of California at Berkeley student and granddaughter of the publishing magnate who built the

86,000-acre opulent estate known as Hearst Castle, shows that anybody has the capacity for violence. Patricia Hearst was ripped at gunpoint from her life of comfort and privilege, kidnapped from her campus apartment by a heavily armed gang of criminals, held in solitary confinement, tormented and sexually assaulted for months by left-wing radicals who fantasized themselves as leaders of a new army, the Symbionese Liberation Army, and recruited into their gang. Her experience shows how anyone under extreme stress can be turned to violence, even extreme criminality including robbery and murder; to depart from her lifelong identity and join a new violent tribe of robbers, terrorists, and murderers who held views completely opposite to her own before her kidnapping.

Stress puts the brain's threat-detection and -response circuitry on a hair trigger. The physiology of this heightened sensitivity to snapping is well understood. It involves stress hormones such as cortisol and adrenaline, neurophysiological changes in cognitive circuits increasing vigilance, emotional states of anger, paranoia, and increased sensitivity, irritability, and heightened sensory reception. Under stress you can hear a pin drop.

Chronic stress literally rewires the rage circuits in your brain, setting the snap response on edge. Research like that described in chapter 3, in which a window was installed into the skulls of mice to enable scientists to see synapses being lost after chronic stress, as well as fMRI studies, show that long-range connections from the prefrontal cortex (the executive control network, involving the dorsolateral prefrontal cortex, the anterior cingulate cortex, and the orbitofrontal cortex) to the amygdala are weakened after chronic stress. This reduces the power of the brakes on the amygdala's threat response that is provided by the prefrontal cortex. The result is snapping in response to even minor provocations.

The depiction in the media of someone who is a normal law-abiding citizen suddenly snapping violently and harming someone is a fiction. The word "snapping," as used in this context, really means that the backstory that would explain the situation that had been producing chronic stress is not known or not reported in the news account. There are always chronic underlying stresses involved when a normally trivial circumstance causes a person to snap violently. Chronic stress is always there somewhere behind the everyday instances of snapping verbally or in a sudden tantrum. The LIFEMORTS trigger mechanism is in place at all times.

And so are the physiological mechanisms that constantly adjust the sensitivity of the LIFEMORTS triggers. A more sensitive threat-detection mechanism and a more rapid threat-response system make sense under chronic stress conditions. It is like putting the military on high alert. The emotion we call stress is the result of all the brain's threat-detection mechanisms and internal status monitoring of our body reaching a conclusion that we are in danger. Your unconscious brain cannot speak to your conscious mind and lay out all the evidence in a logical argument. Even if it could, your conscious mind could not hold such a vast amount of data in working memory to comprehend it all. Your unconscious brain calculates an enormous amount of information exceedingly rapidly and communicates the result the only way it can, with this very specific emotion—stress—telling you that you are in danger. Fear arrives when you realize the danger has become immediate. If you look back to the examples of snapping violently that have been given throughout this book, you are likely to find evidence of the perpetrator enduring chronic stress in most cases; even the mother moose that attacked Myong Chin Ra on the University of Alaska campus had been tormented by students before the attack. If alcohol, steroids, amphetamines, or other drugs are involved in a situation, which interfere with prefrontal cortical control of the amygdala/hypothalamic threat response, snapping is highly likely.

The Navy SEAL, I imagine, could dispatch a robber with a lightning chop to the throat while ordering a pizza on his cell phone with his other hand, but he said he would probably just hand over his wallet. "I never gamble," he says.

But here's the question: Did the neurocircuits of threat detection fail me in Barcelona because they were set on such high alert from the chronic stresses after the previous robbery in Paris? Or did these circuits work exactly as nature intended? The fact is, I had just had my pocket picked while unaware on the Paris Métro. My amygdala had learned. That was not going to happen again if it could be avoided. My amygdala did not discuss this with my cerebral cortex. This is implicit avoidance learning. The amygdala, not the cortex, is where fear conditioning takes place. The fact is that we *were* in heightened danger. I have never had such difficulty with pickpockets while traveling before. Maybe it was because I was an easy mark, traveling with my daughter and more distractible. Maybe it is because I am getting older and street robbers and criminals do prey on the old and vulnerable. Maybe there were just an enormous number of pickpockets working the tourist areas while we were there. I may never know why we were such prey to robbers, but my unconscious threat-detection system knew.

But consider how complicated this situation is. If the pickpocket had been twice my weight, I doubt I would have grabbed him by the neck. I suspect that even though I did not consciously have any idea who was following closely next to me, my unconscious brain knew. It was not an innocent old lady I throttled blindly, and it was also not a 230-pound goon. That the guy was about my size must have been a calculation made instantly and unconsciously. The consequences of a second robbery must also have been weighted into the calculation. My brother would not be able to send us money again. The cash we had just received that morning from the spooky Internet café was extraordinarily precious.

At the same time, if I had seen that there was an entire gang involved, I would not have fought back. I suspect that my unconscious mind would not have triggered the response if it had detected the gang either. It did miss this vital but very-difficult-to-detect information. My point is that we need to appreciate what a truly amazing threat-detection and -response system has evolved in the human brain. Why we snap is an important question, but *how* we are able to snap is fascinating. As with any system, the neurocircuitry of the rage response can and will fail sometimes, but what it achieves in an instant in such a complex and dangerous situation is astonishing. Moreover, this is something that could not be accomplished by all the cognitive power of the cerebral cortex if your life depended on executing the action in a split second.

So in the future I will be more sensitive to the effect of stress on the LIFEMORTS triggers and realize that under stress it is more likely that they will snap inappropriately. An important lesson here is to be aware of chronic stress when you experience it and understand what it does to this circuitry in you and in others. It is essential to ask yourself whether this stress is a signal of credible danger, or if the stress is derived from circumstances in the modern world that your circuitry was never designed to cope with, like being late for work. If circumstances in the modern world are the cause of your stress, be on guard for misfires of your threat-detection and rage circuitry. I will try to remember that the right thing to do is to surrender your wallet. However, it is difficult to know how you will act in a sudden dangerous situation, because there are too many variables. I do know that there are circumstances where people who did give up their wallet were still harmed (although not by pickpockets, usually). Despite all the hours of rational analysis I have applied to this incident, the fact is that I did get my wallet back. The robber was incapacitated. It is essential to understand how the threat-detection and defense circuits in our brain operate, but it is also very important to realize that the neurocircuits of rage operate amazingly well and that we have them for a good reason. Sometimes trusting your rage circuit works, and in that case are we sure we want our threshold triggers set high?

## The Meek Versus the Mighty

On January 12, 2015, a subway train in the Washington, DC, Metro transit system experienced a mechanical failure, stranding it on the tracks inside a dark tunnel beneath the city streets. Many passengers suffered serious injuries, and one person died. Every single person's fate on that train was set by the LIFEMORTS triggers, threat-detection, and unconscious defense circuits in their particular brain. These circuits in some people predispose them toward reacting like lions, and in others the same brain circuits predispose them toward behaving like lambs. In this sudden threat on the subway, which response is the correct move—aggressive or passive? If you were stranded on that ill-fated train, what would your response be—fight, flight, or freeze? Your fate hangs in the balance. The balance is tipped in this dangerous situation not necessarily by your body strength or by the specifics of the situation; it is your predisposition to react instinctively in a certain way to threat and to respond aggressively—your instinctive character in the moment of danger. The unconscious brain circuits that we have been exploring in this book will determine your fate with a snap reaction to that danger.

Imagine that, like others on this subway car, you are returning home from work early at 3:14 p.m. on the Yellow Line Metro. The noise of metal wheels rattling over steel track and the jostling and roar inside the train as it streaks through the dark tunnel suddenly come to an abrupt stop. There is a loud pop and the train is plunged into darkness and silence. A faint smell of smoke begins to waft into your nostrils. The smell of brakes perhaps?

As you sit in silence among the other passengers, you all share the same irritation. Service on Metro is increasingly abysmal. The system is in dire need of repair and better maintenance. Passengers around you begin to make snarky comments. You and fellow Metro riders shake your heads in resignation. The woman across the aisle snaps a cell-phone photo and sends it with a text message to her boss to explain why she will be late for work.

But the smoke rises. Soon the car is filled with a dense fog of acrid smoke as a cacophony of coughing and choking ensues. You and other riders cover your mouths with scarves or coat sleeves to filter out the smoke. Many passengers around you drop to their hands and knees, where the smoke is slightly thinner near the floor. Some people are frantically dialing 911 on cell phones.

Over the loudspeaker the train conductor announces that help is on the way and that everyone should remain calm. The train is not on fire; there is only smoke in the tunnel. Investigators will later determine that the 750 volts of electricity running through the third rail were short-circuiting and melting the rails like an arc welder, generating thick clouds of smoke. The insulation on the power cables was smoldering, releasing toxic halogen gases. The tunnel had turned into a gas chamber. The conductor's voice over the speaker instructs passengers not to force open the doors or windows. The smoke is worse outside the car. Everyone should wait calmly. The fire department is on the way and they will evacuate everyone safely.

The reactions of different people around you span the range of human behavior. Some men become profane, shouting obscenities and banging their fists on the windows in anger. Others are panicked, shaking violently, but stuck to their seats shivering in fear. Some women and children are crying. Some begin to pray. Others sit calmly, either with confidence that they will be rescued or stoically accepting that this is probably the last few minutes of their life.

What are you doing?

The information and directions from an authority, the train operator, carry great weight. The authority is responsible and has the information and resources to guide everyone through the perilous ordeal safely. So you and the other passengers follow his instructions. The human herding instinct for bonding into cooperative groups is on display in force as a neurocircuit that evolved in our brain, and separated us from our Neanderthal brothers, is engaged.

But time passes slowly. As you sit stranded on the tracks in the dark tunnel the smoke inside the car is thickening. Your heart is racing and you are experiencing the dilation of time that always happens in a threatening situation as your mind blazes at maximum speed. Your watch shows that much time is in fact passing, and no one has come to help.

Ten minutes, twenty minutes, thirty minutes, you and the other passengers are still trapped. Many people in the train car are now in serious medical distress; some are slumped unconscious from the lack of oxygen and toxic fumes. You can barely see through the eye-stinging fog of smoke. The train operator, trying to keep everyone calm, suggests that people might want to hug the person next to them.

One woman, sixty-one-year-old Carol Glover, a senior business analyst at DKW Communications, is in serious respiratory distress. She sits calmly in her seat, not crying out for help or panicked, but breathing with great difficulty. Glover is traveling among strangers. Then she melts to the floor. Panting rapidly, she gasps for people to fan her so she can get some air. It is hot inside the smoky car, so a stranger named Jonathan Rogers and a few other passengers come to her aid and try to get her winter coat off. Another man scoops the lady up in his arms and carries her like a baby to what looks like a safer spot on the train. He lays her on the floor as he and others try to help the stranger.

What the passengers could not know was that the firefighters were only eight hundred feet away. They were assembled underground in the L'Enfant station, ready and waiting to enter the tunnel, but they could not get clear confirmation from Metro officials that electricity to the third rail had been shut off. With their masks and respirators donned, the firefighters were fully equipped for smoke, but a step into the water that had puddled all along the tracks would fry them in a flash if 750 volts of electricity was running through it. This danger was an unacceptable risk to the lives of the first responders, and moreover, to the passengers they were desperate to rescue. Forty-five minutes would elapse from the time the train was disabled until firefighters could finally reach the train car stranded only a short walk from the underground station.

"I thought it was more prudent to stay inside the train, especially given the uncertainty of the third rail," one woman who survived the ordeal said later. "But I also figured some of those panicked people would probably have literally stampeded out of there, and I had a good chance of being crushed or pushed into the rail. So, I made the call to stay on the train."

Riding in a different car from where Carol Glover is on the floor in respiratory distress and surrounded by strangers trying to help her, Jeffrey Todd, forty-three years old, feels the same way as the others at first—do as instructed and stay put. But after twenty minutes of waiting, Todd, a trained Navy pilot, is grappling with the powerful urge to take matters into his own hands, pry open the car doors, and escape into the dark tunnel. He knows he would be defying the commands of the train operator to stay inside the car, also risking electrocution, and he might be putting others at risk if indeed the smoke poured into the car when he sprung open the door.

He asks others around him if they think they should try to escape. Most people reply, "We should stay put." A gentleman in glasses suggests they wait three minutes, and if it's worse, go.

Todd gathers other passengers near the door in the center of the car. In his mind his chances of dying were greater if they stayed inside the train.

Three minutes elapse and Todd pulls the escape lever and pushes open the sliding-glass door. Stepping out into the dark tunnel, he crouches low below the smoke and moves toward the front of the train. As he passes in front of the train someone behind him shouts, "Slow down!" Todd turns around to see that all the other passengers except one man, named John, have turned back.

The two men cross the tracks in the black smoke-filled tunnel, and step over the wires of the third rail in darkness.

Carol Glover has lost consciousness. Her breathing stops. Jonathan Rogers and others begin giving her CPR. "It was just like—it wasn't helping," Rogers said afterward. He knew he had to get her fresh air but "We couldn't take her off" because the train operator was insisting that passengers stay on the train.

Taking turns, passengers give the woman CPR even as they cough and gag from the smoke. Working together, they keep giving the unconscious woman CPR for twenty-five minutes. Then a big man picks her up in his arms and carries her motionless body through the doors between train cars toward the back of the train, where the smoke might be thinner. "Nobody followed because they told us to stay on the train," Rogers said to reporters for *The Washington Post* afterward.

Inside the dark and damp tunnel Jeffrey Todd and the stranger named John have found a staircase. They climb to the top and see a steel grate. Todd swings a lever and opens the grate, and the two men emerge into fresh air and daylight.

Sixty people were injured by the toxic smoke in the Metro accident. Carol Glover of Alexandria, Virginia, mother to two sons and grandmother to three, died.

Recalling the moment of decision to pry open the door, Todd said, "At this point, I did not think we were going to die, but I did believe, and stated it, that there was a chance we would die if we remained in the train." It was a risky move—too risky for anyone else on the train except for him and John.

This accident illustrates the reason that different people have different responses to threat regulated by our unconscious neural circuits. There is no correct way to respond in every encounter with sudden danger. Natural selection and life experience have given different individuals different snap responses to threat, because while any individual may take the wrong reaction to a threat and suffer injury or die, the group as a whole will not perish because other individuals in the group will behave differently.

Todd could have been electrocuted the minute he stepped out of the train. He and John could have been trampled in a stampede of panicked passengers. The poisonous smoke could have overcome them and poured into the car, killing other passengers. What is significant is that Todd was a former Navy fighter pilot; Glover was a business analyst. Snapping, like the result of all biological processes, is a matter of genetics, environment, and chance. Todd had to have been someone who would take risks in an instant, with aggressive action, even knowing full well that he could die because of it. He had no doubt experienced threatening situations in the past in his fighter-pilot training that required quick action. Both genetics and training had given Todd's threat-detection and threat-response circuits, spanning from his amygdala to his hypothalamus and prefrontal cortex in his brain, the ability and the propensity to respond aggressively to threat.

No one ever knows how they will react in a moment of danger. Most of us do have a good idea of whether we tend toward the lion or the lamb end of the spectrum. That knowledge can be lifesaving, because it means that while it may be impossible to know whether it is better to react aggressively or remain passive in any given situation, if you fail, in your case it will more likely be because of your propensity to react in one way more readily than the other. Todd is going to be more likely to fail as a result of taking aggressive action because he is more inclined to act aggressively regardless of the stakes. Someone who tends to be passive or fearful will have the opposite proclivity.

Had Carol Glover quickly escaped the car at the first signs of distress, she would have lived. Even if she had snapped hysterically and kicked through the door in a panicked blind rage, she would have lived. Charges have been made that the ventilation system on the Metro cars was not shut off and was sucking the smoke into the passenger cars. Whether this is a fact, or it was a factor in the tragedy, is not yet known. The point here is that there *is* no way of knowing any of this except through 20/20 hindsight after the event. This is why the advice of the man with glasses to wait for three minutes before opening the door was a wise move for Todd, who was about to undertake an aggressive risk. This deliberate pause assured that the prefrontal cortex and conscious thought were overriding the basic instinct of the lion to crash open the door and escape. The situation he was facing, while perilous, did afford an opportunity of delaying his reflexive response, which is designed for instantaneous reaction to sudden threats. Glover could have taken a similar strategy, by giving herself three minutes to resist her natural inclination to remain passive, but if the situation worsened, to go pull the emergency handle on the door to get fresh air.

Certainly, either scenario is dangerous and either action could have proven to be a fatal mistake, but the odds are that Glover would err by being too passive and Todd would err by being too aggressive. That self-realization may be the narrow edge that can beat the odds and allow someone to survive a life-threatening encounter. This is why Navy SEALs are taught to control their rapid and powerful instincts to identify and neutralize a threat. They are not losing that ability by being aware that, in their case, controlling this reaction is going to improve their outcome. They are like racecars with powerful engines. Having brakes does not hamper the power of the engine; the brake pedal permits control of that power for maximum benefit.

It must be made clear that the luxury of this analysis after the fact is not to be taken as critical in any way of the actions of any of the individuals who faced this ordeal. There was a risk of serious injury or death no matter what action each person took. Moreover, while Todd and John survived by a risky evacuation, hundreds of others survived by doing what the train operator directed them to do. They did not stampede in a panic and trample over others, as could have happened in the frightening darkness of the smoky tunnel, and as often does happen in crowded events when smoke rises. But none of these different reactions to the danger taken by every individual on that train would have been possible without the brain's threat-detection and

-response mechanisms and the circuits controlling snap reactions, while other brain circuits promote cooperation among groups of people.

Arland Williams Jr., the middle-aged, balding bank examiner who refused to save himself until all others had been rescued from the crash of the Boeing 737 into the frozen Potomac River, had been educated at the Citadel, the Military College of South Carolina. Although his graduation in 1957 was in the distant past, service to others is a core discipline at the school. "Always take care of your people first," Frank Webster, Williams's roommate at the Citadel said, explaining their training. "That's an unbreakable code. You go last. Your people go first."

Among the passengers who took down the hijackers on United Airlines Flight 93 on September 11, 2001, were Jeremy Glick, a judo champion; Mark Bingham, a rugby player; Tom Burnett, a college quarterback; Louis Nacke, a weightlifter; and William Cashman, a former paratrooper. The background of each of these men reveals experiences—often in organized sport—in dealing with stressful, difficult, and in some cases dangerous situations requiring teamwork, commitment, and individual strong physical action. These experiences better equip a person to respond physically in a threatening situation, but at the same time, individuals who are genetically predisposed toward this snap response will gravitate toward such pursuits and tend to excel in them.

It is not possible to predict how anyone will respond in any given threatening situation, because there are an infinite variety of threats and factors at play. For example, if there is anything in Jasper Schuringa's background that could explain why he alone instantly attacked Umar Farouk Abdulmutallab and subdued the underwear bomber on Flight 253, it is not obvious. Schuringa is a Dutch film director living in Amsterdam. His reaction was instantaneous and instinctive, springing from his hypothalamic/amygdala circuits, while everyone else on the plane panicked.

"I didn't think," Schuringa said afterward.

These subconscious brain circuits of snapping in rage saved his life, and the lives of all 290 people on board.

## A Community Without Rage

The neurocircuitry of anger, aggression, and sudden violence are indelible in the subcortical tissue of the human brain, derived as they are from the tooth-and-nail realities for survival of the fittest in the natural world. But human beings in the last few centuries have separated themselves from the natural world to an extraordinary extent. Violence still boils throughout the man-made environment and permeates our varied social organizations, but there are apparent exceptions. Some cultures renounce violence, anger, and aggression under any circumstance: Quakers, the Amish, and certain Buddhist sects, for example. If the roots of rage are a biological necessity, how is it possible for some groups of people to survive without engaging them?

When one thinks of nonviolence, Mahatma Gandhi is the first name to come to mind. The bespectacled gaunt Indian, wrapped in a white sarong, won independence for India through nonviolent civil disobedience against the brutal violence of the British Empire. In the United States, Martin Luther King Jr. is the icon of nonviolence for his peaceful struggle against racism and brutality in the time of the Ku Klux Klan and government-sanctioned discrimination against black Americans. The peaceful actions of both leaders, who remained strictly nonviolent in response to brutal beatings and killings of their followers, sprang from a deep religious faith. King, a Christian, and Gandhi, a Hindu, both built their nonviolent beliefs on the foundation of an even more ancient religion. King adopted the nonviolent principles of Gandhi, and Gandhi adopted the religious principle of *ahimsa* (doing no harm to other creatures) from Jainism, the predominant religious faith in the Gujarat region of India where he was raised. Remarkably, Jainism began 3,000 years before Christianity. Older than the Egyptian pyramids, Jainism has remained remarkably unchanged for 5,000 years.

Jains believe that all life is sacred and that it is immoral to harm any form of life. Jains are strictly vegetarian, even renouncing milk, honey, and root crops because they are obtained through thievery or violence against another organism. Believing in reincarnation, Jains go to extremes to avoid harming any creature. Some sweep the path in front of them to avoid treading on an insect, or carefully rescue earthworms and ants when digging a foundation for a building. Men in the most extreme sects in India even forgo clothes, but no religion is best understood by studying its extremists. I visited a Jain temple in the United States to learn more about how some people can control the rage circuits in their brains—to meet people who are apparently incapable of snapping.

Early Sunday morning there is no traffic as I turn off of the four-lane avenue near a Safeway store in an empty parking lot. Fierce gusty winds remain after the two-inch blanket of snow overnight, leaving icicle fringes on dirty cars and house gutters. It is bitterly cold. I am dressed in a heavy down coat and wearing gloves even as my car heater works feebly to warm the cold surfaces inside my pickup truck. A side road leads through a quiet suburban residential neighborhood. Then a long driveway comes to an end in front of a yellow, ranch-style house. There are no monks in colorful robes with shaved heads meandering about. No one is here. There is no structure or any visible sign that would suggest that this empty, dark house is anything more than someone's vacant home.

Soon a few cars begin to arrive and park next to mine. I step out into the cold wind and approach two men of Indian descent as they emerge from their car. Both men are dressed casually in jeans, but are bundled in scarves, hats, and gloves. They are speaking in an Indian language, which I later learn is Gujarati. When I introduce myself, both men are friendly and polite. I explain my interest in learning about nonviolence and Jainism, and they generously invite me into the Sunday meeting, which is about to begin.

The night before, I had interviewed Nimesh Shah by telephone, a Jain who lives in New Jersey. Like these gentlemen in the parking lot, Nimesh evinces an extraordinary calm, open, and friendly demeanor. Shah outlined the basic tenets of Jainism for me. Paramount is *ahimsa*, the spiritual inter-dependence and equality among all forms of life, which requires strict non-violence toward all living beings, and thus vegetarianism. Jains accept that humans must eat to live, but Jains respect a hierarchy of life according to the number of senses a being possesses. Human beings have six senses (the five common senses and the ability to think), and violence in thought or action must never be directed toward another human being. Immobile forms of life, such as plants, have the fewest senses and so these can be sacrificed respect-fully for food.

Jains lead ascetic lives, avoiding confrontation and renouncing attach-ment to possessions (*aparigraha*). They will never ask for anything that is not willingly offered (*asteya*). They vow to always speak the truth (*satya*). If speak-ing truth would cause harm or conflict, they resolve to remain silent and non-judgmental. They exercise strict control over their bodily emotions and desires at all times (*brahmacharya*). This includes abstaining from sexual activity and avoiding sensational pursuits, such as watching unwholesome movies, televi-sion, or even using spices in their food. Monks and nuns are celibate, but lay-men refrain from sexual activity to the extent possible and they reject promiscuity. Anger, pride, deceitfulness, and greed are passions of the mind that Jains work forcefully and constantly to conquer. The word "jain" derives from a Sanskrit word meaning "to conquer." Jains do not strive to conquer anyone; they strive to conquer the neurocircuits of rage in their brain.

The goal of Jainism is not to reach heaven. Rather, it is to liberate the soul from worldly attachments and desires. This is achieved through a process of reincarnation, which revolves around the concept of karma. Generally speaking, karma is the balance of good and bad acts and thoughts of a person in this life and in past lives.

There are many differences between Jainism and other religions. In Jainism there are no priests or gods. There are only teachers and deities who can illuminate the way to *nirvana* or *moksha*—the blissful state of a freed pure and everlasting soul. There is no creator in Jainism. "The universe has always existed and always will," another Jain explained to me. "If you have a creator, then you have the problem of explaining who created the creator." Meditation is practiced to achieve a perfect state of calmness and to contemplate the vastness of the universe and the reincarnation of self in the infinite space and time of the universe, but in prayer Jains never ask for anything, except forgiveness. Two examples of Jain prayers posted on the Internet read:

> *If shot by a gun or pistol, or beaten by a stick,*
> *If bound and thrown into a prison,*
> *May I lose my body,*
> *But not my noble forgiveness.*

> *I ask for forgiveness not only if I might have hurt your feelings knowingly or unknowingly, but also during this life or during previous lives.*

It is significant that Nimesh Shah is a pseudonym. The gentle man I interviewed works at a major corporation and his overriding concern, even before I asked my first question, was that in sharing his own religious beliefs and philosophy he wished to cause no friction or conflict for anyone.

"I was born into a Jain family." Nimesh lived in India until he was twenty-six. "Both my father and my mother are Jains and both my grandparents are Jains," he explained. "When you grow up in that kind of an ecosystem you kind of get wired in that way, I guess."

"We have the neurocircuitry for anger and violence," I said. "All animals have defensive and aggressive behaviors and we humans have them wired in our brain too. Do you not experience rage, or do you suppress it? Insult, for example, don't you feel anger if you are insulted?"

"We feel it. I certainly feel it, and I know others feel it too, but there are different ways to handle it," he replied. "One could get angry and smash their fist against the wall, or throw something against the wall, or yell out loud and curse, but the first thing I do when I get angry is try to rationalize and put it into perspective. There are other ways to handle the situation. If it is something personal, you try to understand why that person put you in that spot and you address that issue. If it is something at work, you put your head down and think, How do I make things better so this does not happen again? If it is something like road rage, you say something like, 'It just isn't worth it because nothing good can come out of you reacting in an angry manner.'"

"Yes," I said, "but in these situations there is often not enough time to apply a rational perspective. These are often reflexive responses."

"It comes through time. I don't think it happens overnight. Growing up riding in the back of my father's car I have seen opportunities for my father to get angry and behave in a rash manner or react in a negative way to an incident on the road, but I saw him tackle it differently. I think I have learned that from him. If I am in a situation like that today behind the wheel, I am wired in the way I saw my father deal with those things."

"What if you were pickpocketed and someone took your wallet, how would you react?" I asked provocatively.

"I would call the credit card company and tell them that my wallet was taken."

I pondered how to reformulate my question. . . .

"What can you do?" he asked. "You go into damage-control mode."

"But people in those situations react reflexively, and there is not time to have this rational control. Often they will regret an impulsive act," I said, pushing my point.

"What could be an impulsive act to being pickpocketed?" he asked.

Puzzling over my question he asked again, "What kind of thing? What would it be?"

There was a long silence. . . .

It was astonishing to realize that I was speaking to someone who cannot even conceive of any sort of aggressive reaction to being robbed.

"Well," I said, fumbling shamefully to suggest that there might be another way to react. . . . "Uh . . . get into a tussle with the person to get your wallet back."

This was met only with a very long silence.

The Jains and I enter the yellow house, remove our snow-caked shoes, and place them in a cubby designed to hold perhaps 100 pairs of shoes. We walk in stocking feet into the kitchen and I am immediately overwhelmed with a strong feeling of familiarity. The place reminds me instantly of the Quaker Meeting House in La Jolla, California. Except for the faint cedar smell of incense, the two places are so similar in adapting a home into a place for people to meet and worship together, they are completely interchangeable.

I am guided into a large room, where there is a man, Mr. Pravin Dand, sitting cross-legged on a small stage. The room has a suspended acoustic-tile ceiling with square fluorescent ceiling lights and two white ceiling fans hanging motionlessly. Another rotating fan stands idle on the stage. I can imagine the sweltering heat and humidity in this room filled with people in summer. Today, though, everyone wears their heavy jackets inside, including me.

A brown wall-to-wall carpet runs up over the wall of a knee-high narrow stage where Dand sits with a book resting open on a rosewood stand between his crossed ankles. The pages, yellowed with age, are filled with incomprehensible but beautiful squiggles of Sanskrit. Balding, with close-cropped gray hair, he wears a plain blue V-neck sweater. My host brings me up onstage and introduces me to Mr. Dand as others begin to assemble. I learn that there are no priests or honorary titles in Jainism. The man who is about to lead the group is simply called Mr. Dand or Uncle Dand. He carefully writes down his name, home address, and phone numbers in my notebook and invites me to sit down and observe the proceedings that are about to begin.

Men are seated on one side of the room in folding chairs facing the stage and women are seated on the other side, separated by a central aisle. One woman in a beautiful sunny yellow sarong sits cross-legged on the floor just in front of the first row of folding chairs. I move to sit in one of the separate seats lining the walls, but the men insist that I join them. "Come sit next to me," a man says, waving his hands for me to come. I do so.

The room is unadorned except for some small tapestries, pictures of some of the twenty-four Jain deities, called *tirthankaras,* pinned to the white walls. These, it is later explained to me by Mr. Dand, are not gods but rather enlightened individuals from ancient times who have attained *moksha,* and thus can guide individuals through example in separating their souls from karma. Strands of coin-sized mirrors threaded on strings are hung over pleated tan curtains that cover the back wall of the stage. They twinkle gently like the ornaments of a sleeping belly dancer. Bolts of red, green, and white fabric attached to the acoustic-tile ceiling festoon both sides of the room. This is the only color in the otherwise tan-and-white room.

The proceedings begin as everyone brings their palms together, elbows out in prayer. Reciting the short prayer in unison, I thought it sounded repetitive, like a mantra, but I can't really say because the language is foreign. There is no ceremony, no song, no incense burning. This is not a sermon. The room is not filled with the heavy atmosphere of solemn devotion, but instead with an air of tranquility and friendship. It is more like a lecture, with the teacher explaining and engaging the audience periodically with questions. The men sit with arms crossed listening intently to the man speaking onstage. Several women consult photocopied pages in Sanskrit clipped into three-ring binders. There are no children. I later learn that there are separate Sunday school–like sessions for children.

Mr. Dand's speech is calm and friendly. He is not reading from the book; he is speaking, but in a language that I cannot understand. Sitting relaxed, much like the *tirthankaras* up on the wall who are seated in the lotus position, his right hand gestures gracefully as he speaks. His left hand rests open and relaxed, palm up, in his lap. His gestures evoke the hand motions of Indian dancers, with elegant flicks of his wrist. At times he clutches his heart with his fingertips and then points a single finger to the sky. I cannot understand a word he is speaking, but the calmness and friendliness of his manner is reflected in all those in the room. Members of the audience ask questions occasionally and he replies. Sometimes others in the audience contribute a brief comment to the discussion. Occasionally I pick out the words "karma," followed by "*pap*" and "*punya,*" "*pap* and *punya.*"

After about thirty minutes, the incomprehensible lecture is suddenly interrupted with words directed toward me. "We are talking about our animal brothers," he says, engaging me with eye contact. "The united brotherhood. Every living creature has a soul and must never be harmed—even bacteria and viruses."

I nod and he continues speaking in Gujarati. Another thirty minutes pass. As he speaks, all in the room listen closely and at times giggle politely. I hear the words "Martin Luther King," and about twenty minutes later "security guard." From Mr. Dand's body language it is apparent that he is seeking confirmation from the audience that he has selected the correct word in English, and he says again "security guard." Everyone nods in agreement. I am struck by the realization that there could be a language with no word for "security guard." One more time he breaks spontaneously into English saying not to me but to those in the audience, "It is not easy. Difficult. Difficult."

After speaking for one and a half hours he finishes and then apologizes to me from the stage for having to speak for the last hour and a half in a language that I do not know.

I thank everyone for their generosity in inviting a stranger to join them and ask if I might ask one question of him and the others assembled. He nods.

"You have all found a way to achieve peace and nonviolence," I say. "If there is one thing you could share with others to help them achieve control over their anger and aggressive impulses, what would that be?"

Mr. Dand says, "This is what we have been discussing. We have been discussing the example of the Middle East, the beheadings and burning alive of the prisoner, and how to react. Build your own karma."

He goes on to explain good and bad karma to me, the "*pap*" and "*panya*," as those about me nod in agreement. Karma to Jains means not only acts of good and evil, but also thoughts. "Thinking the bad things builds your own bad karma. Build the good karma! The bad is destroyed then."

He gives me examples of bad karma, such as looking at violent videos or engaging in road rage. In the case of the recent atrocities committed in the Middle East he says that one must stay neutral. The violent act is a result of the bad karma of the perpetrators. "You can't do anything about that, so stay neutral. The middle is the best. Say to yourself, *It is* his *karma not* my *karma*. Stay neutral, don't say he is good or bad. Stay in the middle."

He goes on to emphasize the importance of respecting all life. "You can't kill anything. Who gave you the right to take the body from the soul of any creature?" This includes not only physical but mental killing also, such as verbal abuse, treachery, or deceitfulness. "Nonviolence is three parts: mind, speech, and body [actions]," he explains.

A man seated in my row addresses me to offer his advice. "Control desire. Desires are the root cause of all evil," he says. "Ambition—give it up as much as you can." Mr. Dand nods and all heads in the room nod in agreement.

Another man, with a white mustache drooping like two curved icicles framing his bearded chin, turns around in his seat in front of me and adds, "We say, 'Eat to live, not live to eat.'"

I am struck by how different these people seem. How calm, friendly, gracious, and peaceful. The gentleness and tranquility were infectious. I find myself regarding them as somehow fragile and, like any delicate thing of value, something to be handled carefully.

After the meeting I meet one-on-one with Mr. Dand, talking until everyone else has left the house empty. He leads me to a small bedroom that holds a small shrine of three statues of the *tirthankaras* sitting in the lotus position on a cloth-covered table. He clasps his palms together before his face as we enter the room with incense burning and stand silently for a moment, then leave.

He tells me that it is necessary to work every day at controlling anger and aggression, that it is a lifetime of work. If you react to road rage with anger, he says, "anger will only lead to one thing—a fight." He emphasizes the need to control one's desires. Nonviolence is the way to live, he says.

"It sounds wonderful," I say. "I can see how this can work within a non-violent society, but how can a nonviolent society exist surrounded by a world that is and always has been filled with violence? We all know how Martin Luther King died," I say provocatively, having heard him mention that name in his lecture. "Can a nonviolent society survive the likes of Hitler and others of such brutality and war?"

"You have your empirical answer," he replies. "Jainism is the world's oldest religion."

## Change

Quakers are a Christian religion that is well recognized for its members' commitment to nonviolence. I asked a Quaker, David Connell, how it is possible to have a nonviolent society, considering that we have defensive rage circuits wired into our brain by our evolutionary struggle for survival of the fittest.

By avoiding the LIFEMORTS, he told me. I had just explained the LIFEMORTS concept to him, and he quickly embraced them. Quakers have a mnemonic to encapsulate their core principles: "SPICES."

"Simplicity, peace, integrity, community, equality, stewardship," he explained. "SPICES are the opposite of the triggers you are talking about."

These principles, although stated differently, are essentially the same as those embraced by the Jains to live a life of nonviolence.

Connell is the person who teaches anger-management classes in the Arlington, Virginia, jail, whom we met earlier in this book. I wondered how effective these jailhouse classes could be in changing criminals.

"In your experience, people can change?" I ask.

"Yeah. Definitely. Most of my students are there because they want to be there. They want to self-improve. The older students are more affected by it. After the thirties and beyond they realize what they are doing isn't working. They keep coming back to jail, or their relationships are suffering. When you are young you think you can finesse things. They get into patterns. This isn't working [they say to themselves]. I gotta figure something else out.

"You have been conditioned throughout your entire life to be who you are right now, but if you start to change through daily practice, eventually you will get it," he says.

New research is revealing at a circuit level how the wiring from the prefrontal cortex to the amygdala, fear, and even memories can be changed.

I'll never forget it. They strapped electrodes to my wrist, cranked up a black dial on a frightening electronic device encrusted with switches and knobs, and shocked me repeatedly with jolts of electricity. No, this was not torture, and this memory is not a traumatic one. I was inside the laboratory of Dr. Daniela Schiller, a psychologist at Mt. Sinai Medical School in New York City, experiencing the same treatment that she and her coworkers had used to discover a new way to alter traumatic memories. Research from her team provides a new method for blotting out traumatic memories that are stored in the amygdala—not just suppressing them as current treatments for PTSD do, but altering the fear memory itself.

Just as the participants in her study did, I am watching a computer screen in a dark room while my right wrist is wired up to an electrical stimulator to deliver a painful shock. A second set of sensing electrodes, two black Velcro strips with wires attached, are strapped like rings around the pads of two fingertips on my left hand. These sense the amount of nervous perspiration I produce. Sweaty palms are an involuntary reaction to threat, part of the body's fight-or-flight response that braces the mind and body to defend against an attacker or flee to safety. The heart pounds, stomach churns, muscles twitch with adrenaline-fueled energy, sweat beads up on our forehead, and mental focus sharpens to rev up all systems in the body to survive a potentially deadly danger.

These bodily sensations of fear are the same reactions set off in people who suffer panic attacks and other anxiety disorders. The lifesaving rapid response of fear can become an overwhelming debilitation in people with anxiety disorders or PTSD. The problem is that panic is triggered inappropriately in their brain by stimuli that are not true threats. Terror can grip these people without warning, sometimes crippling their lives. Sleep may become impossible. Others may fear venturing outside or they cannot fly in an airplane. A military veteran from Afghanistan may panic suddenly upon hearing a sound that is connected in their memory to wartime trauma.

Current behavioral methods for treating anxiety disorders utilize exposure therapy. This is based on animal research in which a painful stimulus becomes associated with a second stimulus that is not in itself dangerous. For example, if a rat hears the sound of a bell and then receives a mild electrical shock, it will instantly learn that the bell heralds a nasty jolt of pain to follow. Sound the bell again and the rat freezes in fear even if you do not shock it. This conditioned fear response is how many of us learned as children not to stick hairpins into an electric outlet or to play with matches.

Eventually we lost our fear of matchboxes and electrical outlets after many subsequent experiences with them that were harmless. This is how exposure therapy works. A soldier who survived a harrowing roadside bombing in Afghanistan might develop extreme anxiety about driving a car. Therapists may treat this debilitating fear by having the person drive in a safe environment repeatedly until the terror of the bombing connected in his memory with driving gradually subsides. This can be helpful, but frequently exposure therapy is not effective.

"Some of the bravest people I know are people with PTSD," Schiller tells me as I sit wired up to her experimental apparatus, because unlike individuals who may indeed be fearless, people with PTSD courageously cope with ceaseless terror yet persevere in their daily lives.

Rather than suppress or endure the fear, it would be better to break the connection in memory between the roadside bombing incident experienced in war and the normal experience of riding in a car.

"That memory can change is a natural process that is occurring every day of our lives. We pretty much create a false memory on a daily basis," Schiller says. So rather than suppress the fear, the scientists set out to change the conditioned fear response recorded inside the brain.

Research has shown that when a specific memory is recalled, it becomes vulnerable for a certain window of time to being altered or even eliminated. Recalling a memory is something like pulling a book off the library shelf for review. The book is now subject to alteration or destruction, and it must be put back in the proper place on the shelf. Disrupt a person's attention in the middle of browsing, and the book can be easily misfiled. The process of reshelving a memory immediately after it is recalled is called reconsolidation, and research, including some research in my lab, has uncovered the details of how this works down to the specific molecules in synapses that encode information.

Reconsolidation may sound odd, but it does make sense when you consider what memories are for in the first place. Fundamentally, memories allow us to use past experience to direct our behavior appropriately in the future. This means that memories need to be updated, because things change. Your memory of Barack Obama has certainly changed since the first time you heard the name, for example. The memory has become richer, linked with many other experiences, and separated from others that are no longer relevant and now forgotten.

"In principle, reconsolidation suggests that in order to change memories one must first retrieve them," Schiller explains. The electrodes on my wrist are how Schiller and her team tested the idea that they could break the terrorizing connection with a traumatic memory by first recalling it.

A blue square flashes on the computer screen. Shortly thereafter a purple square appears, followed by a painful jolt of electricity that makes my fingers clench automatically. *Ouch!* Meanwhile, signals from the electrodes that test the perspiration on my fingers trace out a graph on a computer monitor that Schiller and her graduate student are watching. I see the trace being graphed out in real time on the computer monitor shoot up the instant I am shocked. Pain has triggered my body's fight-or-flight response. The sensors on my fingertips have detected the spike in nervous perspiration that is launched automatically from my hypothalamus.

I see the purple box appear on the screen again after several blue ones have passed and instantly the trace showing my perspiration spikes again, but it spikes even before I am shocked! My amygdala has already learned to associate the purple square with the painful shock. If this were a real threat, my body would be that much faster in responding to it because of the fearful memory of my past threatening experience recorded in my amygdala. Just seeing the purple square trips my body's fight-or-flight response the way driving a car would do for the veteran with PTSD who'd survived the road-side bombing in Afghanistan. In contrast to the purple square, when a blue square appears on the screen there is no rise in my perspiration or anxiety. Blue squares are safe. I am just a passive observer into this window into my mind's fear circuitry. I have no conscious control over my sweaty fingertips, frightened by the sight of the purple square. That fear is long lasting. People in these experiments will show the same automatic anxiety reaction to see-ing the purple square even when tested many days later.

This automatic fear can, however, be erased. If Schiller and her col-leagues began to flash the purple square over and over again without giving the electric shock, the stress response to the purple square diminishes with time and ultimately ceases. This is because the prefrontal cortex has learned that bad things don't always happen every time one sees the purple square, and it sends inhibitory signals to the amygdala to suppress its threat response. This process is called extinction. Schiller and her colleagues, including neurobiologist Joseph LeDoux and Elizabeth Phelps of New York University, were able to see this altered brain functioning happening by having the subjects participate in these experiments while inside an fMRI brain scan-ner. The researchers saw that the prefrontal cortex was becoming active in addition to the amygdala during extinction therapy, and the functional con-nections between the prefrontal cortex and the amygdala were growing stronger. However, when these subjects were tested a day later, the finger-tip stress monitor showed that seeing the purple square often triggered the threat and fear reaction again. Exposure therapy helps, but the feared asso-ciation between the purple square and an electric shock is still recorded in memory inside the amygdala.

Next the team tested whether the mechanism of memory reconsolida-tion can be exploited to break the mental association between the purple square and the painful electrical shock. To do this, they simply reminded the person of this connection by flashing the purple square on the screen and delivering the nasty shock. Then they follow up immediately with exposure therapy, by flashing the purple square repeatedly without an electric shock.

Doing this proved to be far more effective in reducing the long-term stress response to the purple square than if they had used extinction therapy without first reminding the participants of the purple-square threat. By monitoring changes in the brain's activity using fMRI, they could see how this was working inside these neural circuits of fear and threat detection.

Two things could explain why extinction therapy during the memory reconsolidation period is more effective: It could be that the prefrontal cortex is now strongly inhibiting the memory of threat associated with the purple square, or alternatively, the mental connection between the purple square and the painful shock, which is stored in the amygdala, could have been diminished. What they found using fMRI was that the prefrontal lobes did not become activated when the purple square was flashed in people who were given extinction therapy during the period of memory reconsolidation. In essence, the fMRI results show that the amygdala had forgotten the association between the painful electric shock and the purple square because the prefrontal cortex was not being activated to inhibit the threat memory.

To put this new finding into a real-life scenario, imagine that you are bullied at the school bus stop by neighborhood thugs John and his delinquent brother Greg. Every time you see John or Greg you become anxious and fearful. If days go by without either brother bothering you, your body's threat response will gradually subside, but you haven't forgotten that the brothers are potential threats. An fMRI of your brain would show that your prefrontal cortex was suppressing the fear response recorded in your amygdala that you have learned from John or Greg attacking you in the past. This is how extinction operates at the level of neural networks.

One morning, John bullies you again when you arrive at the bus stop, but immediately after assaulting you he reverses his behavior and is friendly toward you. Then days go by without any harassment from either brother. The next time you see John and Greg, what happens? Your body does not react in fear to the sight of John, but when Greg approaches your heart races. The reversal of John's hostile behavior immediately after bullying you altered your fearful memory of him. But Greg, who hasn't even done anything mean lately, is now more frightening. An fMRI would show that there is now less neural activity between your prefrontal cortex and your amygdala when you see John as compared to when you see Greg. In fact, your body's defensive response when seeing John is no different from that provoked by seeing his kindly sister Betty, because the fearful memory of John's bullying was replaced when that memory was reshelved right after he acted friendly to you.

"We hope that the reconsolidation window would prove useful for treating PTSD," Schiller tells me. "This would require some modifications in current therapy to specifically target this phase of memory."

Later I ask a professor of neurobiology at the Weizmann Institute in Israel, Yadin Dudai, who was not involved in the study, about this research on erasing fearful memories. He agrees with Schiller, but he emphasizes that more research is needed to adapt this finding into a practical treatment for PTSD. "In real life, PTSD is very persistent; it involves a very dense web of associations and lingers or even becomes intensified over years and decades," Dudai explains. The simple association between a purple square and a mild electric shock is much weaker than the fearful associations people with PTSD must overcome. "I had a conversation last week with a colleague who experienced trauma in combat forty years ago. This still haunts him at nights."

In discussing these studies with B. J. Casey, director of the Sackler Institute at Weill Cornell Medical College, she draws parallels to her own research on phobias and anxiety-related disorders in adolescents. She sees some interesting implications, because the prefrontal cortex is not fully developed in adolescents. Thus teenagers may not be able to suppress threat responses in the amygdala as effectively as adults can. "We have previously shown diminished extinction learning in adolescents due to the maturational changes in the prefrontal cortex," Casey says, meaning that because there is weaker wiring between the prefrontal cortex and the amygdala in children and adolescents, fear and anxiety are more difficult to suppress until adulthood. The new findings using extinction therapy during reconsolidation could be especially effective in treating adolescent anxiety, because the process bypasses the need for regulation of emotional responses by the prefrontal cortex. It erases the fear memory itself inside the amygdala.

As I watched my brain and body respond automatically to the threatening squares, I was astonished by how robotic it all seemed. There was nothing I could do to control it. I felt some sense of what it must be like for people living with anxiety disorders where fear overwhelms them suddenly in ways that are entirely beyond their ability to control. As researchers uncover how our brain encodes memory, updates it, and as the neurocircuits are traced that connect emotional fear and threat reactions to specific triggers, the findings are bringing us closer to developing better therapies. "It is our social responsibility to find treatments for PTSD, especially when putting soldiers and Special Forces at risk," Schiller says.

You don't need to have experienced the horrors of war to suffer serious psychological trauma from uncontrollable anxiety, fear, and aggression. Haley Peckham, the neuroscientist and psychiatric nurse we met in chapter 10 who treats children who cut themselves, also cares for children dealing with drug addiction or uncontrollable aggression for other reasons. The origin of the rage in these children is often traceable to their home environment, but with proper counseling and intervention, such children can be helped to overcome their aggressive and violent impulses.

"Sometimes you will see kids come in with drug issues and their fathers are in gangs," Peckham says. "That's pretty clear."

She continues, "You can have experiences with a child who cannot regulate themselves emotionally at all. They are very hostile and aggressive, and then you can speak to parents on the phone and end up having a conversation in a similar emotional tone, where if you are not very careful in your handling of them then they feel very defensive, and then you end up getting quite a lot of aggression from them about that. So you can see if you've got a very disregulated parent, you've got a very disregulated child."

Deep-brain unconscious emotional responses, threat responses, and violence are under inhibitory control by the prefrontal cortex acting on the amygdala. The strength of these inhibitory inputs is influenced by genetic factors and very much by environmental factors. If you are in a hostile environment, you cannot afford to have a slow reaction to threats and hostility. This puts these violent rage responses on a hair trigger. And because adolescence is the period of life when the prefrontal cortex is developing, the effects of early life experience can be very strong and persist into adulthood. fMRI studies show changes in the brain's wiring in people who suffer sexual abuse as children, severe childhood neglect, or even verbal abuse in the middle school years. The physical changes in the brain persist into adulthood, and correlate with increased risk of drug addiction, depression, anxiety, and other psychological difficulties.

"It's like they have no skin," Peckham says, describing her kids who lack impulse control and who are very aggressive to the point of borderline personality disorder.

"They cannot regulate themselves. Whenever they are fearful, there's no point—it is a *waste* of anyone's time to try to reason with someone who is in that state. They're just locked into that kind of limbic fear, and until they are out of it they cannot think. The first thing you need to do is to get them to feel safe again, which can be quite hard because often it takes a very aggressive form. You are appealing to a part of their brain that isn't developed enough to be able to be engaged into that conversation.

"As a society I think we are shit at managing people who have any kind of emotional turbulence. We don't really know how to help with that. A child who comes in who says, 'I want to die,' or 'I want to hurt myself,' we are so bad at sitting with that risk and letting them express their distress in any way they can that doesn't involve them killing themselves. Because we don't let them do that, we kind of shut it down a little bit [with temporary crisis care and medications], which doesn't resolve it. And if they keep coming back to hospital it turns into almost a game. They *up* the stakes to show us they need to be there, because hospital is validating. You know, 'There is something wrong with me.' You know, 'I do need some help.' So in order to get the validation they need, they end up hurting themselves more and more, but we are not solving anything.

"They need to be validated for the distress they feel before they have to show us it—before they cut themselves. We want to get in there and say, 'Hey, are you struggling? Because it seems like you really are. What's going on?' We need much more preventative, long-term practitioners who are able to sit with the risk and validate those kids' feelings."

When this is done, people do change; brains are rewired. The woman who shared her experience with self-harm earlier in this book is an example of such a success story. She escaped the vicious cycle of self-harm through three years of psychotherapy. Tricia, the blind woman who can see with her fingers, is a dramatic example of how the brain can change according to experience. The purpose of the organ inside our skull is to enable us to interact with our environment. The experience of psychotherapy, like any environmental experience, will change the brain and change the lives of people who, through genetics, environment, or chance, find themselves in trouble, and in need.

# Family

"The steeper for me, the better. I do like to make my heart pump!" says extreme skier Wendy Fisher as we ride the chairlift back up to the starting gates with her five-year-old son Devin for his next practice run before his race starts.

"Mom? Can I have another snack?" Devin asks.

Wendy talks about growing up in a skiing family and entering ski racing with her friend Shane McConkey. She was not a great student in school, but like McConkey, she found success in skiing. McConkey revolutionized big mountain extreme skiing, but his increasingly dangerous feats ultimately took his life.

"The way we ski today is because of him. The skis we ski on today are because of him," Wendy says.

Shane is the person seen in the opening of *The World Is Not Enough*, when James Bond flees down a mountain on skis pursued by a gang of bad guys blazing away with automatic rifles. They drive Bond off the edge of an enormous cliff, seemingly to his horrendous doom, but Bond (that is, Shane) ejects his skis and pops a parasail parachute to glide dramatically to safety.

In reality, Shane's life ended abruptly in a wingsuit BASE-jumping accident, leaving his young daughter and wife to go on without him. After seeing a movie about Shane's life, Wendy's father asked her if she would have ever taken up such a dangerous sport as BASE jumping. She told him yes; if she had grown up in Tahoe, where it was all happening, she "would have totally been a part of it. I would have definitely been persuaded by that kind of energy and excitement."

"Mom?" Devin says.

"I don't know what bag it is in. Have a gummy instead, Devin. Suck it up. I'll buy you a candy somewhere else, OK?"

It is freezing on the chairlift. Ice crystals spewing like fine shrapnel in the wind prick raw, frozen cheeks. Noses run to counter the cold air.

"No, my parents were not thrill seekers at all," Wendy says. "My mom is super mellow and conservative."

We slide off the chairlift and glide over to the starting gate. Devin gets in line with the rest of the competitors. As we chat on the sidelines a man skis up, bundled from head to toe against the cold, and plants himself expertly with skis crossways on the slope.

"Hi, Dad!" Wendy introduces me to her seventy-nine-year-old father.

Her parents were great skiers, she says. Her father was a double-black-diamond skier on the groomed runs, but off terrain he was a disaster. Now after the grandparents moved to Crested Butte, the extended family skis all the difficult terrain off-trail.

"She wouldn't be in skiing if it weren't for her brothers," Wendy's father says.

"I think it is what you grow up with. I was adventurous growing up at Squaw [Ski Resort]. Skied with a lot of guys and the twist of why I do this, which I think is kind of weird, why I do all of this, is because my brother passed away skiing at Squaw and I was there for it."

The traumatic experience of seeing your brother die as a young child while skiing together would have ended skiing for many, but the family persevered in doing what they always had done together.

"I didn't want to ski anymore but my parents said, like, 'Would Mark not want you to cross the street if he got hit by a car?'"

Wendy looks into her dad's eyes. "You guys didn't make it like a scary thing. It wasn't the sport that did it."

Given the pressures of competition in downhill racing and extreme free skiing, I asked Wendy if anger was a help or a hindrance to her.

"Anger? No. My dad made me work on my anger management. He would get mad if I ever showed anger."

She says that she would see other competitors lose it in anger and wrap a ski pole around something, but her father would admonish her. She recalls her dad saying to her when she was a child competitive skier, "Wendy, if I ever see you throw your skis down. If I ever see you whack your pole. That is *not* OK."

"I was always worked on to not show that anger emotion because I had my dad watching over me, making sure I didn't," Wendy says, crediting her father.

"I don't want to see that anger," Wendy says about dealing with her own children now. "I don't want to go as far as my dad, but I don't want my kids to be total assholes either; there's a fine line."

"You know, you watch guys like Jimmy Connors and you say, 'What a jerk,'" her father explains. Connors was a tennis pro, infamous for his atrocious temper tantrums on the courts. "And I just thought some of those kids come down, slam their skis, throw their poles . . ."

"Why did you start telling me not to act that way?" Wendy asks. "Did I start showing that kind of anger?"

"Yes, in the *beginning* you did. And I said, 'You don't need to do that.'"

"Right."

"You see these great athletes; *they* don't do that. Failure is part of life."

"Yeah."

"Everybody who has started a business has always failed. They didn't throw temper tantrums."

"Well, behind closed doors, maybe."

"Yeah! That's fine."

"I feel like it is me—there is no one to blame," Wendy explains. "But on the ski team the girls used to get so mad at their ski reps. 'Oh my God! He waxed the wrong wax! He doesn't tune the skis right!' I never, you know, I just said, 'I sucked today. I didn't ski very good today.' I never blew up, and I always took ownership in that kind of stuff."

Strong leg muscles are essential for skiing, but an inner strength is what separates the most elite athletes from the rest. That early childhood training in self-control strengthened the development of inhibitory connections from Wendy's prefrontal cortex to her amygdala. This is what now allows Wendy today to keep her cool, no matter what. It allows her to persist despite fear, to perform at the highest level despite the emotions and stresses of competition and real danger. To give it all on a treacherous, icy slope in the Olympics and to outwit an avalanche, her greatest fear, coolly in the face of death.

Wendy's parents taught her to suck it up.

I hope that by looking carefully at the human behavior of snapping, which we tend to avoid examining, it will be seen not as an aberration but as a normal and necessary physiological mechanism. It is a lifesaving mechanism and often drives the most heroic acts of self-sacrifice for others.

There is mental illness, psychosis, criminality, and evil in the world, but also much anger and aggression acted upon spontaneously and regrettably every day that cannot be dismissed as psychological dysfunction. Psychology, philosophy, and religion all provide bright beams of illumination into the darker aspects of human behavior, and they are powerful forces of self-control and understanding, but I also hope this new information from neuroscience can be helpful. No one perspective can possibly encompass all the complexities of human behavior, but multiple perspectives on anything bring a more complete picture. Understanding that these rage circuits evolved to help us, not to harm us, and seeing how they can misfire, can help us to understand how people with psychological difficulties or who are dealing with chronic stresses can be even more prone to these circuits of rage misfiring. I hope the LIFEMORTS triggers we have traced in this book and the neurobiology that controls them can be taught to children in school and by parents. I hope people will find the new information helpful in their own lives. That it will help us to avoid and manage danger and conflict, and help us respond in the best possible way in a sudden dangerous situation. My hope is that this bit of neuroscience can help make us better people, but also to work better together within groups, in families, with friends, among coworkers, or as citizens. I hope that understanding how our brain instantaneously distinguishes people by race, class, sex, or other group can help us to see how we are all united by our biology rather than separated by it. I hope that understanding fear and emotion, appreciating the intricate and powerful threat-detection mechanism of the human brain, and marveling at the complex information processing and computation constantly taking place in our unconscious mind will bring you the same sense of awe that it does for me.

These circuits of rage had enabled me to get my wallet back and elude a gang pursuing my daughter and me for two hours after our dangerous incident in Barcelona. The circuits had worked as nature intended, but unease lingered after the victory. The aftermath was disconcerting enough to send me on a four-year quest to understand that moment, which resulted in writing this book. Ambivalence lingers still after having scrutinized the biological mechanism at work and after reliving the many stories in which people reacted in a split second and their lives were changed permanently—either for the better in heroism or for the worse in tragedy.

We have these circuits of rage and aggression for a good reason, just as all animals do, and unfortunately we do need them. But the human brain has evolved another brain region that has become elaborated disproportionately beyond that of any other animal. The frontal cortex of humans comprises one-third of our cerebral cortex. This new brain tissue is what makes us human.

When I returned home after the hour-and-a-half meeting in the Jain temple where the language was incomprehensible to me, I found an email waiting. It was from a man who had been seated a few seats away. While listening to Mr. Dand, the man had thoughtfully transcribed the entire proceedings from Gujarati into English on his cell phone to send it to me.

"Please pardon the mixed use of English/Gujarati," he wrote. "I can try to clarify if you want. I beg your forgiveness for anything written here against the Jain Shastra/text."

The whole sermon was laid out for me, which I could recognize by the landmarks I still remembered. Here are some snippets:

Overcome internal enemies using *Shaant bhav*—not *krodh/kashay.*

*Marganusari*—method to overcome internal enemies—follow the leader—Martin Luther King.

*Samta—chitta baalak*—keep reminding yourself of 12+4 (*dharm bhavna*) *bhavna* medicine and stay focused, else 63 *durdhyaan pisaach* will take control of your mind and will make you do embarrassing things.

All activities have a goal—even a fool does not do any activity when there is no goal/result.

And then this . . . the answer to the question that had launched me on this quest:

You are just a *security guard* for your money.

Recalling the peculiar moment when Mr. Dand said those two words of English in the middle of his hour-and-a-half sermon, a single word flashed into my mind—Barcelona.

# References and Notes

**1. SNAPPING VIOLENTLY**

3 **"Frankly, my dear"**: *Gone with the Wind* (1939), directed by Victor Fleming, movie based on the 1936 novel by Margaret Mitchell. A clip of this scene can be found at https://www.youtube.com/watch?v=YYj0M19Ap34.

9 **"We like it here"**: Author interview with resident of Dumont Oaks neighborhood, Silver Spring, MD, August 2012.

10 **"He is the kind of neighbor"**: M. Laris, "Man Held in Stabbing at MD Post Office," *The Washington Post*, July 27, 2012.

10 **"He was always very polite"**: Ibid.

10 **to complete his paperwork**: "Man Stabbed at Silver Spring Post Office after Another Man Mistakenly Thought He Cut in Line," CBS News, July 26, 2012, http://washington.cbslocal.com/2012/07/26/stabbing-at-post-office-in-silver-spring-one-person-in-custody.

10 **blasted Young with pepper spray**: Crimesider Staff, "Md. Man Stabbed at Post Office in Reported Confrontation over Cutting in Line," CBS News, July 27, 2012, http://www.cbsnews.com/8301-504083_162-57481287-504083.

11 **while officers searched his car**: M. Laris, "Montgomery County Post Office Stabbing Suspect Charged with Attempted Murder," *Crime Scene* (blog), *The Washington Post*, July 27, 2012, http://www.washingtonpost.com/blogs/crime-scene/post/montgomery-county-post-office-stabbing-suspect-charged-with-attempted-murder/2012/07/27/gJQAg1gnDX_blog.html.

11 **"I've never seen him be aggressive"**: Laris, "Man Held in Stabbing."

11 **"He seemed just like a standard"**: Ibid.

11 **"He's as stable a citizen"**: J. Arias, "Silver Spring Man Charged with Attempted Murder After Post Office Stabbing," Gazette.net, July 27, 2012, http://www.gazette.net/article/20120727/NEWS/707279565/1081.

11 **documents in a bright-yellow file folder**: Author reporting from District Court of Maryland, Rockville, MD, August 24, 2012.

11 **"I am not willing"**: Arias, "Silver Spring Man Charged."

11 **"Maybe I should have stayed home"**: Laris, "Man Held in Stabbing."

12 **"Never get out of your vehicle"**: M. Weil, "Fairfax Woman Charged with Hit-and-Run in Shelinda Arrington Death," *The Washington Post*, May 7, 2012.

12 **"It's chilling now"**: L. Sausser, "Dela Rosa an 'Angry Woman,' Psychologist Says," *Vienna Patch*, October 5, 2011, http://patch.com/virginia/vienna/dela-rosa-an-angry-woman-psychologist-says.

13 **"It looked like a jacket"**: L. Sausser, "Update: Grandmother Guilty of First-Degree Murder in Toddler Death Case," *Vienna Patch*, October 6, 2011, http://patch.com/virginia/vienna/grandmother-charged-with-first-degree-murder-in-toddl01d21fb313.

13 **"She'll never be able to"**: J. Jouvenal, "Va. Woman Gets 35 Years for Fatally Tossing Granddaughter off Mall Walkway," *The Washington Post*, January 6, 2012.

13 **"She was basically angry"**: Sausser, "Dela Rosa an 'Angry Woman.'"

13 **"What's going to happen"**: Ibid.

13 **"I'm very sorry"**: Jouvenal, "Va. Woman Gets 35 Years."

14 **"This isn't insanity"**: Sausser, "Dela Rosa an 'Angry Woman.'"

14 **deep into the brain of a cat**: W. R. Hess, *Diencephalon*, Monographs in Biology and Medicine, Vol. 3 (New York: Grune & Stratton, 1954), 1–79.

14 **added one more critical element**: A. H. Rosenfeld, and S. A. Rosenfeld, *Of Cats and Rats: Studies of the Neural Basis of Aggression*, Interview with John P. Flynn, July 1975 (Washington, DC: NIMH, U.S. Government Printing Office, 1976).

16 **"I was defending myself"**: Laris, "Man Held in Stabbing."

17 **Loughner appeared mentally deranged**: J. Hopper, E. Friedman, and D. Adib, "Accused Tuscon Shooter Jared Loughner Smirks in Court, Smiles for Mug Shot," ABC News, January 10, 2011, http://abcnews.go.com/US/story?id=12580344.

17 **receiving psychiatric treatment and medications**: S. D. James, "Newtown Shooter Lanza Had Sensory Processing Disorder," ABC News, February 20, 2013, http://abcnews.go.com/Health/newtown-shooter-adam-lanza-sensory-processing-disorder-controversial/story?id=18532645.

17 **maximum possible under Norway's law**: "Norway Court: Anders Breivik Sane, Going to Prison," CBS News, August 24, 2012, http://www.cbsnews.com/news/norway-court-anders-breivik-sane-going-to-prison.

## 2. NEUROCIRCUITS OF RAGE

20 **"If you imagine the lizard brain"**: D. J. Linden, *The Accidental Mind* (Cambridge: Harvard University Press, 2007). Quoted in J. Hamilton, "From Primitive Parts, a Highly Evolved Human Brain," *The Human Edge*, National Public Radio, August 9, 2010, http://www.npr.org/templates/story/story.php?storyId=129027124.

21 **conceived the triune brain theory**: J. Pearce, "Paul MacLean, 94, Neuroscientist Who Devised 'Triune Brain' Theory, Dies," *The New York Times*, January 10, 2008.

22 **TV series about a serial killer**: *Dexter*, season 5, episode 3, Showtime, directed by Ernest R. Dickerson, original air date October 10, 2010. See the Showtime website http://www.sho.com/sho/dexter/home.

23 **"The greatest language barrier"**: As quoted in Pearce, "Paul MacLean, 94, Neuroscientist."

30 **José Manuel Rodriguez Delgado**: B. Blackwell, "Jose Manuel Rodriguez Delgado," *Neuropsychopharmacology* 37 (2012): 2883–84.

## 3. WHAT ARE THE TRIGGERS?

33 **immediately after the race and embraced him**: J. Steinberg, "London 2012: History Boy Oscar Pistorius Signs Off with a Smile," *The Guardian*, August 5, 2012.

33  **in a fit of rage or by mistake:** A. Topping, "Oscar Pistorius Charged with Murder After Girlfriend Shot Dead," *The Guardian*, February 14, 2013.

33  **Staff Sgt. Robert Bales:** J. Kaminsky, "U.S. Soldier Behind Afghan Massacre Apologizes for 'Act of Cowardice,'" Reuters News Service, August 22, 2013, http://www.reuters.com/article/2013/08/22/us-usa-afghanistan-trial-idUSBRE97L0YV 20130822.

33  **"It didn't make any sense":** A. Ashton, "Robert Bales Takes Stand at Sentencing Trial, Apologizes for Killings, Disgracing Army," *The News Tribune*, August 22, 2013.

34  **"stood out and had a real positive attitude":** J. Hopperstad, "Bales: 'I'm Responsible . . . I'm Truly Sorry, I Murdered Their Family,'" Fox News, August 22, 2013, http://q13fox.com/2013/08/22/bales-takes-the-stand-in-afghan-murder-trial.

34  **"This bastard stood":** J. Van Sant, "Day 2 of Bales' Sentencing Hearing," Fox News, August 21, 2013, http://q13fox.com/2013/08/21/bales-sentencing-enters-day-2.

34  **"I don't think anybody":** Kaminsky, "U.S. Soldier Behind Afghan Massacre."

35  **Some 30 percent of US drivers:** A. Halsey III and B. S. Berkowitz, "Number of drivers who say they feel road rage has doubled, poll finds." *The Washington Post*, September 1, 2013.

36  **"I'm going to give you":** *Fawlty Towers*, "Gourmet Night," season 1, episode 5, directed by John Howard Davies, original air date October 17, 1975. Scene can be viewed at https://www.youtube.com/watch?v=mv0onXhyLlE.

47  **"It's true that we can't react violently":** "Pope Francis on Freedom of Speech: 'One Cannot Make Fun of Faith,'" NBC News, January 15, 2015, http://www.nbcnews.com/storyline/paris-magazine-attack/pope-francis-freedom-speech-one-cannot-make-fun-faith-n286631.

50  **"I have a bad temper":** "Will snapping at my son a lot effect [*sic*] him?" [question posted to online forum by user sarasami], BabyCenter.com, July 30, 2011, http://www.babycenter.com/400_will-snapping-at-my-son-a-lot-effect-him_9611969_959.bc.

50  **"When I come home from work":** J. Warda, "Learn to Manage Stress Before Snapping at Kids," *Chicago Tribune*, August 8, 1999.

51  **"Even me being in anger management":** S. Michels, "Special Courts Take on Criminal Cases of Veterans Struggling with Trauma," *NewsHour*, PBS, February 24, 2015, http://www.pbs.org/newshour/bb/special-courts-take-criminal-cases-veterans-struggling-trauma.

51  **"In the moments":** "Ready to Snap," *PhD in Parenting* (blog), November 19, 2010, http://www.phdinparenting.com/blog/2010/11/19/ready-to-snap.html.

51  **"It seems like I snap at her":** Chrisinohio, "Why Am I Snapping at My Wife Without Even Realizing It?" HealthBoards forum, January 15, 2005, http://www.healthboards.com/boards/anger-management/242875-why-am-i-snapping-my-wife-without-even-realizing.html.

52  **"I snapped out on my husband":** Posted by "Braunwyn," City-Data.com, June 22, 2010, http://www.city-data.com/forum/relationships/1011858-snapping-out-my-spouse.html.

52  **"It seems like lately":** M.S., "Constantly Snapping at My Husband," Mamapedia.com, July 15, 2008, http://www.mamapedia.com/article/constantly-snapping-at-my-husband.

53  **"Anger is a very powerful emotion":** David T. Derrer, MD, "Men and Anger Management," WebMD, http://www.webmd.com/men/guide/anger-management.

53  **"Some Simple Steps":** American Psychological Association, "Controlling Anger Before It Controls You," http://www.apa.org/topics/anger/control.aspx.

56 **"A whistleblower feels vulnerable":** Email interview with Jason Zuckerman of Zuckerman Law, March 10, 2015.

57 **net loss of synaptic connections over time:** C. Liston and W-B. Gan, "Glucocorticoids Are Critical Regulators of Dendritic Spine Development and Plasticity in Vivo," *Proceedings of the National Academy of Sciences of the United States of America* 108, no. 38 (2011): 16074–6079.

57 **prevent them from withering away:** C. Liston et al., "Circadian Glucocorticoid Oscillations Promote Learning-Dependent Synapse Formation and Maintenance," *Nature Neuroscience* 16, no. 6 (2013): 698–705.

57 **"Jail is one of the most incredibly stressful":** Author interview with Mr. David Connell, Washington, DC, February 20, 2015.

**4. REACHING A VERDICT**

61 **"I heard the guy hit the tree":** Author interview with rock-climbing guide Brendan, Silver Spring, MD, January 15, 2014.

61 **"By the time I got there":** Author interview with rock-climbing guide Jennie, Silver Spring, MD, January 15, 2014.

64 **"Talked until blood shot out":** "Foul Play Suspected in Death of 'Carderock Geoff,'" *Dead Point Magazine,* January 14, 2014, http://www.dpmclimbing.com/articles/view/foul-play-suspected-death-%E2%80%9Ccarderock-geoff%E2%80%9D.

64 **"He always seemed more than a bit unusual":** BrianWS, Mountain Project forum, January 14, 2014, http://www.mountainproject.com/v/carderock-geoff-rip/108550580.

65 **Geoff Farrar, AKA Carderock Geoff:** J. Gregory, Potomac Mountain Club forum, December 29, 2013, http://www.potomacmountainclub.org/Climbing?mode=MessageList&eid=1464595&tpg=0.

66 **found next to the body:** M. Mimica, "Man Accused of Killing Fellow Rock Climber with Claw Hammer," NBC News, January 15, 2014, http://www.nbclosangeles.com/news/national-international/David-DiPaolo-Geoffrey-Farrar-Man-Accused-of-Killing-Fellow-Climber-With-Claw-Hammer-240162931.html?akmobile=o&nms=y.

66 **"Pictures that have surfaced":** P. Hermann, "Climber Charged with Killing Friend with Claw Hammer in National Park in Potomac," *The Washington Post,* January 14, 2014.

66 **upstate New York near his father's home:** S. McCabe, "Rock-Climber Accused of Killing Friend with Claw-Hammer Under Bethesda Cliffs," *DC Crime Stories,* January 13, 2014, http://dccrimestories.com/bethesda-area-rock-climber-accused-of-killing-friend-with-claw-hammer.

66 **"I'm sorry this happened":** Hermann, "Climber Charged with Killing Friend."

67 **"I think he kinda":** Email interview of climbing guide Tom Cecil of Seneca Rocks Mountain Guides, January 21, 2014.

71 **a pitchfork rage murder the author had witnessed:** J. Parini, "FILM; Of Bindlestiffs, Bad Times, Mice and Men," *The New York Times,* September 27, 1992. Steinbeck explained the origins of the story in an interview with *The New York Times* in 1937: "I was a bindlestiff myself for quite a spell. I worked in the same country that the story is laid in. The characters are composites to a certain extent. Lennie was a real person. He's in an insane asylum in California right now. I worked alongside him for many weeks. He didn't kill a girl. He killed a ranch foreman. Got sore because the boss had fired his pal and stuck a pitchfork right through his stomach. I hate to tell you how many times. I saw him do it. We couldn't stop him until it was too late."

74 **"I couldn't recognize him":** This and subsequent quotes from an author interview with John Gregory, Carderock, MD, February 1, 2014.

78 **"Chris went to the hospital with Geoff":** Author report from the remembrance for Geoff Farrar, held at Carderock, MD, on December 28, 2014, one year after his death.

80 **tending to real-estate matters:** J. Reid, "Where Hamilton Fell: The Exact Location of the Famous Duelling Ground," *The Hoboken Evening News*, June 10, 1898, http://duel2004.weehawkenhistory.org/Duel-JohnReid.pdf.

83 **"If this devil is not dying":** L. Bui, "Man, 63, Seeks Insanity Finding in Fatal Stabbing of Girlfriend in Beltsville," *The Washington Post*, May 3, 2014.

85 **"Meaning, you cannot in court say":** Author interview with Professor Bruno Maurice Kappes, in his office at University of Alaska, Anchorage, April 19, 2014.

85 **Some have speculated:** S. Helman and J. Russell, *Long Mile Home* (New York: Dutton, 2014); D. Stableford, "Tamerlan Tsarnaev Heard Voices in His Head: Report," *Yahoo! News*, December 15, 2013; S. Jacobs, D. Filipov, and P. Wen, "The Fall of the House of Tsarnaev," *The Boston Globe*, December 15, 2012.

86 **The defense did adopt:** K. Q. Seelye, "Dzhokhar Tsarnaev Is Guilty on All 30 Counts in Boston Marathon Bombing," *The New York Times*, April 8, 2015, http://www.nytimes.com/2015/04/09/us/dzhokhar-tsarnaev-verdict-boston-marathan-bombing-trial.html?_r=0.

86 **"The big choice for the jury":** Ibid.

86 **On May 15, 2015:** A. Goldman, "Boston Marathon Bomber Dzhokhar Tsarnaev Sentenced to Death Despite Appeals for Mercy," *The Washington Post*, May 15, 2015, http://www.washingtonpost.com/world/national-security/jury-weighs-death-penalty-for-boston-marathon-bomber/2015/05/15/30a493b8-fb07-11e4-9ef4-1bb7ce3b3fb7_story.html.

86 **another prime example:** "James Holmes Appears in Court Being Accused of Killing 12 People in Aurora Cinema Shooting," *Belle News*, July 23, 2012, http://www.bellenews.com/2012/07/23/world/us-news/james-holmes-appears-in-court-being-accused-of-killing-12-people-in-aurora-cinema-shooting.

92 **These psychiatrists argue:** C. McKinley, "Aurora Shooting Trial: 10 New Things from 22 Hours of James Holmes Psychiatric Evaluation Interviews," *ABC World News*, June 5, 2015, http://abcnews.go.com/US/aurora-shooting-trial-10-things-22-hours-james/story?id=31540740.

93 **sent him to prison for thirty years:** L. Bui, "Prince George's Jury Rejects Beltsville Man's Insanity Defense in Girlfriend's Fatal Stabbing," *The Washington Post*, May 7, 2014.

93 **"We never would have dreamt":** J. Borden, "Babysitter Guilty in Virginia Toddler's Death," *The Washington Post*, October 21, 2013.

94 **18 to 25 percent of babies:** R. G. Barr, "Preventing Abusive Head Trauma Resulting from a Failure of Normal Interaction Between Infants and Their Caregivers," *Proceedings of the National Academy of Sciences of the United States of America* 109, Supplement 2 (October 2012): 17294–301.

95 **men tend to carry on:** N. De Pisapia et al., "Sex Differences in Directional Brain Responses to Infant Hunger Cries," *Neuroreport* 24, no. 3 (2013): 142–46.

95 **Women in general show greater response:** J. W. Chun et al., "Common and Differential Brain Responses in Men and Women to Nonverbal Emotional Vocalizations by the Same and Opposite Sex," *Neuroscience Letters* 515, no. 2 (2012): 157–61.

95 **this reduces stress and anxiety:** W. Tschacher, M. Schildt, and K. Sander, "Brain Connectivity in Listening to Affective Stimuli: A Functional Magnetic Resonance Imaging (fMRI) Study and Implications for Psychotherapy," *Psychotherapy Research* 20, no. 5 (September 2010): 576–88.

95 **interaction between infants and caregivers:** Barr, "Preventing Abusive Head Trauma."

95  **might account for the paradox:** During sentencing, Judge J. Howe Brown stated in court, "Nothing I can do brings back Elijah or makes [his parents] feel better. Likewise, nothing I can do to punish [Fraraccio] is more important than her memory of what she did. . . . The sentence that I'm going to give is not going to satisfy probably anybody, and people may talk about it for a long time."

Prosecutors had asked for a sentence of fifty years in prison. When Judge Brown read his sentence, gasps swept the courtroom—five years in prison. The Nealeys were shocked. "Not five years. That's definitely not right." Jessica Fraraccio cried as Jennifer Nealey spoke in court of the agony of losing her child. The family was shattered and heartbroken by the murder of their innocent child, whom they entrusted for a short time to the care of Fraraccio. Fraraccio said in court, "I don't know that I'll ever be able to forgive myself—I really did love Elijah."

Quotes from J. Zauzmer, "Babysitter Given 5 Years in Prison for Killing Boy," *The Washington Post,* January 14, 2014.

## 5. TO DO THE RIGHT THING FAST

101  **"Lucky, what are you doing?":** S. Hendrix, "F-16 Pilot Was Ready to Give Her Life on Sept. 11," *The Washington Post,* September 8, 2011.

101  **"This sounds coldhearted":** S. Hendrix, "F-16 Pilot Was Ready to Down Plane Her Father Piloted on 9/11," *The Washington Post,* September 14, 2011.

102  **"Let's roll!":** E. Vulliamy, "Let's Roll . . . ," *The Guardian Observer,* December 1, 2001.

102  **"They were the true heroes":** B. Hewitt, "Because of What They Did, We Didn't Have To," *Parade,* September 6, 2014.

103  **"I believe it's a human instinct":** "Ordinary Heroes—Lenny Skutnik," YouTube video, 7:38, posted by COSMOTOPPER777, January 17, 2009, https://www.youtube.com/watch?v=eXSQ9YGQvOI.

103  **"He seemed sort of middle-aged":** C. Mcdougall, "The Hidden Cost of Heroism," NBCNews.com, November 26, 2007, http://www.nbcnews.com/id/21902983/ns/health-behavior/t/hidden-cost-heroism/#.VXMsG89VhBd.

104  **He had drowned saving:** J. Williams, "A Hero—Passenger Aids Others, then Dies," *The Washington Post,* January 14, 1982.

105  **Christopher instantly rushed:** P. Kotz, "Chris Patino, Mustafa Said, Occupation: Men; Rescue Girl Being Stabbed by Boyfriend," *True Crime Report,* January 31, 2011, http://www.truecrimereport.com/2011/01/chris_patino_mustafa_said_occu.php.

105  **"It was just like, It's now or never":** "I Saw Something Tragic on the News Today," Experience project, www.experienceproject.com/stories/Saw-Something-Tragic-on-the-News-Today/1984756.

106  **"I didn't really think":** K. Sturtz, "Kirkville Teen Saves 93-Year-Old Man from Drowning in Madison County," Syracuse.com, August 18, 2013, http://www.syracuse.com/news/index.ssf/2013/08/kirkville_teen_saves_93-year-old_man_from_drowning_in_madison_county.html.

106  **"This is a neighborhood watering hole":** R. Gazarik and P. Peirce, "Victim in Ligonier Bar: 'I Don't Want . . . Trouble,'" Trib Live News, July 13, 2011, http://triblive.com/x/pittsburghtrib/news/westmoreland/s_746488.html#axzz3cJBvAAeI.

106  **Don Holler was sitting at the bar:** P. Peirce, "Bar-Shooting Heroes Describe During Fromholz Trial the Chaos at a Ligonier Borough Tavern," Trib Live News, August 29, 2012, http://triblive.com/home/2493629-74/fromholz-haldeman-holler-bar-ligonier-ledgard-rifle-shot-jurors-testified#axzz3cJBvAAeI.

107 **"I don't want any trouble"**: Gazarik and Peirce, "Victim in Ligonier Bar: 'I Don't Want . . . Trouble.'"

107 **"Within thirty seconds"**: Peirce, "Bar-Shooting Heroes Describe During Fromholz Trial the Chaos at a Ligonier Borough Tavern."

108 **"We look out and there was blood"**: Associated Press, "Hawaii Shark Attack Victim's Rescuer Recalls Daring Ordeal," Fox News, August 17, 2013, http://www.foxnews .com/us/2013/08/17/shark-attack-victim-rescuer-describes-ordeal.

108 **The children's grandmother, fifty-four-year-old Virginia Grogan**: L. King, "Victims in Fatal Gloucester County Fire Identified," *The Virginian-Pilot*, January 17, 2013, http://hamptonroads.com/2013/01/victims-fatal-gloucester-county-fire-identified.

108 **"save my baby!"**: E. Kurhl, "Teen Recounts Saving Boy from Certain Death at Yosemite Waterfall," *San Jose Mercury News*, April 15, 2013, http://www.sltrib.com/sltrib/ world/56160061-68/alec-smith-family-yosemite.html.csp.

109 **"I have forty dead contacts in my cell phone"**: Telephone interview with former member of SEAL Team 6, March 26, 2014.

## 6. THE FLAVORS OF THREATS

111 **"You know that you would have to possibly lay your life down"**: Author interview with Secret Service agent Scott Moyer, Washington, DC, December 17, 2012.

114 **"Absolutely!"**: All quotes in this section from a telephone interview with former member of SEAL Team 6, March 26, 2014.

118 **out of competition for the rest of the season**: Associated Press, "American Skis Race of Her Life," *Lodi News-Sentinel*, February 12, 1992.

118 **All of them were badly injured**: Ibid.

118 **"Noodles is sharp"**: "American Wendy Fisher, Austrian Are Injured in Skiing Accidents," Philly.com, February 12, 1992, http://articles.philly.com/1992-02-12/sports/ 26042259_1_megan-gerety-sabine-ginther-fis-rules.

118 **"She must have flown"**: Ibid.

119 **"I stopped liking skiing"**: Author interview with Wendy Fisher, Olympic skier, on Paradise Run at Crested Butte Ski Resort, February 28, 2014.

120 **"So-called choking"**: Telephone interview with Christopher S. Ahmad, head team physician of the New York Yankees, February 7, 2014.

120 **"There's nothing harder"**: A. Kilgore, "For Washington Nationals' Jayson Werth, Hitting Is a Journey Without End," *The Washington Post*, March 18, 2014.

121 **"Players are under enormous pressure"**: Telephone interview with Dr. Charley Maher, team psychologist for the Cleveland Indians, February 5, 2014.

122 **"And not only do they have more of it"**: Telephone interview with Christopher S. Ahmad, head team physician of the New York Yankees, February 7, 2014.

123 **"I have had multiple friends die"**: Telephone interview with a Navy SEAL, 2014.

123 **The book describes a heroic moment**: M. Owen, *No Easy Day* (New York: Dutton, 2012), 236.

123 **"I guarantee he thought through every single step"**: Telephone interview with a Navy SEAL, 2014.

126 **"A batter gets hit by a pitch"**: Telephone interview with Christopher S. Ahmad, head team physician of the New York Yankees, February 7, 2014.

126 **Canandaigua Motorsports Park**: J. Sutton and S. Almasy, "NASCAR's Tony Stewart Hits, Kills Driver at Dirt-Track Race in New York," CNN, August 11, 2014, http://www.cnn.com/2014/08/10/justice/tony-stewart-hits-driver.

127 **demonstrates this convincingly**: T. G. Lee, R. S. Blumenfeld, and M. D'Esposito, "Disruption of Dorsolateral but Not Ventrolateral Prefrontal Cortex Improves

Unconscious Perceptual Memories," *The Journal of Neuroscience* 33, no. 32 (August 2013): 13233–37.

129 **suicide, drug abuse, and violent crime:** L. Bevilacqua et al., "A Population-Specific HTR2B Stop Codon Predisposes to Severe Impulsivity," *Nature* 468, no. 7327 (December 2010): 1061–66.

129 **Dopamine is an important neurotransmitter:** A. Björklund and S. B. Dunnett, "Dopamine Neuron Systems in the Brain: An Update," *Trends in Neuroscience* 30, no. 5 (May 2007): 194–202.

130 **the pons and mesencephalon:** I. P. Dorocic et al., "A Whole-Brain Atlas of Inputs to Serotonergic Neurons of the Dorsal and Median Raphe Nuclei," *Neuron* 83, no. 3 (August 2014): 663–78.

131 **more likely to join gangs:** "'Warrior Gene' Linked to Gang Membership, Weapon Use," *Science Daily*, June 8, 2009, http://www.sciencedaily.com/releases/2009/06/090605123237.htm; See also Beaver et al., "Monoamine Oxidase A Genotype Is Associated with Gang Membership and Weapon Use," *Comprehensive Psychiatry* 51, no. 2 (March–April 2010): 130–34.

131 **The statistics show that five to ten percent:** J. Tiihonen et al., "Genetic Background of Extreme Violent Behavior," *Molecular Psychiatry* 20 (June 2015): 786–92.

131 **associated with depression and other mental disorders:** T. C. Eley et al., "Association Analysis of MAOA and COMT with Neuroticism Assessed by Peers," *American Journal of Medical Genetics Part B: Neuropsychiatry Genetics* 120B, no. 1 (July 2003): 90–96.

132 **On July 31, 2012, there was a near disaster:** A. Halsey III, "Two Planes Taking off from National Put on Collision Course with Plane Trying to Land," *The Washington Post*, August 1, 2012.

134 **researchers A. H. Olivier and colleagues:** A. H. Olivier et al., "Minimal Predicted Distance: A Common Metric for Collision Avoidance During Pairwise Interactions Between Walkers," *Gait and Posture* 36, no. 3 (July 2012): 399–404.

135 **Researchers from Princeton University:** G. J. Stephens, L. J. Silbert, and U. Hasson, "Speaker–Listener Neural Coupling Underlies Successful Communication," *Proceedings of the National Academy of Sciences of the United States of America* 107, no. 32 (August 2010): 14425–30.

135 **with a short lag of about one second:** R. D. Fields, "Of Two Minds: Listener Brain Patterns Mirror Those of the Speaker," *Scientific American Online*, July 27, 2010, http://blogs.scientificamerican.com/guest-blog/2010/07/27/of-two-minds-listener-brain-patterns-mirror-those-of-the-speaker.

135 **"Communication is a joint action":** Email interview with Uri Hasson, Assistant Professor, Princeton University Department of Psychology, August 2010.

136 **could it be affecting deeper mental processes:** R. D. Fields, "The Power of Music: Mind Control by Rhythmic Sound," *Scientific American Online*, October 19, 2012, http://blogs.scientificamerican.com/guest-blog/2012/10/19/the-power-of-music-mind-control-by-rhythmic-sound.

136 **Rhythmic sound "not only coordinates the behavior":** Author interview with Dr. Annette Schirmer presenting her research at the Society for Neuroscience meeting in New Orleans, October 15, 2012.

138 **"But they, I am sure":** Edward Everett, *Address of Hon. Edward Everett at the Consecration of the National Cemetery at Gettysburg, 19th November, 1863, with the Dedicatory Speech of President Lincoln, and the Other Exercises of the Occasion* (Boston: Little, Brown and Company, 1864), 82.

140 **Much higher ratings:** S. Dolcos et al., "The power of a Handshake: Neural Correlates of Evaluative Judgments in Observed Social Interactions," *Journal of Cognitive Neuroscience* 24, no. 12 (December 2012): 2292–305.

## 7. EXTRASENSORY PERCEPTION?

143 **The researchers studied people:** C. Bertini, R. Cecere, and E. Làdavas, "I Am Blind, but I 'See' Fear," *Cortex* 49, no. 4 (April 2013): 985–93.

145 **"threat signals trigger defensive responses":** S. Pichon, B. de Gelder, and J. Grèzes, "Threat Prompts Defensive Brain Responses Independently of Attentional Control," *Cerebral Cortex* 22, no. 2 (February 2012): 274–85.

147 **"I tend to be hyperaroused":** Author interview with Tricia in her home, December 13, 2014.

152 **"Absolutely nothing!":** Author interview with Dr. Manzar Ashtari, Penn Station, Philadelphia, PA, September 25, 2014.

## 8. HEROES AND COWARDS

174 **"Anybody who's been in combat":** Author interview with former member of SEAL Team Six, 2014.

174 **a shabby man, looking "drugged-out":** J. S. Feinstein et al., "The Human Amygdala and the Induction and Experience of Fear," *Current Biology* 21, no. 1 (January 2011): 34–38.

175 **"I'm going to *cut* you, bitch!"** Quoted in supplemental information to Feinstein et al., "The Human Amydala."

176 **SM's medical condition:** J. S. Feinstein, "Lesion Studies of Human Emotion and Feeling," *Current Opinion in Neurobiology* 23, no. 3 (June 2013): 304–9.

176 **"Every time I was on top of a run":** Author interview with Wendy Fisher, Olympic skier, on Paradise Run at Crested Butte Ski Resort, February 28, 2014.

176 **"Fear is a force":** M. Luttrell, *Service: A Navy SEAL at War* (New York: Little, Brown and Company, 2012), 127.

176 **"It is quite remarkable":** Feinstein, "Lesion Studies."

176 **activity in the amygdala increases:** M. B. Stein et al., "Increased Amygdala and Insula Activation During Emotion Processing in Anxiety-Prone Subjects," *American Journal of Psychiatry* 164, no. 2 (February 2007): 318–27.

179 **a low resting heart rate correlates:** L. C. Wilson and A. Scarpa, "Baseline Heart Rate, Sensation Seeking, and Aggression in Young Adult Women: A Two-Sample Examination," *Aggressive Behavior* 39, no. 4 (July–August 2013): 280–89.

179 **below 66 beats per minute:** T. A. Armstrong et al., "Low Resting Heart Rate and Antisocial Behavior," *Criminal Justice and Behavior* 36, no. 11 (November 2009): 1125–40.

179 **men tend to show more direct aggression:** Wilson and Scarpa, "Baseline Heart Rate."

180 **Heart-rate monitors indicated:** Dr. Zebra, "Medical History of Spacefarers," http://www.doctorzebra.com/drz/s_medhx.html.

183 **Lynne Isbell, professor of anthropology:** Q. Van Le et al., "Pulvinar Neurons Reveal Neurobiological Evidence of Past Selection for Rapid Detection of Snakes," *Proceedings of the National Academy of Sciences of the United States of America* 110, no. 47 (November 2013): 19000–19005.

184 **"The results show that the brain has special neural circuits to detect snakes":** Quoted in "Snakes on the Brain: Are Primates Hard-Wired to See Snakes?," University of California, Davis, press release, October 28, 2013.

184 **Susan Mineka's research at Northwestern:** A. Öhman and S. Mineka, "The Malicious Serpent: Snakes as a Prototypical Stimulus for an Evolved Module of Fear," *Current Directions in Psychological Science* 12, no. 1 (February 2003): 5–9.

184 **A subsequent study on humans:** V. LoBue, D. H. Rakison, and J. S. DeLoache, "Threat Perception Across the Lifespan: Evidence for Multiple Converging Pathways," *Current Directions in Psychological Science* 19, no. 6 (December 2010): 375–79.

185 **"One is that we detect them quickly":** Quoted in "People Aren't Born Afraid of Spiders and Snakes: Fear Is Quickly Learned During Infancy," Association for Psychological Science, press release, January 24, 2011.

185 **"I don't like heights":** Telephone interview with former Navy SEAL, 2014.

186 **"Let this be the time I don't feel fear":** Author interview with Wendy Fisher, Crested Butte, Colorado, February 15, 2014.

186 **"In the case of good self-controllers":** Quoted in Caltech press release, April 30, 2009, describing the study: T. A. Hare, C. F. Camerer, and A. Rangel, "Self-Control in Decision-Making Involves Modulation of the vmPFC Valuation System," *Science* 324, no. 5927 (May 2009): 646–48.

186 **psychological phenomenon called "regulatory depletion":** W. M. Hedgcock, K. D. Vohs, and A. R. Rao, "Reducing Self-Control Depletion Effects Through Enhanced Sensitivity to Implementation: Evidence from fMRI and Behavioral Studies," *Journal of Consumer Psychology* 22, no. 4 (January 2012): 486–95.

187 **Dr. Josef Parvizi of the Department of Neurology:** J. Parvizi et al., "The Will to Persevere Induced by Electrical Stimulation of the Human Cingulate Gyrus," *Neuron* 80, no.6 (December 2013): 1359–67.

187 **"That few electrical impulses":** M. B. O'Leary, quoted in "Electrical Brain Stimulation May Evoke a Person's 'Will to Persevere,'" Cell Press, press release, December 5, 2013.

188 **Neuroscientist Florent Meyniel:** F. Meyniel et al., "Neurocomputational Account of How the Human Brain Decides When to Have a Break," *Proceedings of the National Academy of Sciences of the United States of America* 110, no. 7 (February 2013): 2641–46.

189 **"We are currently testing drugs on this paradigm in our lab":** Email interview with Dr. Mathias Pessiglione, January 21, 2013.

189 **"snakes on a brain-imaging machine":** U. Nili et al., "Fear Thou Not: Activity of Frontal and Temporal Circuits in Moments of Real-Life Courage," *Neuron* 66, no. 6 (June 2010): 949–62.

190 **"Our findings delineate":** C. Genova, quoted in "Brave Brains: Neural Mechanisms of Courage," Cell Press, press release, June 23, 2010.

191 **"These are really interesting questions":** Author interview with bush pilot Drake Olson as a passenger in his airplane over the Davidson Glacier in Alaska, July 13, 2013.

192 **with 56.1 fatalities:** T. Korch, "10 of the Most Dangerous Jobs in the U.S.," *Yahoo! Finance*, May 19, 2013, http://finance.yahoo.com/news/10-of-the-most-dangerous-jobs-in-the-u-s--191643548.html.

193 **"Well, that was exciting":** Author interview with bush pilot Drake Olsen, landing on Glacier Point next to Chilkat Inlet, July 8, 2013.

195 **"The bravest people I know":** Author interview with Dr. Daniela Schiller in her laboratory at the Mt. Sinai School of Medicine, New York, November 4, 2013.

195 **"Andretti would come over":** Author interview with bush pilot Drake Olson in his airplane hangar, July 14, 2013.

196 **an interest in mastering complex tasks:** Wilson and Scarpa, "Baseline Heart Rate."

197   Using an fMRI scanner: B. C. Wittmann et al., "Striatal Activity Underlies Novelty-Based Choice in Humans," *Neuron* 58, no. 6 (June 2008): 967–73.

197   "Increased novelty-seeking may play a role": Quoted in "Adventure—It's All in the Mind, Say UCL Neuroscientists," Wellcome Trust Centre, press release, June 25, 2008, https://www.ucl.ac.uk/news/news-articles/0806/08062502.

197   Other studies also show: N. Hiroi and S. Agatsuma, "Genetic Susceptibility to Substance Dependence," *Molecular Psychiatry* 10, no. 4 (April 2005): 336–44.

197   both behavioral activities are also linked: D. T. Chau, R. M. Roth, and A. I. Green, "The Neural Circuitry of Reward and its Relevance to Psychiatric Disorders," *Current Psychiatry Reports* 6, no. 5 (October 2004): 391–99.

197   In fact, a well-known side effect: D. Weintraub and M. J. Nirenberg, "Impulse Control and Related Disorders in Parkinson's Disease," *Neurodegenerative Diseases* 11, no. 2 (2013): 63–71.

197   sensation seeking in alcoholic males: F. Limosin et al., "Association Between Dopamine Receptor D1 Gene Ddel Polymorphism and Sensation Seeking in Alcohol-Dependent Men," *Alcoholism: Clinical and Experimental Research* 27, no. 8 (August 2003): 1226–28.

197   1,591 adolescent twin pairs: J. R. Koopmans et al., "A Multivariate Genetic Analysis of Sensation Seeking," *Behavioral Genetics* 25, no. 4 (July 1995): 349–56.

197   Another study showed that boredom: L. C. Wilson and A. Scarpa, "Baseline Heart Rate, Sensation Seeking, and Aggression in Young Adult Women: A Two-Sample Examination," *Aggressive Behavior* 39, no. 4 (July–August 2013): 280–89.

198   "Not everyone who's high on sensation": Quoted in "A Thirst for Excitement Is Hidden in Your Genes," Association for Psychological Science, press release, October 5, 2010, in reference to J. Derringer et al., "Predicting Sensation Seeking from Dopamine Genes: A Candidate-System Approach," *Psychological Science* 21, no. 9 (September 2010): 1282–90, https://www.psychologicalscience.org/index.php/news/releases/a-thirst-for-excitement-is-hidden-in-your-genes.html.

198   In fact, another study by Matthew Cain: "Study Links Personal, Corporate Risk-Taking," University of Oregon, press release, August 9, 2011, www.eurekalert.org/pub_releases/2011-08/uoo-slp080911.php. Article by M. D. Cain and S. B. McKeon forthcoming in the *Journal of Financial and Quantitative Analysis.*

198   "These adventures are physically challenging": J. Preston, "Richard Branson: My Fascination with Adventure," Virgin News, undated, http://www.virgin.com/news/richard-branson-my-fascination-adventure.

198   "These individuals take on higher leverage": Quoted in "Study Co-Authored by UO Finance Professor Links Personal, Corporate Risk-Taking," University of Oregon, press release, August 9, 2011.

199   "The reason that teenagers take risks": Quoted in "Teenagers Programmed to Take Risks," University College London, press release, March 25, 2010, https://www.ucl.ac.uk/news/news-articles/1003/10032503.

199   Functional MRI studies show that: N. Steinbeis, B. C. Bernhardt, and T. Singer, "Impulse Control and Underlying Functions of the Left DLPFC Mediate Age-Related and Age-Independent Individual Differences in Strategic Social Behavior," *Neuron* 73, no. 5 (March 2012): 1040–51.

199   Men tend to score higher: C. P. Cross, D. M. Cyrenne, and G. R. Brown, "Sex Differences in Sensation-Seeking: A Meta-Analysis," *Scientific Reports* 3 (2013): 2486.

199   Salivary testosterone levels: B. C. Campbell et al., "Testosterone Exposure, Dopaminergic Reward, and Sensation-Seeking in Young Men," *Physiology & Behavior* 99, no. 4 (March 2010): 451–56.

199  "Some will experience sex": Quoted in "New Study Suggests that a Propensity for 1-Night Stands, Uncommitted Sex Could Be Genetic," Binghamton University, press release, November 30, 2010. The study is published in *PLOS ONE* : J. R. Garcia et al., "Associations Between Dopamine D4 Receptor Gene Variation with both Infidelity and Sexual Promiscuity," *PLOS ONE* 5, no. 11 (November 2010): e14162.

199  when high sensation seekers view arousing photographs: J. E. Joseph et al., "Neural Correlates of Emotional Reactivity in Sensation Seeking," *Psychological Science* 20, no. 2 (February 2009): 215–23.

200  Risk takers have lower rates of Parkinson's disease: A. H. Evans et al., "Relationship Between Impulsive Sensation Seeking Traits, Smoking, Alcohol and Caffeine Intake, and Parkinson's Disease," *Journal of Neurology, Neurosurgery & Psychiatry* 77, no. 3 (March 2006): 317–21.

200  "The Hero of the Hudson": B. Hutchinson, "Chesley (Sully) Sullenberger, Hero of US Airways Flight 1549, Breaks Silence About Emergency Landing," *Daily News*, February 3, 2009.

201  "Cactus 1529, turn right": B. Evans, "Hero Pilot Sullenberger's Exchange with Air-Traffic Control," *Information Week*, February 5, 2009, http://www.information week.com/it-leadership/ hero-pilot-sullenbergers-exchange-with-air-traffic -control/d/d-id/1076362?

201  "He was the last one off": K. Burke, P. Donohue, and C. Siemaszko, "US Airways Airplane Crashes in Hudson River—Hero Pilot Chesley Sullenberger III Saves All Aboard," *Daily News*, January 16, 2009.

201  Thirty-two people were killed: B. L. Nadeau, H. Yan, and G. Botelho, "Costa Concordia Captain Convicted in Deadly Shipwreck," CNN, February 11, 2015, http:// www.cnn.com/2015/02/11/world/costa-concordia-trial.

201  "Captain Coward": E. Ide, "Costa Concordia Captain's Trial Resumes in Italy," Fox News, July 17, 2013, http://www.foxnews.com/world/2013/07/17/costa-concordia -captain-trial-resumes-in-italy.

202  "At around 2:30 a.m.": B. Forer, "Cruise Ship Captain Cried Like a Child, Chaplain Says," ABC News, January 20, 2012, http://abcnews.go.com/International/cruise-ship -captain-cried-baby-reaching-shore/story?id=15403207.

202  wasn't "lucid": Ibid.

202  tripped and fell into a lifeboat: Ibid.

202  "Go back to your cabins": Associated Press, "Costa CEO Says Captain Misled Company, Crew," ABC News, January 20, 2012, http://abclocal.go.com/story?section-news /national_world&id=8513099.

202  an innocent scapegoat: B. L. Nadeau, "Costa Concordia Captain: I'm a Scapegoat for Carnival Cruise Lines," *The Daily Beast*, March 11, 2014, http://www.thedailybeast .com/articles/2014/03/11/costa-concordia-captain-i-m-a-scapegoat-for-carnival -cruiselines.html.

202  In February 2015, Captain Schettino: C. Pleasance, "Costa Concordia's 'Captain Coward' Sentenced to 16 Years in Jail on Manslaughter Charges over 2012 Cruise Ship Disaster which Killed 32," *Daily Mail*, February 11, 2015, http://www.daily mail.co.uk/news/article-2949670/Costa-Concordia-s-Captain-Coward-sentenced -16-years-jail-manslaughter-charges.html.

202  "The frontal cortex developed": J. Thayer, "Neurovisceral Integration: Implications for Cognition, Emotion, and Health," NIH Director's Seminar, December 21, 2001, http://videocast.nih.gov/summary.asp?Live=1169&bhcp-1.

203  variability decreases as the servicemen are preparing: A. L. Hansen, B. H. Johnsen, and J. F. Thayer, "Relationship Between Heart Rate Variability and Cognitive

Function During Threat of Shock," *Anxiety, Stress, & Coping* 22, no. 1 (January 2009): 77–89.

203 **"They are able to take a non-routine play":** Telephone interview with Dr. Charley Maher, team psychologist for the Cleveland Indians, February 5, 2014.

204 **increased blood phobia:** Thayer, "Neurovisceral Integration."

204 **either susceptible to fear or resistant to fear:** J. L. McGuire et al., "Traits of Fear Resistance and Susceptibility in an Advanced Intercross Line," *European Journal of Neuroscience* 38, no. 9 (2013): 3314–24.

204 **"I didn't think":** R. Schapiro, "Jasper Schuringa, Dutchman Who Subdued Umar Farouk Abdulmutallab on Flight 253, Recalls Moment," *Daily News*, December 26, 2009.

205 **"I didn't think of anything":** H. Hong and S. Jiang, "Outrage in China After Toddler Run Over, Ignored," CNN, October 18, 2011, http://www.cnn.com/2011/10/17/world/asia/china-toddler-hit-and-run.

205 **"I knew the situation":** A. D. Stein, "Ph.D. Alumnus Awarded Top Honor by Human Factors and Ergonomics Society," Naval Postgraduate School, October 14, 2010. http://www.nps.edu/About/News/Ph.D.-Alumnus-Awarded-Top-Honor-by-Human-Factors-and-Ergonomics-Society.html.

206 **"Low resting heart rate appears":** J. Ortiz and A. Raine, "Heart Rate Level and Antisocial Behavior in Children and Adolescents: A Meta-Analysis," *Journal of the American Academy of Child and Adolescent Psychiatry* 43, no. 2 (February 2004): 154–62.

206 **A 2009 study confirms these findings:** T. A. Armstrong et al., "Low Resting Heart Rate and Antisocial Behavior: A Brief Review of Evidence and Preliminary Results from a New Test," *Criminal Justice and Behavior* 36, no. 11 (November 2009): 1125–40.

## 9. THE BEST DEFENSE

207 **circuit that underlies all types:** M. S. Fanselow, "Neural Organization of the Defensive Behavior System Responsible for Fear," *Psychonom Bulletin & Review* 1, no. 4 (December 1994): 429–38.

208 **As seen captured on video:** Dr. Bruno M. Kappes, "Moose," January 9, 1995, http://afbmk.uaa.alaska.edu/media.html.

209 **"There were people standing around":** P. S. Goodman, "Witnesses Say Students Harassed Cow, Calf for Hours," *Anchorage Daily News*, January 11, 1995.

209 **"I think my response was a normal panic response":** S. Toomey, "Psychotherapist Turns Patient after Encounter with UAA Moose," *Anchorage Daily News*, January 1995, http://hosting.uaa.alaska.edu/afbmk/psychotherapistTurnsPatient.jpg.

210 **cameras in the London Underground system:** Associated Press, "Harrowing Video Shows Moment Woman Rescues Child After Wind Blows Stroller onto London Tube Tracks," *National Post*, August 11, 2014, http://news.nationalpost.com/news/harrowing-video-shows-moment-woman-rescues-child-after-wind-blows-stroller-onto-london-tube-tracks.

216 **He still had his rifle:** M. Luttrell, *Lone Survivor* (New York: Little, Brown and Company, 2007), 77.

216 **Mikey worked his way:** Ibid., 84.

216 **Mark Owen's book *No Easy Day*:** Mark Owen, *No Easy Day* (New York: Dutton, 2012).

218 **"momma bears":** S. C. Motta et al., "Ventral Premammillary Nucleus as a Critical Sensory Relay to the Maternal Aggression Network," *Proceedings of the National Academy of Sciences of the United States of America* 110, no. 35 (August 2013): 14438–43.

219  **This spot resides in a general region:** A. Siegel, *The Neurobiology of Aggression and Rage* (Boca Raton, FL: CRC Press, 2004), 1–79; W. R. Hess, *Diencephalon: Autonomic and Extrapyramidal Functions* (New York: Grune & Stratton, 1954), 1–79; A. H. Rosenfeld and S. A. Rosenfeld, *Of Cats and Rats: Studies of the Neural Basis of Aggression* (Washington, DC: NIMH, US Government Printing Office, 1976). This is a summary of John P. Flynn's research.

224  **neuroanatomy book:** L. W. Swanson, "The Amygdala and Its Place in the Cerebral Hemisphere," *Annals of the New York Academy of Sciences* 985 (April 2003): 174–84; P. H. Janak and K. M. Tye, "From Circuits to Behavior in the Amygdala," *Nature* 517, no. 7534 (January 2015): 284–92.

224  **circuits are in different modules inside the amygdala:** C. T. Gross and N. S. Canteras, "The Many Paths to Fear," *Nature Reviews Neuroscience* 13, no. 9 (September 2012): 651–58.

225  **researchers compared a rat's brain responses to a predator:** S. C. Motta et al., "Dissecting the Brain's Fear System Reveals the Hypothalamus Is Critical for Responding in Subordinate Conspecific Intruders." *Proceedings of the National Academy of Sciences of the United States of America* 106, no. 12 (March 2009): 4870–75.

225  **Electrical stimulation of this region in humans elicits panic attacks:** W. B. Wilent et al., "Induction of Panic Attack by Stimulation of the Ventromedial Hypothalamus," *Journal of Neurosurgery* 112, no. 6 (June 2010): 1295–98.

225  **Life-or-limb and Environment trigger circuits are separate:** B. A. Silva et al., "Independent Hypothalamic Circuits for Social and Predator Fear," *Nature Neuroscience* 16, no. 12 (December 2013): 1731–33.

225  **"We have also some findings":** Email interview with Newton Canteras, April 26, 2014.

226  **"I think the guy who reacted that way":** Telephone interview with former Navy SEAL, 2014.

227  **a laser illuminates the fiber-optic cable and the mouse freezes instantly:** P. Tovote, "Periaqueductal Grey Circuits Mediating Freezing and Flight," and A. Adhikari, "Optogenetic Dissection of the Role of the BNST in Anxiety," lectures at the Society for Neuroscience meeting, San Diego, CA, November 13, 2013.

## 10. SEX . . . AND LOVE

234  **"Your training kicks in":** "Voices of Valor: An American Hero: Army Woman Earns Silver Star and Makes History," Women in Military Service for America Memorial Foundation, Inc., http://www.womensmemorial.org/Education/WHM 08KitUSA.html.

235  **"For my brothers it was easy":** M. Husain, "Malala: The Girl Who Was Shot for Going to School," BBC News, October 7, 2013, http://www.bbc.com/news/maga zine-24379018.

235  **"I had a terrible dream":** Ibid.

236  **youngest recipient of the Nobel Peace Prize:** G. Witte, "Shared Honor Rekindles Hope for Peace," *The Washington Post*, October 11, 2014.

236  **While the world media awaited:** R. Leiby and K. Adam, "Malala Yousafzai Says She Yearns to Be 'Normal,' Despite Fame—and Now Nobel," *The Washington Post*, October 11, 2014.

236  **"It gives a message":** Ibid.

236  **"One [Nobel recipient] is from Pakistan":** Witte, "Shared Honor."

236  **"Some people are silent":** Ibid.

237  **compared with about 14 percent:** R. Eglash, "Israel's Women Warriors," *The Washington Post*, September 20, 2014.

238 **"Wish List":** Wish list for donations from the House of Ruth, Maryland, http://www.hruth.org/wish-list.asp.

238 **30 percent of the female population:** A. Travis, "Domestic Violence Experienced by 30% of Female Population, Survey Shows," *The Guardian*, February 13, 2014.

238 **dragging her unconscious body:** TMZ, "Ray Rice Knocked Out Fiancée," https://www.google.com/webhp?sourceid=chrome-instant&rlz=1C1ASUT_enUS501US502&ion=1&espv=2&ie=UTF-8#safe=off&q=ray%20rice%20video.

239 **"I love my husband":** A. M. Steiner, "Ray Rice: Wife Janay Rice Defends Him on Instagram—'This Is Our Life,'" *Hollywood Life*, September 9, 2014, http://hollywoodlife.com/2014/09/09/ray-rice-wife-janay-rice-defends-husband-instagram-statement.

239 **"It was unbelievable":** L. Logan, "Lara Logan Breaks Silence on Cairo Assault," *60 Minutes,* May 1, 2011, http://www.cbsnews.com/news/lara-logan-breaks-silence-on-cairo-assault.

240 **"I'm screaming":** Ibid.

240 **"The atmosphere was one of jubilation":** N. Smith, "Please God. Please Make It Stop," Blog post by Natasha Smith, June 26, 2012, https://natashajsmith.wordpress.com/2012/06/26/please-god-please-make-it-stop.

241 **"In a split second, everything changed":** "'Please God, Make It Stop!' British Female Journalist, 21, Describes Horrific Sexual Assault in Egypt's Tahrir Square After Election Result," *Daily Mail*, June 27, 2012, http://www.dailymail.co.uk/news/article-2165445/British-journalist-Natasha-Smith-22-recalls-horrific-sexual-assault-Egypts-Tahrir-Square.html.

241 **forty-six cases of sexual assault:** N. Smith, "One Year On," July 4, 2013, https://natashajsmith.wordpress.com.

241 **Amnesty International believes the actual number is much higher:** D. Eltahawy, *Amnesty International, Livewire*, July 3, 2013, http://livewire.amnesty.org/2013/07/03/sexual-assaults-on-women-protestors-continuing-amid-the-political-turmoil.

241 **at least ninety-one women were raped:** "Egypt: Epidemic of Sexual Violence," Human Rights Watch, July 3, 2013, http://www.hrw.org/news/2013/07/03/egypt-epidemic-sexual-violence.

241 **"Yasmine El Baramawy":** Ibid.

241 **99.3 percent of women:** N. Mayen, "Women in Egypt Suffer More Sexual Violence Under Islamist Rule," *Al Arabiya News*, June 2, 2013, http://english.alarabiya.net/en/perspective/features/2013/06/02/Women-in-Egypt-suffer-more-sexual-violence-under-Islamist-rule-.html.

241 **a bus in Delhi, India:** G. Harris, "Charges Filed Against 5 over Rape in New Delhi," *The New York Times*, January 3, 2013.

241 **"Burn them alive":** A. Singh and K. Bajeli-Datt, "Burn Them Alive: Delhi Gang-Rape Victim's Last Words," *India Times*, September 11, 2013.

243 **"It seems to me to be clear":** C. Darrow, *Crime: Its Cause and Treatment* (New York: Thomas Y. Crowell Company, 1922), 274.

243 **30 percent of all Internet traffic is pornography:** "Pornography Statistics," Enough Is Enough, http://www.internetsafety101.org/Pornographystatistics.htm.

243 **Porn sites have more visitors:** Staff Reporter, "Porn Sites Get More Visitors Than Netflix, Amazon and Twitter Combined," *The Huffington Post*, May 4, 2013, http://www.huffingtonpost.com/2013/05/03/internet-porn-stats_n_3187682.html.

244 **10 percent of viewers report uncontrollable addiction:** Safe Families, "Statistics on Pornography, Sexual Addiction and Online Perpetrators," http://www.safefamilies.org/sfStats.php.

244 **activate the release of the neurotransmitter dopamine:** F. Hoeft et al., "Gender Differences in the Mesocorticolimbic System During Computer Game-Play," *Journal of Psychiatric Research* 42, no. 4 (March 2008): 253–58.

244 **become active during video gaming:** M. J. Koepp et al., "Evidence for Striatal Dopamine Release During a Video Game," *Nature* 393, no. 6682 (May 1998): 266–68.

244 **someone threatened a mass shooting:** H. Tsukayama, "Trade Group Responds to 'Gamergate,'" *The Washington Post*, October 16, 2014.

244 **"There's a toxicity":** S. T. Collins, "Anita Sarkeesian on GamerGate: 'We Have a Problem and We're Going to Fix This,'" *Rolling Stone*, October 17, 2014.

245 **violent aggression can be addictive:** J. Bryner, "Humans Crave Violence Just Like Sex," *LiveScience*, January 17, 2008, http://www.livescience.com/2231-humans-crave-violence-sex.html.

245 **"Aggression occurs among virtually all vertebrates":** Ibid.

245 **serotonin is involved in both sexual gratification and violence:** R. J. Nelson and S. Chiavegatto, "Molecular Basis of Aggression," *Trends in Neurosciences* 24, no. 12 (December 2001): 713–19.

245 **surged 50 percent in one year:** C. Whitlock, "Sex Assaults Reported by U.S. Service Members Surged 50 Percent Last Year," *The Washington Post*, May 2, 2014.

245 **fifty-five colleges and universities:** N. Anderson, "Colleges Scrutinized on Sex Assault Cases," *The Washington Post*, May 2, 2014.

246 **"Colleges and universities can no longer":** N. Anderson and K. Zezima, "White House Targets Assaults on Campus," *The Washington Post*, April 30, 2014.

246 **"I have cut myself only once":** Author interview with a woman treated for self-harm who wishes to remain anonymous, 2014.

247 **1 in 12 young people:** K. Kelland, "One in 12 Teenagers Self Harm, Study Finds," Reuters News Service, November 17, 2011, http://www.reuters.com/article/2011/11/17/us-self-harm-idUSTRE7AG02520111117.

248 **"We have a huge amount":** Author interview with Dr. Haley Peckham, Bethesda, MD, November 21, 2014.

250 **2012 crime statistics reveal:** M. O'Malley, A. G. Brown, and M. L. Brown, "Crime in Maryland: 2012 Uniform Crime Report," Maryland State Police, Central Records Division, Ida J. Williams, Division Director, http://www.goccp.maryland.gov/msac/documents/2012-Crime-In-Maryland.pdf.

253 **in the presence of others:** N. B. Hirschinger et al., "A Case-Control Study of Female-to-Female Nonintimate Violence in an Urban Area," *American Journal of Public Health* 93, no. 7 (July 2003): 1098–103.

253 **18.3 percent of women have been raped:** "Anti-Violence Resource Guide: National Intimate Partner and Sexual Violence Survey, 2010," Feminist.com, http://www.feminist.com/antiviolence/facts.html.

253 **challenge requiring a response to save face:** D. Baskin and I. Sommers, *Casualties of Community Disorder* (Boulder, CO: Westview Press, 1997). Quoted in N. B. Hirschinger et al., "A Case–Control Study of Female-to-Female Nonintimate Violence": 1098.

254 **a drunken, violent outburst:** C. Young, "The Surprising Truth About Women and Violence," *Time*, June 25, 2014.

254 **"too fat and overweight":** C. Boren, "Hope Solo Arrest: Nephew Says She Was Drinking, Called Him 'Too Fat to Be an Athlete,'" *The Washington Post*, June 24, 2014.

255 **"That's why I have a degree":** A. Stapleton and S. Almasy, "ESPN Reporter Britt McHenry Suspended After Berating Towing Company Clerk," CNN, April 20, 2015, http://www.cnn.com/2015/04/16/us/espn-reporter-britt-mchenry-tirade.

255 **"In an intense and stressful moment":** B. McHenry (@BrittMcHenry), April 16, 2015, 1:18 p.m., https://twitter.com/brittmchenry/status/588798710684987393.

255 **In fact, a month later:** P. Sullivan, "Predatory Tow Companies Targeted in Proposal from Md. and Va. Lawmakers," *The Washington Post*, May 21, 2015, http://www.wash ingtonpost.com/local/virginia-politics/predatory-tow-companies-targeted -in-proposal-from-md-and-va-lawmakers/2015/05/21/d510203a-ffde-11e4-805c -c3f407e5a9e9_story.html.

256 **276 schoolgirls in Nigeria:** P. Constable, "Nigerian Girls Inspire International Action," *The Washington Post*, May 7, 2014.

257 **to service the military men** A. Olivo, "Homage to WWII Comfort Women Puts Fairfax in a Delicate Situation," *The Washington Post*, May 31, 2014.

257 **most survivors were left infertile:** A-M. de Brouwer, *Supranational Criminal Prosecution of Sexual Violence* (Cambridge, UK: Intersentia, 2005), 8.

257 **"In the 'comfort station'":** J. R. O'Herne, witness to United States House of Representatives congressional hearing, "Protecting the Human Rights of Comfort Women," February 15, 2007, US Government Printing Office, http://archives.repub licans.foreignaffairs.house.gov/110/33317.pdf.

257 **raped by Japanese soldiers:** K. Hirano, "East Timor Former Sex Slaves Start Speaking Out," *Japan Times*, April 28, 2007.

257 **those who fought back were executed:** Ibid.

257 **twenty-five to thirty-five men a day:** Nelson, "Consolation Unit" also in *The Journal of Pacific History* 43, no. 1 (2008): 1–28.

257 **as many as sixty to seventy times:** E. Heineman, "The Hour of the Woman: Memories of Germany's 'Crisis Years' and West German National Identity," *The American Historical Review* 101, no. 2 (April 1996): 354–95.

258 **"Females changed their diet":** M. Kim, "Studies Offer Insights into Evolution of Monogamy," *The Washington Post*, July 31, 2013.

258 **"Infanticide is a real problem":** Ibid.

258 **There is something about the apes:** R. Wrangham and D. Peterson, *Demonic Males:* (New York: Houghton Mifflin, 1996), 135.

259 **"[She] fought the ape with":** Ibid., 137.

260 **"What I call the four F's":** D. J. Anderson, "The Neural Circuitry of Sex and Violence," lecture at the Society for Neuroscience meeting, San Diego, CA, November 11, 2013.

260 **"The boy at school":** C. Martin, "Signs That a Guy Is About to Ask You Out," Match.com Dating Tips, http://datingtips.match.com/signs-guy-ask-out-13443341 .html.

262 **Dayu Lin, while working in Anderson's laboratory:** D. Lin et al., "Functional Identification of an Aggression Locus in the Mouse Hypothalamus," *Nature* 470, no. 7333 (February 2011): 221–26.

264 **"I know players who":** Author interview with anonymous major league baseball team member, 2014.

265 **"Sex differences exist":** L. Cahill, "Why Sex Matters for Neuroscience," *Nature Reviews Neuroscience* 7 (June 2006): 477–84.

266 **wiring of a child's brain is permanently altered:** R. D. Fields, "Sticks and Stones— Hurtful Words Damage the Brain," *Psychology Today*, October 2010, http://www .psychologytoday.com/blog/the-new-brain/201010/sticks-and-stones-hurtful -words-damage-the-brain; R. D. Fields, "Rudeness Is a Neurotoxin," *The Huffington Post*, January 5, 2011, http://www.huffingtonpost.com/dr-douglas-fields/rudeness-is -a-neurotoxin_b_765908.html.

266  **followed a large group of children:** R. J. Herringa et al., "Childhood Maltreatment Is Associated with Altered Fear Circuitry and Increased Internalizing Symptoms by Late Adolescence," *Proceedings of the National Academy of Sciences of the United States of America* 110, no. 47 (November 2013): 19119–24.

267  **"This could very well be adaptive":** Email interview with R. J. Herringa, November 1, 2013.

268  **"There is this pattern":** Author interview with Dr. Larry Cahill, Bethesda, MD, October 20, 2014.

272  **"Skiing deep in the trees":** Author interview with Dr. John Petraitis, in his office, University of Alaska, Anchorage, April 19, 2014.

272  **mock electronic dating service:** J. Petraitis et al., "Sex Differences in the Attractiveness of Hunter-Gatherer and Modern Risks," *Journal of Applied Social Psychology* 44, no. 6 (June 2014): 442–53.

275  **"Help me! Help me!":** K. L. Alexander, "Testimony of 'Good Samaritan' Sways Jury in Sex Assault Case," *The Washington Post*, November 22, 2014.

275  **"I got nieces":** Ibid.

276  **Sharperson was sentenced to thirteen years in prison:** K. L. Alexander, "Fort Washington Man Sentenced to 13 Years in Prison for 2014 Sex Assault," *The Washington Post*, March 13, 2015, http://www.washingtonpost.com/local/crime/fort-washington-man-sentenced-to-13-years-in-prison-for-2014-sex-assault/2015/03/13/c758aa2e-c9ca-11e4-b2a1-bed1aaea2816_story.html.

276  **"I do not expect":** "History of the Carengie Hero Fund Commission," Carnegie Hero Fund Commission, http://www.carnegiehero.org/about-the-fund/history.

276  **Nearly one in four Carnegie Hero medals:** C. McDougall, ed., "The Hidden Cost of Heroism," *Men's Health*, November 2007.

277  **Valeri Bezpalov, Alexei Ananenko, and Boris Baranov:** "The Chernobyl Nuclear Power Plant Disaster," Friends of Chernobyl's Children, http://www.foccwestlothian.com/chernobyl.html.

277  **"You can't sit next to":** A. Cooper, "Should Male Passengers Be Allowed to Sit Next to Unaccompanied Children?" CNN, August 15, 2012, http://www.cnn.com/2012/08/14/travel/unaccompanied-children-flights/index.html.

278  **Males prevent violent crime:** N. Barber, "Countries with Fewer Males Have More Violent Crime: Marriage Markets and Mating Aggression," *Aggressive Behavior* 35, no. 1 (January–February 2009): 49–56.

279  **"fall into a hole or trap":** "Why Do We Say 'to Fall in Love'? Is It Something Unwished For?," English Language & Usage Stack Exchange, http://english.stackexchange.com/questions/138876/why-do-we-say-to-fall-in-love-is-it-something-unwished-for.

281  **interpersonal relationships constitute the foundation:** J. E. Swain et al., "Parenting and Beyond: Common Neurocircuits Underlying Parental and Altruistic Caregiving," *Parenting: Science and Practice* 12, no. 2–3 (2012): 115–23.

282  **spiritual beings and agents in their life:** D. Kapogiannis et al., "Brain Networks Shaping Religious Belief," *Brain Connectivity* 4, no. 1 (February 2014): 70–79.

283  **Structural differences in some brain regions:** A. D. Owen et al., "Religious Factors and Hippocampal Atrophy in Late Life," *PLOS ONE* 6, no. 3 (March 2011).

283  **researchers observed less atrophy:** R. D. Hayward et al., "Associations of Religious Behavior and Experiences with Extent of Regional Atrophy in the Orbitofrontal Cortex During Older Adulthood," *Religion, Brain and Behavior* 1, no. 2 (2011): 103–18.

283  **meditation activates the precuneus network:** Z. Josipovic, "Neural Correlates of Nondual Awareness in Meditation," *Annals of the New York Academy of Sciences* 1307 (January 2014): 9–18.

283 **role for meditation in memory consolidation:** M. Engström et al., "Functional Magnetic Resonance Imaging of Hippocampal Activation During Silent Mantra Meditation," *Journal of Alternative and Complementary Medicine* 16, no. 12 (December 2010): 1253–58.

283 **"mind wandering" and focused attention:** A. Manna et al., "Neural Correlates of Focused Attention and Cognitive Monitoring in Meditation," *Brain Research Bulletin* 82 no. 1–2 (April 2010): 46-56.

283 **mindfulness-based stress reduction intervention:** B. K. Hölzel et al., "Stress Reduction Correlates with Structural Changes in the Amygdala," *Social Cognitive and Affective Neuroscience* 5, no. 1 (March 2010): 11–17.

284 **pain-regulating brain processes:** U. Schjoedt et al., "Highly Religious Participants Recruit Areas of Social Cognition in Personal Prayer," *Social, Cognitive and Affective Neuroscience* 4, no. 2 (June 2009): 199–207.

284 **Brain imaging in Carmelite nuns:** M. Beauregard and V. Paquette, "Neural Correlates of a Mystical Experience in Carmelite Nuns," *Neuroscience Letters* 405, no. 3 (September 2006): 186–90.

284 **supernatural resistance to the pain and horror:** W. G. Berry, ed., *Foxe's Book of Martyrs* (New York: Eaton and Mains, 1907; orig. published in 1563).

285 **"Indeed, there is a case":** K. Armstrong, *A History of God* (New York: Alfred A. Knopf, 1993), xix.

## 11. A WORLD OF TROUBLE

287 **human beings have killed off *half*:** C. Ingraham, "We've Killed Off Half the World's Animals," *The Washington Post*, September 30, 2014.

288 **7.4 people in 1,000:** R. Walmsley, *World Prison Population List (10th edition)*, International Centre for Prison Studies, November 21, 2013, http://www.prisonstudies .org/sites/default/files/resources/downloads/wppl_10.pdf.

288 **25 percent of the world's prisoners:** J. Holland, "Land of the Free? US Has 25 Percent of the World's Prisoners," *Moyers & Company*, December 16, 2013, http:// billmoyers.com/2013/12/16/land-of-the-free-us-has-5-of-the-worlds-population -and-25-of-its-prisoners.

288 **One-fifth of our national wealth:** B. Plumer, "America's Staggering Defense Budget, in Charts," *Wonkblog, The Washington Post*, January 7, 2013, http://www.washing tonpost.com/news/wonkblog/wp/2013/01/07/everything-chuck-hagel-needs -to-know-about-the-defense-budget-in-charts.

289 **than on Germany and Japan combined:** M. Schudel, "Exposed U.S. Bombing Campaign Inside Laos," *The Washington Post*, October 5, 2014.

290 **"The bird is a creature of the air":** H. G. Wells, *Mind at the End of Its Tether* (London: William Heinemann Ltd., 1945), xii, 25.

291 **"Their semi-erect attitude":** Ibid., 62.

291 **"Families and tribes":** Ibid., 40.

291 **"Ordinary man is at the end":** Ibid., 62.

293 **Ötzi the Iceman of the Alps:** J. Owen, "5 Surprising Facts About Otzi the Iceman," *National Geographic*, October 18, 2013, http://news.nationalgeographic.com/news/ 2013/10/131016-otzi-ice-man-mummy-five-facts; B. Yirka, "Study Finds Modern Relatives of Otzi Alive and Well in Austria," Phys.org, October 14, 2013, http://phys .org/news/2013-10-modern-relatives-otzi-alive-austria.html.

294 **all other *Homo* species except *Homo sapiens*:** "Human Evolution," from an exhibit entitled *Understanding Evolution*, University of California Museum of Paleontology, UC Berkeley, http://evolution.berkeley.edu/evosite/evo101/IIE2cHumanevop2.shtml.

295  **a 20 percent population decrease:** C. Mooney, "A Staggering 400 Million Birds Have Vanished from Europe Since 1980," *The Washington Post*, November 3, 2014, http://www.washingtonpost.com/news/wonkblog/wp/2014/11/03/20-percent-of -european-birds-have-vanished-since-1980.

295  **168 species of amphibians:** "Worldwide Amphibian Declines: How Big Is the Problem, What Are the Causes and What Can Be Done?," AmphibiaWeb, February 13, 2013, http://amphibiaweb.org/declines/declines.html.

296  **clutched as if by talons beneath its wings:** "P-8A Poseidon Maritime Surveillance Aircraft, United States of America," Naval Technology, http://www.naval-technology .com/projects/mma.

296  **"very much continuing to push":** C. Whitlock, "China and Russia Push Limits of Airspace," *The Washington Post*, October 5, 2014.

297  **In August 2015, Finland fired:** G. Taylor, "Finland Fires on Suspected Russian Submarine in Waters off Helsinki," *The Washington Times*, April 28, 2015, http://www .washingtontimes.com/news/2015/apr/28/finland-fires-suspected-russian -submarine-waters-h.

297  **a major fishing area for Japan and China:** P. Goble, "Moscow Closes Okhotsk Sea to Outsiders," The Jamestown Foundation, April 29, 2014, http://www.jamestown .org/single/?tx_ttnews%5Btt_news%5D=42282&no_cache=1#.VDFcXvldWSo.

299  **"We are not about to send":** S. Mintz and S. McNeil, "LBJ," Digital History, http:// www.digitalhistory.uh.edu/disp_textbook.cfm?smtID=2&psid=3461.

299  **"These new documents":** P. Paterson, "The Truth About Tonkin," U.S. Naval Institute, *Naval History Magazine* 22, no. 1 (February 2008), http://www.usni.org/mag azines/navalhistory/2008-02/truth-about-tonkin.

300  **"The ultimate responsibility":** The United States Senate Joint Committee on the Investigation of the Pearl Harbor Attack, June 20, 1946, http://www.senate.gov/ artandhistory/history/common/investigations/PearlHarbor.htm.

301  **an example of a "just war":** K. Burns and L. Novick, *The War: The Declaration of War Against Japan and Just War Theory*, PBS, https://www.pbs.org/thewar/down loads/just_war.pdf.

301  **A man suspected of communicating:** This and subsequent quotes from J. Steinbeck, *Steinbeck in Vietnam: Dispatches from the War*, Thomas E. Barden, ed. (Charlottesville, VA: University of Virginia Press, 2012), 35. The Letter to Alicia dated January 7, 1967/Saigon was first published in *Newsday* as a series of reports from Viet Nam, that were written in the style of a letter home.

301  **At about 10 o'clock in the evening:** Ibid., 44. Originally published in *Newsday* as a Letter to Alicia, dated January 21, 1967/Can Tho.

302  **Last week in a remote village:** Ibid., 114. Originally published in *Newsday*, as a Letter to Alicia, dated March 4, 1967/Bangkok.

302  **"John changed his mind totally about Vietnam":** Ibid., 134.

303  **"We're the redcoats in this":** T. Barden, in ibid., 135.

304  **Pakistani officer kills man:** Reuters News Service, "Pakistani Officer Kills Man over Alleged Blasphemy," *The Washington Post*, November 8, 2014.

304  **November 2014 the body of journalist Par Gyi:** N. Noreen, "Par Gyi Tortured before Death," DBV News, November 5, 2014, https://www.dvb.no/news/par-gyi -tortured-before-death-myanmar-burma/45665.

304  **Hakan Yaman sustained serious injuries:** H. Eissenstat, "Tweet Hakan Yaman the Birthday Gift He Deserves: Justice," *Human Rights Now*, Amnesty International, October 10, 2013, http://blog.amnestyusa.org/europe/tweet-hakan-yaman-the-birth day-gift-he-deserves-justice.

304  **7,000 complaints of torture:** "14 Facts about Torture in Mexico," Amnesty International, October 22, 2014, http://www.amnestyusa.org/news/news-item/14-facts-about-torture-in-mexico.

305  **forty-three students at a teaching school:** J. Partlow, "Bodies Believed to Be Those of Some of the 43 Missing Mexican Students," *The Washington Post*, November 7, 2014.

305  **Amnesty International survey:** Amnesty International 2011 Report on Torture, May 2012, http://www.amnestyusa.org/sites/default/files/pdfs/air12-facts-and-figures.pdf

305  **On June 28, 2010:** "United Nations Committee Against Torture, Convention Against Torture, Periodic Report of the United States of America," United States Department of State, http://www.state.gov/documents/organization/213267.pdf.

306  **"I was too young to really understand":** "George Takei on the Japanese internment camps during WWII," YouTube video, 7:00, posted by TVLEGENDS, November 29, 2011, youtube: https://www.youtube.com/watch?v=yogXJl9H9z0.

307  **"It would be perfect for Obama":** C. Harlan, and Z. A. Goldfarb, "U.S. Criticizes N. Korea's Racist Obama Rant," *The Washington Post*, May 10, 2014.

308  **"Although prejudice stems from a mechanism":** D. M. Amodio, "The Neuroscience of Prejudice and Stereotyping," *Nature Reviews Neuroscience* 15 (2014): 670–82.

308  **"In view of substantial doubt as to the subject's loyalty":** T. Fensch, ed., *The FBI Files on John Steinbeck* (Santa Teresa, NM: New Century Books, 2002), 37.

309  **The central nucleus of the amygdala is critical for classical fear conditioning:** S. Ciocchi, "Encoding of Conditioned Fear in Central Amygdala Inhibitory Circuits," *Nature* 468 (November 2010): 277–82.

309  **When Chinese and Caucasians view:** X. Xu et al., "Do You Feel my Pain?: Racial Group Membership Modulates Empathic Neural Responses," *The Journal of Neuroscience* 29, no. 26 (July 2009): 8525–29.

310  **There is more activity in the mPFC:** L. T. Harris and S. T. Fiske, "Dehumanizing the Lowest of the Low: Neuroimaging Responses to Extreme Out-Groups," *Psychological Science* 17, no. 10 (October 2006): 847–53.

310  **"Most forms of implicit learning are resistant":** Amodio, "The Neuroscience of Prejudice and Stereotyping."

311  **"Anzor had a job within three days":** Author interview with Mr. Ruslan Tsarni, Montgomery Village, MD, January 4, 2014.

311  **"He had a heart of gold":** S. Jacobs, D. Filipov, and P. Wen, "The Fall of the House of Tsarnaev," *The Boston Globe*, December 15, 2012.

311  **On September 11 of that same year:** "Dzhokhar and Tamerlan: A Profile of the Tsarnaev brothers," *48 Hours*, April 23, 2013, http://www.cbsnews.com/news/dzhokhar-and-tamerlan-a-profile-of-the-tsarnaev-brothers.

312  **"Why the Boston Marathon?"** Author interview with Mr. Ruslan Tsarni, Montgomery Village, MD, January 4, 2014.

314  **Substance abuse during these critical years:** D. M. Anglin et al., "Early Cannabis Use and Schizotypal Personality Disorder Symptoms from Adolescence to Middle Adulthood," *Schizophrenia Research* 137, nos. 1–3 (May 2012): 45–49.

314  **can lead to depression and leave marks on the brain:** M. H. Teicher et al., "Hurtful Words: Association of Exposure to Peer Verbal Abuse with Elevated Psychiatric Symptom Scores and Corpus Callosum Abnormalities," *American Journal of Psychiatry* 167, no. 12 (December 2010): 1464–71.

315  **Neural correlates of impaired emotional processing:** I. Dogan et al., "Neural Correlates of Impaired Emotion Processing in Manifest Huntington's Disease," *Social Cognitive and Affective Neuroscience* 9, no. 5 (May 2014): 671–80.

315 **brain regions implicated in introspective processes:** Z. A. Englander et al., "Neural Basis of Moral Elevation Demonstrated Through Inter-Subject Synchronization of Cortical Activity During Free Viewing," *PLOS ONE* 7, no. 6 (June 2012): e39384.

315 **brain circuits that process physical pain:** J. E. Beeney et al., "I Feel Your Pain: Emotional Closeness Modulates Neural Responses to Empathically Experienced Rejection," *Social Neuroscience* 6, no. 4 (2011): 369–76.

315 **"Unmet need for social bonding":** C. A. Pedersen, "Biological Aspects of Social Bonding and the Roots of Human Violence," *Annals of the New York Academy of Sciences* 1036 (December 2004): 106–27.

316 **"He put a shame":** Associated Press, "Ruslan Tsarni 'Losers' Interview Video: Uncle Says Dzhozkar, Tamerlan Tsarnaev Were 'Losers' on NBC," NBC News, April 20, 2013, http://www.wptv.com/dpp/news/national/ruslan-tsarni-losers-interview-video -uncle-says-dzhozkar-tamerlan-tsarnaev-were-losers-on-nbc#ixzz2pXuQX9NZ; M. Martinez, "Uncle Calls Boston Marathon Bombers 'Losers,'" CNN, April 19, 2013, http://www.cnn.com/2013/04/19/us/marathon-suspects-uncle.

317 **"The funeral was dear to me":** Author interview with Mr. Ruslan Tsarni, in Montgomery Village, MD, January 4, 2014.

323 *Charlie Hebdo*: "Charlie Hebdo Attack: Three Days of Terror," BBC News, January 14, 2015, http://www.bbc.com/news/world-europe-30708237.

323 **cyberattack on Sony Movie Studio:** B. Barnes and N. Perlroth, "Sony Films Are Pirated, and Hackers Leak Studio Salaries," *The New York Times*, December 2, 2014, http://www.nytimes.com/2014/12/03/business/media/sony-is-again-target -of-hackers.html.

323 **shooting of Michael Brown:** L. Buchanan et al., "What Happened in Ferguson?," *The New York Times*, November 25, 2014, http://www.nytimes.com/interactive/ 2014/08/13/us/ferguson-missouri-town-under-siege-after-police-shooting.html.

323 **Eric Garner by the New York Police:** D. Ford, G. Botelho, and B. Brumfield, "Protests Erupt in Wake of Chokehold Death Decision," CNN, December 8, 2014, http:// www.cnn.com/2014/12/04/justice/new-york-grand-jury-chokehold.

323 **murders of two New York police officers:** B. Mueller and A. Baker, "2 N.Y.P.D. Officers Killed in Brooklyn Ambush; Suspect Commits Suicide," *The New York Times*, December 20, 2014, http://www.nytimes.com/2014/12/21/nyregion/two-police-officers -shot-in-their-patrol-car-in-brooklyn.html.

323 **Tribalism arises from human herding behavior:** R. M. Raafat, N. Chater, and C. Frith, "Herding in Humans," *Trends in Cognitive Sciences* 13, no. 10 (October 2009): 420–28.

324 **"In thirty-four years of law enforcement":** M. Pearce and M. Hennessy-Fiske, "9 Killed, At Least 170 Arrested After Waco Biker Gang Shootout," *Los Angeles Times*, May 18, 2015.

324 **"Not only did he fail to protect and serve":** "Assault Trial to Begin for Undercover NYPD Detective Embroiled in Biker Melee with SUV Driver," Associated Press, May 18, 2015.

324 **Much of the melee was captured on film:** C. Curry, "Motorcyclists Pull Driver from Car in NYC, Beat Him After Collision," ABC News, September 30, 2013, http:// abcnews.go.com/US/video-shows-motorcyclists-pull-driver-car-beat/story?id= 20419705.

325 **Gray's mortal injury was inflicted:** S. G. Stolberg, "Baltimore Enlists National Guard and a Curfew to Fight Riots and Looting," *The New York Times*, April 27, 2015, http://www.nytimes.com/2015/04/28/us/baltimore-freddie-gray.html?_r=0.

325 **"Can we all get along?":** "King: 'Can We All Get Along?,'" ABC News, June 17, 2012, http://abcnews.go.com/US/video/rodney-king-16589937.

327 **"She knows her son":** J. Levs, A. C. Stapleton, and S. Almasy, "Baltimore Mom Who Smacked Son at Riot: I Don't Play," CNN, April 29, 2015, http://www.cnn.com/2015/04/28/us/baltimore-riot-mom-smacks-son.

327 **"That's my only son":** Ibid.

328 **"Seinfeld shutdown":** M. A. Thiessen, "The Seinfeld Shutdown," *The Washington Post*, October 4, 2013, http://www.washingtonpost.com/opinions/marc-thiessen-the-seinfeld-shutdown/2013/10/04/b600d434-2cf2-11e3-97a3-ff2758228523_story.html.

328 **"Mind reading" (from chapter 7):** C. M. Heyes and C. D. Frith, "The Cultural Evolution of Mind Reading," *Science* 344, no. 6190 (June 2014), http://www.sciencemag.org/content/344/6190/1243091.

329 **Studies of the neuroscience of human cooperation:** J. R. Gray, J. A. Bargh, and E. Morsella, "Neural Correlates of the Essence of Conscious Conflict: fMRI of Sustaining Incompatible Intentions," *Experimental Brain Research* 229, no. 3 (September 2013): 453–65.

329 **Serotonin enhances fairness:** M. J. Crockett et al., "Serotonin Modulates Striatal Responses to Fairness and Retaliation in Humans," *The Journal of Neuroscience* 33, no. 8 (February 2013): 3505–13.

329 **Oxytocin differentially modulates:** F. S. Ten Velden et al., "Oxytocin Differentially Modulates Compromise and Competitive Approach but not Withdrawal to Antagonists from Own vs. Rivaling Other Groups," *Brain Research* 1580 (September 2014): 172–79.

329 **Another study published in 2013:** C. C. Ruff et al., "Changing Social Norm Compliance with Noninvasive Brain Stimulation," *Science* 342, no. 6157 (October 2013): 482–84.

329 **Another study found that testosterone:** E. R. Montoya et al., "Testosterone Administration Modulates Moral Judgments Depending on Second-to-Fourth Digit Ratio," *Psychoneuroendocrinology* 38, no. 8 (August 2013): 1362–69.

## 12. BEYOND THE CIRCUIT

331 **Brilliant, menacing flashes:** Author interview with Rusty on the Potomac River, May 13, 2014.

335 **the wild chaos of that biker bar:** *Full Throttle Saloon*, TruTV, season 3 episodes 3–5, (episode 3 aired December 13, 2011; episode 4 aired December 20, 2011; episode 5 aired December 27, 2011).

337 **"I've seen people in different states of that":** Telephone interview with former member of SEAL Team Six, Spring 2014.

340 **"I got my own carbine":** P. C. Hearst, *Every Secret Thing* (New York: Doubleday, 1982), 147–48.

341 **weakened after chronic stress:** C. Liston, B. S. McEwen, and B. J. Casey, "Psychosocial Stress Reversibly Disrupts Prefrontal Processing and Attentional Control," *Proceedings of the National Academy of Sciences of the United States of America* 106, no. 3 (January 2009): 912–17.

346 **"I thought it was more prudent":** A. Halsey III, "Variety of Factors Shaped Stranded Metro Riders' Behavior," *The Washington Post*, January 19, 2015.

347 **"We should stay put":** L. Aratani, P. Sullivan, and M. E. Ruane, "Dying Among Strangers, Metro Victim Found Fellow Riders Who Tried to Save Her Life," *The Washington Post*, January 13, 2015.

350 **"Always take care of your people":** C. McDougall, ed. "The Hidden Cost of Heroism," *Men's Health*, October 4, 2007, http://www.menshealth.com/best-life/heroes-and-self-sacrifice?fullpage=true.

353 **"The universe has always existed":** Author interview with Mr. Pravin Dand in the Jain Temple, Silver Spring, MD, February 15, 2015.

353 *If shot by a gun or pistol:* BBC Religions, September 14, 2009, http://www.bbc .co.uk/religion/religions/jainism/worship/worship_1.shtml.

353 *I ask for forgiveness: JainNet Gazette,* Yahoo! Groups member comment by Premchand Gada, September 21, 2002, https://groups.yahoo.com/neo/groups/jainlist/ conversations/topics/2993.

353 **"I was born into a Jain family":** Telephone interview with Mr. "Nimesh Shah," February 14, 2015.

356 **"We are talking about our animal brothers":** Author interview with Mr. Dand, February 15, 2015.

359 **"Simplicity, peace, integrity":** Author interview with Mr. David Connell, Washington, DC, February 20, 2015.

360 **"Some of the bravest people I know":** Author interview with Dr. Daniela Schiller in her laboratory at Mount Sinai Medical School, New York, November 11, 2013.

361 **how Schiller and her team tested the idea:** D. Schiller et al., "Extinction During Reconsolidation of Threat Memory Diminishes Prefrontal Cortex Involvement," *Proceedings of the National Academy of Sciences of the United States of America* 110, no. 50 (2013): 20040–45.

364 **"In real life, PTSD is very persistent":** Email interview with Dr. Yadin Dudai, Weizmann Institute of Science, Rehovot, Israel, November 22, 2013.

364 **"We have previously shown diminished extinction":** Email interview with B. J. Casey, November 22, 2013.

365 **"Sometimes you will see kids":** Author interview with Haley Peckham, Bethesda, MD, November 21, 2014.

366 **"The steeper for me":** Author interview with Wendy Fisher, Olympic skier, on Paradise Run at Crested Butte Ski Resort, February 28, 2014.

370 **"Please pardon the mixed use of English/Gujarati":** Email to the author sent from a member of the Silver Spring Jain Temple, February 15, 2015.

# Acknowledgments

I would like to thank, first and foremost, all of the wonderful people I interviewed for this book who shared their personal experiences and insights. One of the most rewarding aspects of writing a book is the opportunity to meet such a wide range of remarkable people. In every interview I was struck by the openness and intense interest of people in generously contributing their personal stories. I feel this book is very much a collective effort to explore and understand sudden rage and aggression, which in one form or another touches us all.

Creating a book is the mountaineering of writing. It is all-consuming—impossible without the support of friends and family. No one does it alone. This book was forged and polished by close collaboration with my gifted literary agents Jeff Kellogg and Andrew Stuart. Their insightful criticisms and creative contributions to this work were absolutely essential. My editor at Dutton, Stephen Morrow, was a delight to work with, and it is a privilege to have had his expert guidance and creative collaboration. As every rock climber knows, any achievement on the sharp end of the rope is made possible only by knowing that there is a steady and skillful hand on the belay. I thank my copyeditor, Rachelle Mandik, for her diligent work to make my text as clear, correct, and presentable as possible. I am grateful to my friends, notably Dan Coyle, Tom Cecil, John Gregory, Drake Olsen, Manzar Ashtari, and Pamela Hines, and especially every member of my family for reading snippets of text as it emerged. I am grateful to the Helen R. Whiteley Center at the University of Washington at Friday Harbor Laboratories for the precious opportunity for concentrated research and writing in the summers of 2013 and 2014. Those days of refuge, looking out the window of the writer's studio at sailboats rocking in San Juan Harbor, of uninterrupted study and writing punctuated by picnic lunches with Melanie on a sunny bluff overlooking the ocean, and evenings of scholarship with writers, artists, musicians, scientists, and students, will never leave me.

It is often difficult to trace back to the moment of inspiration that sparked a new idea, but in the case of this book, that moment is crystal clear. One evening I related my experience with pickpockets in Barcelona to friends in a congested noisy cocktail reception at the annual meeting of the National Association of Science Writers in New Haven, Connecticut. John Rennie, science writer and former editor of *Scientific American*, exclaimed above the din, "That's your next book!" Brainstorming from that moment, I scribbled in my notebook the entire train ride home notes that became the outline for this book. Remarkably, that creative burst of energy produced the complete structure and scope of this book, which is essentially unchanged from those notes. Such fleeting chance events are one of life's great delights. They become plot points in one's personal life story. Had I shared a beer that night with someone else in that crowded room, this book might never have happened. But when you are blessed with so many extraordinary people sharing their lives with you—close family, good friends, talented scientists, writers, and editors—such moments have little to do with chance.

# Index